CAMBRIDGE LIBRARY COLLECTION

Books of enduring scholarly value

Physical Sciences

From ancient times, humans have tried to understand the workings of the world around them. The roots of modern physical science go back to the very earliest mechanical devices such as levers and rollers, the mixing of paints and dyes, and the importance of the heavenly bodies in early religious observance and navigation. The physical sciences as we know them today began to emerge as independent academic subjects during the early modern period, in the work of Newton and other 'natural philosophers', and numerous sub-disciplines developed during the centuries that followed. This part of the Cambridge Library Collection is devoted to landmark publications in this area which will be of interest to historians of science concerned with individual scientists, particular discoveries, and advances in scientific method, or with the establishment and development of scientific institutions around the world.

The Life of James Clerk Maxwell

James Clerk Maxwell (1831–79) was a Scottish physicist well-known for his extensive work with electromagnetism, colour analysis, and kinetic theory. Considered by many to be a giant in his field with significant influence on the physicists who would follow, Maxwell spent time as a professor at Aberdeen University, King's College, London, and Cambridge. This 1882 *Life* by his friend Lewis Campbell and natural philosopher William Garnett represents an important – and lengthy – investigation into Maxwell's life and thought. Part I is concerned with biographical matters while the second section focuses upon his scientific mind. A third part contains Maxwell's poetry, so included because the poems are 'characteristic of him' and have 'curious biographical interest'. At nearly 700 pages, the *Life* represents an important starting point for those curious about the state of theoretical physics and the person in whom it reached its culmination in the nineteenth century.

Cambridge University Press has long been a pioneer in the reissuing of out-of-print titles from its own backlist, producing digital reprints of books that are still sought after by scholars and students but could not be reprinted economically using traditional technology. The Cambridge Library Collection extends this activity to a wider range of books which are still of importance to researchers and professionals, either for the source material they contain, or as landmarks in the history of their academic discipline.

Drawing from the world-renowned collections in the Cambridge University Library, and guided by the advice of experts in each subject area, Cambridge University Press is using state-of-the-art scanning machines in its own Printing House to capture the content of each book selected for inclusion. The files are processed to give a consistently clear, crisp image, and the books finished to the high quality standard for which the Press is recognised around the world. The latest print-on-demand technology ensures that the books will remain available indefinitely, and that orders for single or multiple copies can quickly be supplied.

The Cambridge Library Collection will bring back to life books of enduring scholarly value (including out-of-copyright works originally issued by other publishers) across a wide range of disciplines in the humanities and social sciences and in science and technology.

The Life of James Clerk Maxwell

With a Selection from his Correspondence and Occasional Writings and a Sketch of his Contributions to Science

LEWIS CAMPBELL
WILLIAM GARNETT

CAMBRIDGE
UNIVERSITY PRESS

CAMBRIDGE UNIVERSITY PRESS

Cambridge, New York, Melbourne, Madrid, Cape Town, Singapore,
São Paolo, Delhi, Dubai, Tokyo

Published in the United States of America by Cambridge University Press, New York

www.cambridge.org
Information on this title: www.cambridge.org/9781108013703

© in this compilation Cambridge University Press 2010

This edition first published 1892
This digitally printed version 2010

ISBN 978-1-108-01370-3 Paperback

LIFE OF JAMES CLERK MAXWELL

James Clerk Maxwell.

THE LIFE

OF

JAMES CLERK MAXWELL

WITH A SELECTION FROM HIS CORRESPONDENCE
AND OCCASIONAL WRITINGS

AND

A SKETCH OF HIS CONTRIBUTIONS TO SCIENCE

BY

LEWIS CAMPBELL, M.A., LL.D.

PROFESSOR OF GREEK IN THE UNIVERSITY OF ST. ANDREWS

AND

WILLIAM GARNETT, M.A.

LATE FELLOW OF ST. JOHN'S COLLEGE, CAMBRIDGE
PROFESSOR OF NATURAL PHILOSOPHY IN UNIVERSITY COLLEGE, NOTTINGHAM

WITH THREE STEEL PORTRAITS, COLOURED PLATES, ETC.

London

MACMILLAN AND CO.

1882

PREFACE.

IN a work which has more than one author, it is right to distinguish as far as possible what has been contributed by each.

The first part of this book, then, has been mainly composed by the present writer (who has acted as editor), and the second by Mr. Garnett; but it is due to Mr. Garnett to add that, while he had the chief share of the labour of collecting materials for the whole biography, and the entire burden of the account here given of Maxwell's contributions to science, the substance of Chapters XI. XII. XIII. is also largely drawn from information obtained through him. The matter of whole pages remains almost in his very words, although for the sake of uniformity and simplicity the first person has still been used in speaking of my own reminiscences.

The narrative of Maxwell's early life has been facilitated—(1) by a diary kept by Maxwell's father from 1841 to 1847, and often referred to in these pages as "the Diary;" (2) by two albums containing a series of water-colour drawings by Maxwell's first cousin, Mrs. Hugh Blackburn (*née* Isabella Wedderburn), the value of which may be inferred from the

outlined reproductions of a few of them prepared by
Mrs. Blackburn herself as illustrations for this book.
They are literally *bits out of the past*, each containing an exact representation, by a most accurate
observer and clever draughtswoman, of some incident
which had just happened when the sketch was made.
The figures, even as outlined, bring back the persons
with singular vividness to the memory of those who
knew them.

For these and many other advantages the thanks
of the authors are due to Maxwell's relatives, but
above all to his widow, not only for the free access
she has given them to various documents, such as
those mentioned above, but for the generous confidence she has reposed in them throughout, and for
many important suggestions made by her during the
progress of the work.

Their thanks are also due in an especial manner
to Professor G. G. Stokes of Cambridge for his kindness in reading most of Part II. in MS., and for many
valuable suggestions both with regard to the subject
matter and the mode of treating it, which were received from him ; and to Professor P. G. Tait of the
University of Edinburgh for his zealous and able
assistance in many ways.

The book owes much, of course, to those who
entrusted the letters here published to the authors'
care. Their names are duly mentioned in the course
of the work. Our task has also been lightened by
the help of those friends whose contributions are
inserted with their names. In describing the life of

one who was so many-sided, it is no small advantage to be thus enabled to register the impression which he made on different men. Even should this occasion slight repetitions and discrepancies, the reader may thus form a fuller and, on the whole, a truer image than could be conveyed by a single narrator. Attention is here particularly directed to the statements in Chapter XIII. by Dr. Paget, the Rev. Dr. Guillemard (of Little St. Mary's, Cambridge), and Professor Hort.

As a general rule, no attempt has been made to weave the correspondence into the narrative. The facts relating to each period have been grouped together, and the letters have been appended to these in chronological order.

A word should be said respecting Part III. Whatever may be the judgment of critics as to the literary merits of Maxwell's occasional writings in verse, there can be no doubt of their value for the purpose of the present work. Like everything which he did, they are characteristic of him, and some of them have a curious biographical interest. Maxwell was singularly reserved in common life, but would sometimes in solitude express his deepest feelings in a copy of verses which he would afterwards silently communicate to a friend. Again, he shrank from controversy. But his active mind was constantly playing on contemporary fallacies, or what appeared so to him, and his turn for parody and burlesque enabled him to give humorous expression to his criticism of mistaken methods. Of the later pieces here reproduced, several appeared in *Nature* with the signature $\frac{dp}{dt}$ (which happens to be the analytical equivalent of the thermo-dynamical formula JCM); and one (the "Notes on the President's Address") was published in *Black-wood's Magazine* for December 1874. The greater number are now printed for the first time.

The juvenile verses and translations have been included for the same reason which has led to the prominence given to the early life in Part I. If we are right in our estimate of Maxwell, it must be interesting to watch the unfolding of such a mind and character from the first, and this not only for the psychological student, but for all those who share Wordsworth's fondness for "days" that are "linked each to each with natural piety."

While the last sheets were being revised for the press the sad news arrived that Maxwell's first cousin, Mr. Colin Mackenzie, had died on board the *Bosnia*, on his way home from America. There was no one whose kind encouragement had more stimulated the preparation of this volume, or whose pleasure in it would have been a more welcome reward. But he, too, is gone before his time, and this book will be sent into the world with fewer good wishes. He deserves to be remembered with affection wherever the name of James Clerk Maxwell is honoured or beloved.

LEWIS CAMPBELL.

August 1882.

CONTENTS.

PART I.—BIOGRAPHICAL OUTLINE.

CHAPTER I.

CHAPTER II.

CHAPTER III.

CHAPTER IV.

CHAPTER V.

CHAPTER VI.

PART II.—CONTRIBUTIONS TO SCIENCE.

PART III.—POEMS.

LIST OF ILLUSTRATIONS.

STEEL PLATES.

LITHOGRAPHS.

COLOURED PLATES.

PART I.

CHAPTER I.

INTRODUCTORY—BIRTH AND PARENTAGE.

ONE who has enriched the inheritance left by Newton and has consolidated the work of Faraday,—one who impelled the mind of Cambridge to a fresh course of real investigation,—has clearly earned his place in human memory.

But there was more in James Clerk Maxwell than is implied in any praise that can be awarded to the discoverer, or in the honour justly due to the educational reformer,—much, indeed, which his friends feel they can but partly estimate, and still less adequately describe.

We have, notwithstanding, undertaken this imperfect Memoir of him, in which the purpose of this First Part will be to trace the growth from childhood to maturity, and to record the untimely death, of a man of profound original genius, who was also one of the best men who have lived, and, to those who knew him, one of the most delightful and interesting of human beings.

If I can bring before the reader's mind, even in shadowy outline, the wise and gentle but curiously

B

blended influences which formed the cradle of his
young imagination, the channels through which ideas
reached him from the past, the objects which most
challenged his observation and provoked his invention, his first acquaintance with what permanently
interested him in contemporary speculation and
discovery, and the chief moments of his own intellectual progress in earlier years,—such record
should have a right to live. And it may be that
a congenial spirit here and there may look with
me into the depths of this unique personality,
and feel the value of the impulses, often seemingly wayward, and strange even to himself, with
which the young eagle "imped his wings" for flight,
or taught his eyes to bear the unclouded light. And
many to whom modern Science is a sealed book may
find an interest in observing the combination of
extraordinary gifts with a no less remarkable simplicity and strength of character.

James Clerk Maxwell was born at No. 14 India
Street, Edinburgh, on the 13th of June 1831. His
parents were John Clerk Maxwell, one of the Clerks
of Penicuik, in Midlothian, and Frances, daughter of
R. H. Cay, Esq. of N. Charlton, Northumberland.
Excepting a daughter, Elizabeth, who had died in
infancy, James was their only child.

Edinburgh was at this time the natural meeting-
place for the best spirits of the North. How much of
intellect and individuality, of genuine though often
eccentric worth, of high thinking and plain living,
then forgathered in Auld Reekie, and found ample

scope and leisure there, is known to the lovers of Sir
W. Scott and to the readers of Lord Cockburn—both
prominent figures in the Edinburgh of 1824-1831.
And the *agora* of "Modern Athens" was the
Parliament House. There the heir-presumptive could
while away his time of waiting for "dead men's
shoon"; there the laird's brother might qualify for
some berth hereafter to be provided for him ; and
the son of those whose ancestral estates had been
impaired by rashness or misfortune, and who had
perchance sought the asylum of the Abbey, might
hope through honourable industry to restore the
fallen house, or even to win new lustre for an ancient
name.

When John Clerk Maxwell, after leaving the
University, first sought those purlieus of the law,
he was already a laird, although a younger brother.
For he had inherited the estate of Middlebie, which,
by the conditions of the entail under which it had
descended from the Maxwells,[1] could not be held together with Penicuik, and was therefore necessarily
relinquished by Sir George Clerk in favour of his
brother John. This arrangement had been completed
when the two brothers were boys together at the High
School, and were living in George Square with their
mother and their sister Isabella. Their father, James
Clerk,[2] who died before his elder brother, Sir John,
was a naval captain in the H.E.I.C.S., but retired

---

[1] See below, pp. 16-23.

[2] He is said to have played well on the bagpipes, and a set of pipes
was until recently preserved at Glenlair, of which the following singular

early, and married Miss Janet Irving, who thus became the mother of Sir George Clerk and of John
Clerk Maxwell. When Sir George had come of age,
and taken up his abode at Penicuik, John Clerk Maxwell continued living with his mother, Mrs. Clerk, in
Edinburgh. About 1820, in order to be near Isabella,
Mrs. Wedderburn, they "flitted" to a house in the
New Town, No. 14 India Street, which was built by
special contract for them. Mrs. Clerk died there in
the spring of 1824.

The old estate of Middlebie had been considerably
reduced, and there was nothing in what remained of
it to tempt its possessor, while a single man, to leave
Edinburgh, or to break off from his profession at the
Bar. There was not even a dwelling-house for the
laird. Mr. Clerk Maxwell therefore lived in Edinburgh until the age of thirty-six, pacing the floor of
the Parliament House, doing such moderate business
as fell in his way, and dabbling between-whiles in
scientific experiment. In vacation time he made
various excursions in the Highlands of Scotland and
in the north of England, and kept a minute record of
his observations.

But when, after his mother's death, he had married
a lady of tastes congenial to his own and of a sanguine
active temperament, his strong natural bent towards
a country life became irresistible. The pair soon con-

---

story was told :—Captain James Clerk was wrecked in the Hooghly
and swam ashore, using the bag of his pipes for a float ; and when he
gained the shore he " played an unco' fit," whereby he not only cheered
the survivors, but frightened the tigers away.

ceived a wish to reside upon their estate, and began to form plans for doing so; and they may be said to have lived thenceforth as if it and they were made for one another. They set themselves resolutely to the work of making that inheritance of stony and mossy ground to become one of the habitable places of the earth. John Clerk Maxwell had hitherto

appeared somewhat indolent; and there was a good deal of *inertia* in his composition. But the latent forces of his character were now to be developed.

He was one of a race [1] in whom strong individuality had occasionally verged on eccentricity. For two centuries the Clerks had been associated with all that was most distinguished in the Northern kingdom, from Drummond of Hawthornden to Sir Walter Scott. Each generation had been remarkable for the talents and accomplishments of some of its members; and it was natural that a family with such ante-

---

[1] The note appended to this chapter contains a sketch of the family history of the Clerks and Maxwells, which those who believe in heredity, as Maxwell did, will do well to read.

cedents should have acquired something of clannish-
ness. But any narrowing effect of such a tendency
was counteracted by a strong intellectual curiosity,
which kept them *en rapport* with the world, while
they remained independent of the world. And as
each scion of the stock entered into new relations,
the keen mutual interest, instead of merely narrowing, became an element of width. I speak now of
the generation preceding our own. No house was
ever more affluent in that *Coterie-Sprache*, for which
the Scottish dialect of that day afforded such full
materials. It would be pleasant, if possible, to recall
that humorous gentle speech, as it rolled the cherished
vocables like a sweet morsel on the tongue, or minced
them with a lip from which nothing could seem coarse
or broad,—caressing them as some Lady Bountiful may
caress a peasant's child,—or as it coined *sesquipedalia
verba*, which passed current through the stamp of
kindred fancy. This quaint freemasonry was un-
consciously a token not only of family community,
but also of that feudal fellowship with dependents
which was still possible, and which made the language
and the manners of the most refined to be often racy
of the soil. But Time will not stand still, and neither
the delicate "couthy" tones, nor that which they
signified, can be fully realised to-day. But we can
still in part appreciate the playful irony which
prompted these humorous vagaries of old leisure,
wherein true feeling found a modest veil, and a naïve
philosophy lightened many troubles of life by making
light of them.

Mr. John Clerk Maxwell's own idiosyncrasy, as has been said, was well suited for a country life. But to give a true idea of him it is necessary to be more precise. His main characteristic, beyond a warm, affectionate heart, the soundest of sound sense, and absolute sincerity, was a persistent practical interest in *all useful processes*.[1] When spending his holidays at Penicuik as a boy from the Edinburgh High School (as well as long afterwards), he took delight in watching the machinery of Mr. Cowan's paper-mill, then recently established in that neighbourhood. And Mr. R. D. Cay remembers him, when still a young man living in India Street with his mother (about 1821-24), to have been engaged, together with John Cay, who was after-

---

[1] He never lost an opportunity of inspecting manufactures, or of visiting great buildings, ecclesiastical or otherwise, and he impressed the same habit upon his son. The " works " they " viewed " together were simply innumerable, but it will be sufficient to cite one crowning instance. When James Clerk Maxwell was in the midst of his last year's preparation for the Cambridge Tripos, he proposed to spend the few days of Easter vacation which the pressure of his work allowed to him, in a visit to a friend at Birmingham. His father had seen Birmingham in his youth, and gave him the following instructions, which were mostly carried out :—" View, if you can, armourers, gunmaking and gunproving — swordmaking and proving — *Papier-mâchée* and japanning — silver-plating by cementation and rolling—ditto, electrotype—Elkington's works—Brazier's works, by founding and by striking out in dies—turning—spinning teapot bodies in white metal, etc. —making buttons of sorts, steel pens, needles, pins, and any sorts of small articles which are curiously done by subdivision of labour and by ingenious tools—glass of sorts is among the works of the place, and all kinds of foundry works—engine-making—tools and instruments —optical and [philosophical], both coarse and fine. If you have had enough of the town lots of Birmingham, you could vary the recreation by viewing Kenilworth, Warwick, Leamington, Stratford-on-Avon, or such like." James began with the glassworks.

wards his brother-in-law, in a series of attempts to make a bellows that should have a continuous even blast. We can readily imagine, therefore, how closely he must have followed every step in the gradual application of steam to industry, and the various mechanical improvements which took place in his youth and early manhood.[1]

His practical thoroughness was combined with a striking absence of conventionality and contempt for ornament. In matters however seemingly trivial— nothing that had to be done was trivial to him—he considered not what was usual, but what was best for his purpose. In the humorous language which he loved to use, he declared in favour of doing things with *judiciosity*. One who knew him well describes him as always balancing one thing with another—exercising his reason about every matter, great or small. He was fond of remarking, for example, on the folly of coachmen in urging a horse to speed as soon as they saw the top of a hill, when, by waiting half a minute until the summit was really attained, they might save the animal. "A sad waste of work," he would say. Long before the days of "anatomical" bootmaking, he insisted on having ample room for his feet. His square-toed shoes were made by a country shoemaker under his direction on a last of his own and out of a piece of leather chosen by himself. This

---

[1] In 1831 he contributed to the *Edinburgh Medical and Philosophical Journal* (vol. x.) a paper entitled, "Outlines of a Plan for combining Machinery with the Manual Printing Press." His acme of festivity was to go with his friend John Cay (the "partner in his revels"), to a meeting of the Edinburgh Royal Society.

is only one example of the manner in which he did everything. It was thought out from the beginning to the end, and so contrived as to be most economical and serviceable in the long run. In his Diary (1841) we find him cutting out his own and his son's shirts, while planning the outbuildings which still exist at Glenlair. And he not only planned these, but made the working plans for the masons (1842) with his own hand.[1] This habitual careful adaptation of means to ends was the characteristic which (together with profound simplicity) he most obviously transmitted to his son. Its effect, heightened by perfect science, is still apparent in the construction and arrangement of the Cavendish Laboratory at Cambridge.

While thus unostentatious and plain in all his ways, he was essentially liberal and generous. No one could look in his broad face beaming with kindliness and believe otherwise. But his benevolence was best known nearest home. And in caring for others, as in providing for his own house, his actions were ruled; not by impulse, but persistent thoughtfulness. By his ever-wakeful consideration, he breathed an atmosphere of warm comfort and quiet contentment on all (including the dumb animals) within his sphere. Whoever had any claim upon his affectionate heart, whether as an old dependant, or as a relation or

---

[1] The following entry from the Diary (1842) will be appreciated by those who are interested in the country life of a past generation :—
" Wrote to Nanny about check of the yarn of the dead Hogs, to make trowser stuff or a plaid."

friend, might command from him any amount of
patient thought, and of pains given without stint
and without complaint. His *"judiciosity"* was used
as freely for them as for himself.[1] And where need
was he could be an effectual peacemaker.

He was assiduous also in county business (road
meetings, prison boards, and the like), and in his own
quiet way took his share in political movements, on
the Conservative side.

There was a deep unobtrusive tenderness in him,
which in later years gave a touching, almost femi-
nine, grace to his ample countenance, and his portly,
even somewhat unwieldy, frame.[2] He was a keen
sportsman (unlike his son in this), and an excellent shot;
but it was observed that he was above all careful never
to run the risk of wounding without killing his game.

His temper was all but perfect; yet, as " the best
laid schemes o' mice and men gang aft agley," the
minute care with which he formed his plans sometimes exposed him to occasions which showed that
his usual calm self-possession was not invulnerable.
At such times he would appear not angry, only somewhat discomposed or " vexed," and, after donning his
considering-cap for a little while, would soon resume
his benign equanimity.

---

[1] Mr. Colin Mackenzie says :—" He was the confidential friend
of his widowed sister Mrs. Wedderburn's children, who were in the
habit of referring to him in all their difficulties in perfect confidence
that he would help them, and regarded him more as an elder brother
than anything else." This is abundantly confirmed by various entries
in the Diary.

[2] Entry in Diary, Nov. 9, 1844.—Weighed 15 st. 7 lbs.

G.J.Stodart

M.ʳ John Clerk Maxwell. Aet. 66.

Engraved by G.J.Stodart after Sir John Watson Gordon, R.A.

An interesting trait is revealed to us by the Diary. Minute as the entries are for day after day, things which, if mentioned, might reflect unfavourably upon others, are invariably omitted. They have come down through other channels, but in this scrupulous record they have left no trace.

His otherwise happy life was crossed with one deep, silent sorrow,—but was crowned with one long comfort in the life of his son. They were bound together by no ordinary ties, and were extremely like in disposition, in simplicity, unworldliness, benevolence, and kindness to every living thing. Those who knew Maxwell best will be least apt to think irrelevant this somewhat lengthy description of his father.

The portrait of Mr. John Clerk Maxwell by Watson Gordon is a faithful representation of a face which returns more vividly than most others to the eye of memory, but no portrait can restore " the busy wrinkles round his eyes," or give back to them their mild radiance—

> " Gray eyes lit up
> With summer lightnings of a soul
> So full of summer warmth, so glad,
> So healthy, sound and clear and whole,
> His memory scarce can make me sad."

He lived amidst solid realities, but his vision was neither shallow nor contracted. And his sense of things beyond, if inarticulate, was, in later life at least, not the less serious and profound. Yet those who shall compare this likeness with the study of James Clerk Maxwell's head by Mrs. Blackburn, may at once detect something of the difference between the father and the son.   In the one there is a grave and placid acquiescence in the nearer environment, the very opposite of enthusiasm or mysticism; in the other, the artist has succeeded in catching the unearthly look which often returned to the deep-set eyes under the vaulted brow, when they had just before been sparkling with fun,—the look as of one who has heard the concert of the morning stars and the shouting of the Sons of God.

James himself has said to me that to have had a wise and good parent is a great stay in life, and that no man knows how much in him is due to his progenitors.   And yet the speculative ideal element which was so strong in him—the struggle towards the infinite through the finite—was not prominent in either of his parents.   Mrs. Clerk Maxwell was, no doubt, a good and pious (not bigoted) Episcopalian; but, from all that appears, her chief bent, like that of her husband, must have been practical and matter-of-fact.   Her practicality, however, was different from his. She was of a strong and resolute nature,—as prompt as he was cautious and considerate,—more peremptory, but less easily perturbed.   Of gentle birth and breeding, she had no fine-ladyisms, but with blunt determination entered heart and soul into that rustic life. It is told of her that when some men had been badly hurt in blasting at a quarry on the estate, she personally attended to their wounds before a surgeon could be brought, and generally that wherever help was needed she was full of courage and resource.   She was very intelligent and ingenious, played well on the organ, and composed some music, but in other respects was less "accomplished" than most of her family, except in domestic works, and above all in knitting, which in those days was an elegant and most elaborate pursuit.

Her father, R. Hodshon Cay, Esq., of N. Charlton, is thus spoken of in Lockhart's *Life of Scott* (p. 86 of the abridged edition, 1871):—"I find him" (Scott) "further nominated in March 1796, together with Mr. Robert Cay,—an accomplished gentleman, afterwards Judge of the Admiralty Court in Scotland,—to put the Faculty's cabinet of medals in proper arrangement." Mr. Cay at one time held the post of Judge-Admiral and Commissary-General, and while thus dignified in his profession used to reside for part of the year on his hereditary estate of Charlton, which had been freed from certain burdens [1] upon his coming of age.

He married Elizabeth Liddell, daughter of John Liddell, Esq., of Tynemouth, about the year 1789. The eldest son, John, has been already mentioned as an early companion of John Clerk Maxwell's, and both his name and those of Jane and R. D. Cay will reappear in the sequel. Between Frances (Mrs. Clerk Maxwell) and her sister Jane, who was never married, there existed a very close affection. There is a picture of them both as young girls (a three-quarter length in water-colours) done by their mother, who was an accomplished artist. Her gift in

---

[1] Incurred by his father in successfully resisting some manorial claims. These debts had brought the family to Edinburgh.

this way, which was very remarkable, and highly
cultivated for an amateur, was continued in Jane and
Robert, and has been transmitted to the succeeding
generation. Miss Jane Cay was one of the warmest
hearted creatures in the world ; somewhat wayward
in her likes and dislikes, perhaps somewhat warm-
tempered also, but boundless in affectionate kindness
to those whom she loved. Mr. R. D. Cay, W.S.,
married a sister of Dyce the artist, and, after acting
for some time as one of the Judges' clerks, proceeded
in 1844 to Hong-Kong, where he had an appointment.
His wife joined him there in 1845, and died in 1852.
In two of their sons, besides the artistic tastes which
they inherited through both parents, there was developed remarkable mathematical ability. It should
be also noticed that Mr. John Cay, the Sheriff of
Linlithgow, though not specially educated in mathematics, was extremely skilful in arithmetic and fond
of calculation as a voluntary pursuit. He was a great
favourite in society, and full of general information.
We have already seen him assisting at experiments
which might have led to the invention of " blowing
fans," but seem to have produced no such profitable
result. And we shall find that his interest in practical
Science was continued late in after life.[1]

In speaking of the Cay family it has been necessary
to anticipate a little, in order to advert to some particulars which, although later in time, seemed proper

---

[1] It should be remembered that in the early years of the century
considerable interest in experimental science had been awakened in
Edinburgh through the teaching of Professors Playfair and Hope.

to an introduction. Having departed so far from the order of events, I may before concluding this chapter make explicit mention of the loss which coloured the greater part of James Clerk Maxwell's existence, by leaving him motherless in his ninth year. Mrs. Clerk Maxwell died on the 6th of December 1839. There was extant until after Professor Maxwell's death a memorandum or diary kept at the time by her husband, describing the heroic fortitude which she had shown under the pain of her disease, and of the operation by which they had attempted to save her. Anæsthetics were then unknown. She had nearly completed her forty-eighth year, having been born on the 25th of March 1792, and married at the age of 34 (October 4, 1826). Mr. Maxwell was aged fifty-two at the time of his wife's death. He did not marry again.

We now return from this sad record to the birth of the son and heir, which was the more welcome to the parents after the loss of their first-born child. At this joyful epoch Mr. and Mrs. Clerk Maxwell, though retaining the house in India Street, had been already settled for some years in their new home at Glenlair.

NOTE.

THE CLERKS OF PENICUIK AND MAXWELLS OF MIDDLEBIE.

MISS ISABELLA CLERK, of 3 Hobart Place, London, has kindly furnished me with the following statement, to which I have added some annotations. These are chiefly derived from a book of autograph letters, which was long kept at Glenlair, and is now in the possession of Mrs. Maxwell.

"The Clerks of Penicuik are descended from John Clerk, of Kilhuntly, in Badenoch, Aberdeenshire, who attached himself to the party of Queen Mary, and had to leave that part of the country in 1568 during the troubles. His son, William Clerk, was a merchant in Montrose; he lived in the reigns of Mary and James the Sixth, and died in 1620.

"John Clerk, his son, was a man of great ability. He went to Paris in 1634, and having acquired a large fortune there in commerce, returned to Scotland in 1646, and bought the barony of Penicuik, and also the lands of Wright's Houses. He married Mary, daughter of Sir William Gray, of Pittendrum. This lady brought the necklace of Mary Queen of Scots into the family, through her mother, Mary Gillies, to whom it was given by Queen Mary before her execution. He died in 1674. His eldest son John was created a Baronet of Nova Scotia in 1679."

This Sir John Clerk (the first baronet) records a singular affray with sixteen robbers who attacked the house of Penicuik in 1692. He kept them at bay until the neighbouring tenants came to his relief. His pluck, sagacity, presence of mind, good feeling, and piety, are conspicuous in the narrative.

"He served in the Parliament of Scotland, and acquired the lands and barony of Lasswade. He married Elizabeth, daughter of Henry Henderson, Esq., of Elvington, and grand-daughter of Sir William Drummond, of Hawthornden, the poet. Sir William Drummond had only two daughters, the younger of whom married W. Henderson. Their daughter Elizabeth was wonderfully talented and accomplished, and had a special gift for music. Sir John died in 1722. He was succeeded by his eldest son, Sir John Clerk, a man of great learning, who was appointed in 1707 one of the Barons of the Exchequer in Scotland, which judicial employment he retained during the remainder of his life. He was also one of the Commissioners of the Union."

The Baron's precepts to his son George when at school in Cumberland are worthy of Polonius. He advises him to be kind to his companions, as he could not tell what they might be able to do for him thereafter, and to be sure to take the opportunity of "learning the English language," in which the Baron himself regrets his own deficiency. "You have nothing else to depend on but your being a scholar and behaving well." He is described in the autumn of his days as "humming along and stuffing his pipe in order to whiff it away for half an hour," while "my lady is engaged reading over some newspaper. Miss Clerk is labouring with great industry at a very pretty open muffler. Mrs. Dean is winding three fine white clews upon a fourth one," etc. "My lady's" letters to her son George dwell much on family matters, and are full of old-fashioned piety.

"He was highly accomplished in music, painting, and languages. At the age of nineteen he went abroad for three years to finish his education. He studied one year under the celebrated Dr. Boerhaave at Leyden, where he was a pupil of William Mieris in drawing. He afterwards went to Florence and Rome, where he had lessons in music from Antonio Corelli, and in painting from Imperiale. He wrote an opera which was performed in Rome. His brother, William Clerk, married Agnes Maxwell, heiress of Middlebie, in Dumfriesshire."

William also was at Leyden before going into business as a
lawyer, and kept a journal of his tour in Holland, which, like
other writings of the Clerk family, is furnished with pen and
ink sketches of what he saw. His letters are interesting from
the combination of earnest, Covenanting piety, with a gay and
chivalrous bearing in what was evidently the one serious love-
passage of his life. His letters to his wife in the years after
their marriage are as full of tenderness as that in which he
makes his first proposal is instinct with old-world gallantry.

"They left an only daughter, Dorothea, who married her
first cousin, afterwards Sir George Clerk Maxwell."

This George Clerk Maxwell probably suffered a little from
the world being made too easy for him in early life. Such a misfortune was all but inevitable, and the Baron seems to have done
his best to obviate it by good counsel; but the current of circumstances was too strong. George was in the habit of preserving
letters, and from those received by him before succeeding to
Penicuik it is possible to form a tolerably full impression of the
man. In some respects he resembled John Clerk Maxwell, but
certainly not in the quality of phlegmatic caution. His imagination seems to have been dangerously fired by the "little
knowledge" of contemporary science which he may have picked up
when at Leyden with his elder brother James (see p. 19, ll. 18, 25).
We find him, while laird of Dumcrieff, near Moffat, practically
interested in the discovery of a new "Spaw," and humoured in this
by his friend Allan Ramsay, the poet :—by and by he is deeply
engaged in prospecting about the Lead Hills, and receiving
humorous letters on the subject from his friend Dr. James
Hutton, one of the founders of geological science, and author of
the *Theory of the Earth*.[1] After a while he has commenced active
operations, and is found making fresh proposals to the Duke of
Queensberry. Then to the mines there is added some talk of

---

[1] See "Biographical Account of the late Dr. James Hutton, F.R.S.,
Edinburgh," read by Mr. Playfair, in *Transactions of the Royal Society
of Edinburgh*, vol. v. He made a tour in the north of Scotland in
1764, "in company with Commissioner, afterwards Sir George, Clerk,
in which Dr. Hutton's chief object was mineralogy, or rather geology,
which he was now studying with close attention."

a paper manufactory, and the Duchess playfully "congratulates" his "good self on every new sprouting up manufacture by means of so good a planter and planner." But by this time it has been found advisable to add to his studies some more certain source of income, and he applies for the Postmastership, and (through the Duke of Q.) obtains an office in the Customs. In this, as in all relations of life, he seems to have won golden opinions. And ere he succeeded to Penicuik, the loss of Middlebie proper and Dumcrieff had doubtless taught the lesson of prudence which his father the Baron had vainly tried to impress upon his youth. The friendship of Allan Ramsay and the affectionate confidence of the "good Duke and Duchess of Queensberry," sufficiently indicate the charm which there must have been about this man.

"Sir John Clerk (the Baron of Exchequer) married Janet Inglis of Cramond, and was succeeded by his eldest son, Sir James Clerk, in 1755. Sir James died without children in 1782, and was succeeded by his next brother, Sir George Clerk, who had married his cousin, Dorothea Maxwell, heiress of Middlebie."

The betrothal of George Clerk and Dorothea Clerk Maxwell is said by the Baron (in a special memorandum) to have been in accordance with her mother's dying wish. (Dorothea was seven years old when Agnes Maxwell died!) They were married privately when he was twenty and she was seventeen, but "as they were too young to live together," he was sent to join his brother James at Leyden. After a short interval the marriage was declared, and they lived very happily at Dumcrieff.

"Sir George Clerk Maxwell was a Commissioner of Customs, and a trustee for the improvement of the fisheries and manufactures of Scotland. His brother, John Clerk of Eldin, was the author of the well-known work on *Naval Tactics*. He was the father of John Clerk, a distinguished lawyer in Edinburgh, afterwards Lord Eldin, a Lord of Session."

The following account of Mr. Clerk of Eldin is from the above-mentioned biographical notice of Dr. James Hutton :—[1]

---

[1] Edinburgh Royal Society's *Transactions*, vol. v., part iii., p. 97.

" Mr. Clerk of Eldin was another friend with whom, in the formation of his theory, Dr. Hutton maintained a constant communication. Mr. Clerk, perhaps from the extensive property which his family had in the coal-mines near Edinburgh, was early interested in the pursuit of mineralogy. His inquiries, however, were never confined to the objects which mere situation might point out, and through his whole life has been much more directed by the irresistible impulse of genius, than by the action of external circumstances. Though not bred to the sea, he is well known to have studied the principles of naval war with unexampled success ; and, though not exercising the profession of arms, he has viewed every country through which he has passed with the eye of a soldier as well as a geologist. The interest he took in studying the surface no less than the interior of the earth ; his extensive information in most branches of natural history ; a mind of great resource and great readiness of invention ; made him to Dr. Hutton an invaluable friend and coadjutor. It cannot be doubted that, in many parts, the system of the latter has had great obligations to the ingenuity of the former, though the unreserved intercourse of friendship, and the adjustments produced by mutual suggestion, might render those parts undistinguishable, even by the authors themselves. Mr. Clerk's pencil was ever at the command of his friend, and has certainly rendered him most essential service."

In another place the same writer says :—

" Some excellent drawings were made by Mr. Clerk, whose pencil is not less valuable in the sciences than in the arts." [1]

John Clerk, Esq., of Eldin, fifth son of Sir John Clerk of Penicuik, and brother of Sir George, the grandfather of John Clerk Maxwell, was the author of certain suggestions on Naval Tactics, which had the credit of contributing to the victory

---

[1] See the collection of his etchings printed for the Bannatyne Club in 1855, with a striking portrait of him.

gained over the French by Lord Rodney's fleet off Dominique in 1782. It is fair to add that Sir C. Douglas claimed to have independently hit on the new method of "breaking the enemy's line." But John Clerk never relinquished his pretensions to the merit of the discovery, and his book on Naval Tactics in the edition of 1804 [1] is a thick octavo volume, with many woodcut illustrations and diagrams. He used to sail his mimic navies, trying various combinations with them, on the fish-ponds at Penicuik.[2] His eldest son, John Clerk, afterwards Lord Eldin of the Court of Session, made a vivid and lasting impression on his contemporaries, and many anecdotes of legal shrewdness and caustic wit still live about his name.[3] A graphic description of him as the Coryphæus of the Scottish Bar occurs in *Peter's Letters to his Kinsfolk*, Letter xxxii.[4] Quite recently he has appeared again in Carlyle's *Reminiscences*,[5] as the one figure in the Edinburgh law-courts, which remained clearly imprinted on the historian's imagination.

Lord Eldin's younger brother, William Clerk, was the trusted

---

[1] There were in all five issues, viz., in 1781, 1790, 1797, 1804, and (with notes by Lord Rodney) in 1827.

[2] Miss Isabella Clerk is my authority for this.

[3] It has been often told how John Clerk, in an appeal before the House of Lords, when twitted for his pronunciation by Lord Mansfield,—"Mr. Clerk, do you spell *water* with a double *t?*"—replied, "Na, my laird, we spell *wăter* wi' yae *t*, and we spell mănners wi' twa *n's*." The following is less known :—A young advocate in the Court of Session had ventured, in the vehemence of pleading, to exclaim against some deliverance of the Court, "My lords, I am astonished," whereupon my lords contracted their judicial brows. But Clerk, who by this time was one of them, is reported to have said, "Dinna be ill at the laddie! it's just his inexperience. Gin he had kenned ye, my lairds, as lang as I hae, he wad no been astonished at onything ye might dae." It was his father who, on the occasion of Johnson's visit to the Parliament House, clapped a shilling in Boswell's hand with "Here's till ye for the sight o' your Bear!"

[4] "The very essence of his character is scorn of ornament."

[5] Vol. ii. p. 7. "The only figure I distinctly recollect and got printed on my brain that night was John Clerk, there veritably hitching about, whose grim, strong countenance, with its black, far-projecting brows and look of great sagacity, fixed him in my memory. . . . Except Clerk, I carried no figure away with me."

friend and comrade of young Walter Scott. And it may be
observed here, by the way, that the friend of both, John Irving,
was half-brother to Mrs. Clerk, John Clerk Maxwell's mother.[1]

"Sir George Clerk Maxwell died in 1784, and was succeeded by his eldest son, Sir John, who married Mary Dacre,
but died without children in 1798. His brother James
was a lieutenant in the East Indian Navy, and married Janet
Irving of Newton. He died in 1793, leaving two sons and
a daughter. The elder son, afterwards the Right Honourable
Sir George Clerk,[2] succeeded to Penicuik on the death of his
uncle, Sir John Clerk, in 1798; and the younger son John
succeeded to the property of Middlebie, which descended to
him from his grandmother, Dorothea, Lady Clerk Maxwell,
and took the name of Maxwell. He married in 1826
Frances, daughter of Robert Cay of Charlton, and had one
son, James Clerk Maxwell, born in July 1831, and died in
Nov. 1879."

The Maxwells of Middlebie are descendants of the seventh
Lord Maxwell, and, in the female line, of the Maxwells of
Spedoch in Dumfriesshire. Their early history, like that of
many Scottish families, is marked with strong lights and shadows,[3] and John, the third laird, on coming into his inheritance,

---

[1] Many notices both of William Clerk and of John Irving occur in
Mr. John C. Maxwell's Diary. The following are of some interest :—
"1846, Sa., Nov. 5.—Called on William Clerk, and sat a good while
with him. He was in bed, and not able to raise himself, but spoke
freely and well, though very weak and thin. His appetite quite good,
and eats plenty, but gains no strength. Dec. 12, Sa.—Called on W.
Clerk, and sat 45 minutes with him. Weak, but hearty to talk.
1847, Jan. 7.—Wm. Clerk died early this morning."

[2] For many years Member for Midlothian, and one of the most
prominent politicians of his day. He held office under Sir Robert
Peel, who had great confidence in the breadth and soberness of his
judgment. He was noted for his command of statistics, and of questions of weights and measures, etc. He had also considerable skill in
music, and was a good judge of pictures.

[3] The famous feud between the Maxwells and Johnstones is
celebrated in the old ballad, entitled "Lord Maxwell's Good-night."

found it considerably burdened. He made a great effort to
liberate the estate from encumbrances, and in 1722 was able to
entail his landed property, and make it inseparable from the
name of Maxwell. But he was by no means ultimately freed
from debts, and he transmitted these, together with the estate,
to his daughter Agnes, who married William Clerk. (See p. 17.)
Their daughter, Dorothea, was left an orphan at the age of seven,
and became the ward of Baron Clerk of Penicuik, who (in accordance, as he declares, with her mother's wish), betrothed her
to his second son George. They were married, privately at
first, in 1735 ; and, in order to simplify their affairs, the Baron
obtained a decree of the Court of Session, followed by an Act of
Parliament, by which (notwithstanding the entail of 1722)
Middlebie proper was sold to liquidate the Maxwell debts, and
was bought in for George Clerk, whose title was thus freed from
burdens, and, so far, from the conditions of the entail. At the
same time it was arranged that Middlebie should not be held
together with Penicuik. Whatever crafty views the Baron may
have had in this curious transaction were frustrated by the im-
prudence of George Clerk Maxwell. His mining and manu-
facturing speculations involved him so deeply that Middlebie
proper, which, by his father's act, was now at his individual
disposal, was sold in lots to various purchasers (of whom the
Duke of Queensberry was chief), and the hereditary estate was
reduced to something like its present limited area. This accounts
for the fact that Glenlair, the present seat of this branch of the
Maxwell family, is so far distant from the well-known town of
Middlebie, in Dumfriesshire.

CHAPTER II.

GLENLAIR—CHILDHOOD—1831 TO 1841.

THAT part of the old estate of Middlebie which remained
to the heirs of Maxwell was situate on the right or
westward bank of the Water of Orr, or Urr, in Kirk-
cudbrightshire, about seven miles from Castle-Douglas,
the market-town, ten from Dalbeattie, with its granite
quarries, and sixteen from Dumfries. It consisted
chiefly of the farm of Nether Corsock, and the moor-
land of Little Mochrum. But, before building, Mr.
Clerk Maxwell by exchange and purchases had
added other lands to these, including the farm of
Upper Glenlair. The site chosen for the house was
near to the march of the original estate, where a
little moor-burn from the westward falls into the Urr.
The two streams contain an angle pointing south-east,
opposite the heathery brae which hides the village of
Kirkpatrick Durham. There, on a rising ground above
the last descent towards the river and the burn, a
mansion-house of solid masonry, but of modest dimensions, had been erected. It was built of dark-gray
stone, with a pavement and a "louping-on-stane" of
granite before the front door. On the southward
slope, towards the burn, was a spacious garden-ground
and a plantation beyond it, occupying the den or

dingle on either side the burn, and coming round to westward of the house and garden, where it ended in a shrubbery, by which the house was approached from the north. On the eastward slope, towards the Water of Urr, was a large undivided meadow for the "kye" and the ponies. To the northward was a yard with a duck-pond, and some humble "offices" or farm-buildings, which were displaced by the new erection of 1843. At the foot of the meadow, near the mouth of the burn, was a ford with stepping-stones, where the bridge was afterwards to be built, and the regular approach to the completed house was to be constructed. But this was far in the future, for in his building projects the laird would not trench upon the resources that were needed for the land. At the foot of the garden a place was hollowed out in the bed of the burn, which has often proved convenient for bathing. The rocky banks of the Urr, higher up, were fringed with wood, and on the upland, on either side the moor, there were clumps of plantation, giving cover to the laird's pheasants, and breaking the force of the winds coming down from the hill of Mochrum (N.W.). Glenlair was the name ultimately appropriated to the "great house" of Nether Corsock.

Every detail of these arrangements had been planned by the laird himself, and may be said to have been executed under his immediate supervision. The house was so placed and contrived as to admit of enlargement; but, in the first erection of it, space was economised as in the fitting up of a ship. And while it was building, the owners were contented with still

narrower accommodation, spending one if not two
whole summers in what was afterwards the gardener's
cottage. For the journey from Edinburgh was no
light matter, even for so experienced a traveller as
John Clerk Maxwell. Coming by way of Beattock, it
occupied two whole days, and some friendly entertain-
ment, as at the Irvings of Newton (his mother's half-
sisters),[1] had to be secured on the way. Carriages,
in the modern sense, were hardly known to the Vale
of Urr. A sort of double-gig with a hood was the
best apology for a travelling coach, and the most
active mode of locomotion was in a kind of rough
dog-cart, known in the family speech as a "hurly."
A common farmer's cart has been seen carrying the
laird to church, or to a friend's hall-door.

Glenlair was in the parish of Parton, of which the
kirk is by Loch Ken. Mr. Clerk Maxwell was one
of the elders there. (The little church of Corsock,
about three miles up the Urr, was not yet thought
of; for it was built in 1838, and not completely
endowed until 1862.) About seven miles of hilly
road lead from the Urr to Loch Ken and the Dee,
which is reached at a point near Cross-Michael, about
half-way between New Galloway in one direction and
Kirkcudbright and St. Mary's Isle[2] in the other. On
the high ground between the Urr and the Dee is Loch
Roan, a favourite point for expeditions from Glenlair.

Æt. 2 yrs.   In a letter dated "Corsock, 25th April 1834" (the
10 mo.   Boy being then aged two years and ten months),

---

[1] See above, p. 4.
[2] The seat of the Earl of Selkirk.

Mr. Clerk Maxwell writes to Jane Cay, his sister-in-law, in Edinburgh :—

> To MISS CAY, 6 Great Stuart Street, Edinburgh.
>
> *Corsock, 25th April 1834.*
>
> This has been a great day in Parton. Your humble servant and his better half, Ned,[1] and Davie took their departure an hour and a half after screigh [2] for Parton, to appropriate the seats of the new kirk, which was successfully atchieved in four hours' work to the satisfaction of all concerned, by a grand assemblage of the Magnates of the Parochin, who adjourned to a feast at the Manse. Master James is in great go, but on this subject I must surrender the pen to abler hands to do justice to the subject.

Instead of attempting to paraphrase a mode of speech which must be studied in and for itself, I will now give the continuation, written on the same sheet of paper by the "abler pen."

> He is a very happy man, and has improved much since the weather got moderate ; he has great work with doors, locks, keys, etc., and "Show me how it doos" is never out of his mouth. He also investigates the hidden course of streams and bell-wires, the way the water gets from the pond through the wall and a pend or small bridge and down a drain into Water Orr, then past the smiddy and down to the sea, where Maggy's ships sail. As to the bells, they will not rust; he stands sentry in the kitchen, and Mag runs thro' the house ringing them all by turns, or he rings, and sends Bessy to see and shout to let him know, and he drags papa all over to show him the holes where the wires go through. We went to the shop and ordered hats and bonnets, and as he was freckling with the sun I got him a black and white straw till the other was ready, and as an apology to Meg

---

[1] The laird's horse.

[2] *i.e.* Daybreak ; but in the laird's vocabulary, 9 A.M.

said it would do to toss about; he heard me, and acts accordingly. His great delight is to help Sandy Frazer with the water barrel. I sent the fine hat to Mrs. Crosbie [1] for her babe, which is well bestowed. . . . You would get letters and violets by a woman that was going back to her place; the latter would perhaps be rotten, but they were gathered by James for Aunt Jane.

Not much dragging was needed, either then or afterwards, to get Mr. Clerk to explain any mechanism to "boy," and "show him how it doos." Henceforward this was the chief pleasure of his life, until the order was reversed, and the son took hardly less delight in explaining Nature's Mechanics to the father.

Before seeing this letter, I had been told by his cousin, Mrs. Blackburn, that throughout his childhood his constant question was, "What's the go o' that? What does it do?" Nor was he content with a vague answer, but would reiterate, "But what's the *particular* go of it?" And, supported by such evidence, I may hope to win belief for a reminiscence which I might else have shrunk from mentioning. I distinctly remember his telling me, during his early manhood, that his first recollection was that of lying on the grass before his father's house, and looking at the sun, and *wondering*. To which may be added the following anecdote, which has been communicated to me by Mrs. Murdoch,[2] the "Meg" of the preceding letter. "When James was a little boy of two years and a half old, I had given him a new tin plate to play with.

---

[1] Wife of the minister at Parton.
[2] She was distantly connected with the Liddells (above, p. 13).

Mrs Clerk Maxwell and "Boy".

Engraved by G. J. Stodart after William Dyce, R.A.

It was a bright sunny day; he held it to the sun, and the reflection went round and round the room. He said 'Do look, Maggy, and go for papa and mamma.' I told them both to come, and as they went in James sent the reflection across their faces. It was delightful to see his papa; he was delighted. He asked him, ' What is this you are about, my boy?' He said, ' It is the sun, papa ; I got it in with the tin plate.' His papa told him when he was a little older he would let him see the moon and stars, and so he did." Methinks I see the laird in those happiest days, standing on some moonlit night on the pavement at the door, pointing skywards with one hand, while the small astronomer is peering from a plaid upon the other arm, and the glad wife is standing by.[1]

In Mr. Dyce's picture of the mother and child we see this open-eyed loving intercourse with the visible universe already begun. And in the accompanying woodcut (p. 30), taken from a contemporary sketch of a " barn ball " at the Harvest-home of 1837, the boy of six years old (instead of looking at the dancer) is totally absorbed in watching the bow of the " *violino primo*," unshakably determined to make out " the go of that " some day or other. The spirit which after-

---

[1] There has been preserved amongst the Glenlair papers a chart of the celestial globe, cut out into constellations, which fit each other like the pieces of a child's puzzle. A round hole is made in the cardboard for every star, differing in size so as to show the magnitude of each. The whole is executed with laborious neatness, and it seems probable that we have here the means (whether purchased or made at home) whereby the configuration of the starry heavens was still further impressed on " Jamsie's " mind.

wards welcomed the acoustic discoveries of Helmholz was already at work.[1]

Kirn 1837

In still earlier childhood, when he returned from walking with his nurse, she had generally a lapful of curiosities (sticks, pebbles, grasses, etc.) picked up upon the paths through the wood, which must be stored upon the kitchen dresser till his parents had told him all about each one.   In particular, she re-

[1] From his earliest manhood his ear for music was remarkably fine and discriminating, a fact (though in strict accordance with " heredity ") which surprised some of those who had known him as a boy.   He has told me that he remembered a time when it was exquisitely painful to him to hear music.   This time must clearly have been subsequent to 1837.   The truth seems to be that his naturally keen perception of sounds was interfered with by a tendency to inflammation in the ear, which came to a crisis in his sixteenth year, but that having outgrown this together with other signs of delicacy, his powers in this respect also were developed with striking rapidity.   On the other hand, his shortsightedness seems hardly to have been noticed till he was fourteen or fifteen.

members his interest in colours—" that (sand) stone is red; this (whin) stone is blue." " But how d'ye know it's blue?" he would insist. He would catch insects and watch their movements, but would never hurt them. His aunt, Miss Cay, used to confess that it was humiliating to be asked so many questions one could not answer " by a child like that."

But the child was not always observing, or asking questions. Ever and anon he was engaged in *doing*, or in *making*, which he liked better still. And here his inventiveness soon showed itself. He was not long contented with " tossing his hat about," or fishing with a stick and a string (as in an early picture of Miss Cay's); but whenever he saw anything that demanded constructive ingenuity in the performance, that forthwith took his fancy, and he must work at it. And in the doing it, it was ten to one but he must give it some new and unexpected turn, and enliven it with some quirk of fancy. At one time he is seated on the kitchen table, busily engaged in basket-making, in

which all the domestics, probably at his command, are also employed. At another he is "making seals"[1] with quaint devices, or improving upon his mother's knitting. For he must early have attained the skill, of which an elaborate example still exists, in "Mrs. Wedderburn's Abigail," which will be described in the next chapter, and was worked by him in his twelfth year.

Of his education in the narrower sense during this period little is known, except that his mother had the entire charge of it until her last illness in 1839, and that she encouraged him to "look up through Nature to Nature's God."[2] She seems to have prided herself upon his wonderful memory, and it is said that at eight years old he could repeat the whole of the 119th Psalm. His knowledge of Scripture, from his earliest boyhood, was extraordinarily extensive and minute; and he could give chapter and verse for almost any quotation from the Psalms. His knowledge of Milton also dates from very early times. These things were not known merely by rote. They occupied his imagination, and sank deeper than anybody knew.

But his most obvious interests were naturally out of doors. To follow his father "sorting" things about the farm, or "viewing" recent improvements; by and

---

[1] As mentioned in his letter to Miss Cay of 18th January 1840.

[2] When James, being eight years old, was told that his mother was now in heaven, he said, "Oh, I'm so glad! Now she'll have no more pain." Already his first thought was for another.

by (at ten years old) to ride his pony after his father's
phaeton ; to learn from the men " how to pickfork the
sheaves into the cart," [1] to witness a ploughing-match,
to slide on the Urr in time of frost (January 7, 1841),
to leap ditches, to climb trees " of sorts," to see them
felled and " have grand game at getting upon them
when falling," [1] to take wasps' nests on hot days in
July; [2] to blow soap bubbles and marvel at their changing hues ; to scramble up the bed of the eddying stream
that " flowed past the smiddy to the sea," and mark
the intricate tracery of holes and grooves which, in
rolling the shingle, it had worn and carved in the hard
rock; or to watch the same river in a " spate " rushing
and whirling over those " pots " which it had wrought,
and piling up the foam into mimic towers, like the
cumuli of the sky; or to gaze into the wan water
when in a milder mood, and drink in the rich brown
colour tinged with green reflections from the trees;—
such were some of his delights, as I may confidently
infer from what he loved to show me afterwards. For
in that constant soul the impression once made was that
which remained and went on deepening—" as streams
their channels deeper wear."   I well remember with
what feeling he once repeated to me the lines of Burns—

> " The Muse, nae poet ever fand her,
> Till by himsel he learn'd to wander ;
> Adown some trottin' burn's meander,
> An' no think lang." [3]

---

[1] From a letter of J. C. M. to his aunt, Miss Cay, of 18th January
1840 (*Æt.* 8½).

[2] At ten years old this *animosus infans* took four in one day.

[3] From the Epistle to William Simpson, the schoolmaster of Ochil-
tree, May 1785.

There were living companions of his solitude, besides those at home, for no live thing escaped his loving observation. And chief among these was that "child of the mossy pool," the frog,—nay, humbler still, the tadpole. The marvel of that transformation has engrossed many a child ; but in none, unless in some great naturalist, has it awakened such a keen, continuous interest. And it may be here observed, as a trait not to be dissociated from his intimacy with Nature, as one of her familiars, that Maxwell never had a " horror " of any creature. " Clean dirt " was a favourite expression, though no one was ever more cleanly. He would pick up a young frog, handle him tenderly, as loving him (not " as if he loved him ! "), listen for his scarcely audible voice ("hear him squeak !" he would say), put him into his mouth, and let him jump out again ! The movements of the frog in swimming were long a favourite study, and to jump like a frog was one of the pranks with which he astonished his companions, when he " put an antic disposition on " at school ; but of these there will be a time to speak hereafter. It was also at a later time that he was told of Galvani's discoveries ; but the recital had the more vivid interest for him because of this childhood fancy.

With eminent " judiciosity," Mr. Clerk Maxwell had furnished his son with a leaping-pole. This long staff, which appears in many of the early drawings, had at least one excellent effect. Few civilised men have had such perfect use of hands and arms as Maxwell always had.—His hand was the model of a hand, at once effective and refined-looking.—Thus equipped, he went across country anywhere and everywhere, with an eye for all he saw, and pluck enough to meet any emergency. One who knew him as a child, and who is fond of animals, says that he was extraordinarily "game." His endurance, both physical and mental, was always most remarkable.

He was sometimes taken to share in the simple daylight festivities of the neighbourhood (perhaps also to the New Year's gathering at Largnane, where the gifts were dispensed by a Fairy from her grotto), and it is still remembered, how at an archery picnic,[1] when an elaborate pie from Glenlair was being opened by the member of another house, the sturdy scion of Middlebie, who had not yet learned the meaning of Erănos,[2] and had doubtless been at the making and baking, bounded over the cloth just laid upon the turf, and laid his hand upon the dish, crying eagerly, "That's oor's!"   "That's oor pie!"

His resources on wet days were—first, reading voraciously every book in the house, except what his mother kept out of his reach; and secondly, drawing, which was begun at a very early age. Not that he ever showed the highest order of artistic talent (though his young performances are full of spirit); but he had great accuracy of eye, and any singular arrange-

---

[1] About 1838. This was graced by the presence of three persons then in the fulness of life, who were not destined to outlive the next fifteen years—Mrs. Clerk Maxwell, Miss Dyce, called the Pentland Daisy (Mrs. Robert Cay), and Isabella Wedderburn (Mrs. Mackenzie).

[2] Ἔρανος, i.e. a feast to which all contribute, and which all share. See below, chap. xii.

ment either of form or colour had always a fascination for him. Besides his mother's knitting, already
mentioned, his Aunt Jane's Berlin-wool work, and her
landscape drawings, early set him inventing curious
patterns and harmonising colours. And there were
two other frequent visitants at Glenlair, whom it is
now time formally to introduce. These were Mr.
Clerk Maxwell's sister Isabella, the widow of James
Wedderburn, Esq., at one time Solicitor-General for
Scotland, and her daughter Jemima, who was still a
young girl, though somewhat older than her cousin.

Mrs. Wedderburn had been an ornament of Edinburgh society in the days of her youth, combining
beauty of an elegant and piquant kind with great
sprightliness and originality, and the staunchest
loyalty to her kin. In spite of her early widowhood
and of some long illnesses, she retained much of her
spirit, together with her erect, lightsome figure, to the
last, and danced a reel at James's wedding with the
utmost sprightliness though at the age of seventy.
Her daughter, now Mrs. Hugh Blackburn, was only
eight years older than her cousin James; but her rare
genius for pictorial delineation, especially of animals,
was already manifest. It is obvious how this companionship of genius must have influenced the child's
indoor pursuits.

A scientific toy had recently come into vogue, an
improvement on the thaumatrope, called variously
by the names "phenakistoscope," "stroboscope," or
"magic disc." Instead of turning on its diameter, as
in the thaumatrope, it was made to revolve on a

transverse axis, before a mirror, at which the eye
looked through apertures cut at equal distances near
the rim of the disc. And the figures drawn upon it
were so contrived, by being placed in carefully
graduated positions, as thus to produce the impression
of a continuous movement.[1] This was a source of
endless amusement to the two cousins, the younger
generally contriving, and in part executing, the elder
giving life and spirit to the creatures represented.
Through Mrs. Maxwell's kindness, I have in my possession some of these early works, in which the
ingenuity of the contriver is everywhere manifest,
the hand of the artist only here and there. The cow
jumping over the waxing and waning moon, the dog
pursuing the rat in and out of his hole, the circus
horse, on which the man is jumping through the hoop,
have the firmness and truth of touch, the fulness of
life, familiar to the many admirers of J. B. The
tumbler under the horse's feet; the face in which the
pink and white, drawn separately, are made to blend;
the tadpole that wriggles from the egg and changes
gradually into a swimming frog; the cog-wheels moved
by the pendulum, and acting with the precision of
clockwork (showing, in fact, the working of an escape-
ment);—these display, with less power of execution,
the quaint fancy and observation, and the constructive

---

[1] This was afterwards developed into the " zoëtrope " or " wheel
of life,"—how far through suggestions of Clerk Maxwell's, or otherwise,
I am unable to say. The lenses which perfect the illusion were
certainly added by Maxwell himself. See Problem xx. of Cambridge
Tripos Examination, 1869, Thursday, January 7.

ingenuity of the young Clerk Maxwell.   There are
also intricate coloured patterns, of which the hues
shift and open and close as in a kaleidoscope.   I would
not venture to affirm that all these belong to the very
earliest period.[1]   But I believe that the magic disc was
in full operation before 1839, and that it has a real
connection not only with the "wheel of life," but also
with the "colour-top" of after years.

Another playmate and partaker of his whims must
be remembered here.   This was a terrier of the
"Mustard" kind, called Toby, Tobin, Tobs, or Tobit,
according to the moment's humour.   Toby was always

[1] As late as Feb. 27, 1847, on a Saturday half-holiday, when kept
in by weather, he was employed with his father in making a magic disc
for a young cousin.

learning some new trick (performed for his wages of
home-made biscuit after dinner), and neither he nor
James were ever tired of repeating the old ones. To
mention this is not mere trifling, for his power over
animals and perception of their ways[1] was a permanent
characteristic, and he found a scientific use for it at
a later time in inspecting the eyes of dogs with a
view to certain optical investigations.[2]

He does not seem to have been particularly fond of
riding at an early age, though in later life it was his
favourite recreation; but at ten years old, as has been
said, he used to ride his pony behind his father's
"phaeton."

Lastly, amongst the constant surroundings of the
boy's early years, the "vassals" must not be forgotten. Davie M'Vinnie and his family—Sandy Fraser
the gardener, and his—the Murdochs, who were the
kindred of Maggie, James's nurse,—were the objects of
a continuous kindly interest and friendly companionship, which had a genial effect on the heart of the
child. And by those of them whom I have been able
to see, he is still remembered, notwithstanding many
years of inevitable silence, with undoubting affection.
The very names of the places where they lived are
suggestive of quaintness and singularity, as were
most things in the Galloway of that day, where it

---

[1] "He seemed to get inside them more than other people."—J. B.

[2] "Coonie," a favourite terrier in later years, had a trick of howling unmercifully whenever the piano was played. He was completely
cured of this, and Maxwell told a friend, in his grave way, that he
had taken "Coonie" to the piano and explained to him how it went.
That was all.

was supposed that the devil had come after the Creation, with the riddlings of the universe, and had begun
" couping his creels " at Screels, till creel and all fell at
Criffel. Tor-holm, Tor-knows, Tor-brae, Paddock Hall,
Knock-vinney, the Doon of Urr, High Craigs of
Glenlair—such were some of the immediately surrounding names, sorting well with the homely, yet
unusual scenery, and with the picturesque Gallowegian dialect, which, like everything that struck the
boy's fancy, laid a strong and lasting hold upon his
mind.

In speaking of his own childish pursuits, it is
impossible not to recall the ready kindliness with
which, in later life, he would devote himself to the
amusement of children. There is no trait by which
he is more generally remembered by those with whom
he had private intercourse ; and, indeed, in this also
it appears that the boy was father to the man. Behold
him at the age of twelve, with his father's help, good-

naturedly guiding the constructive efforts of a still younger boy.[1]

As the months went on after Mrs. Maxwell's death, the question of education began to press. The experiment of a tutor at home, which had been tried in the autumn preceding that event, was continued until November 1841, but by that time had been pronounced unsuccessful. The boy was reported slow at learning, and Miss Cay after a while discovered that the tutor was rough. He was probably a raw lad, who having been drilled by harsh methods had no conception of any other, and had failed to present the Latin grammar in such a way as to interest his pupil. He had, in short, tried to coerce Clerk Maxwell. Not a promising attempt! Meanwhile the boy was getting to be more venturesome, and needed to be—not driven, but led. And it may be conceded for the tutor's behoof that, when once taken the wrong way, his power of provocation must have been, from a certain point of view, "prodigious." A childhood without some naughtiness would be unnatural. One evening at Glenlair, just as the maid-servant was coming in with the tea-tray, Jamsie blew out the light in the narrow passage, and lay down across the doorway.

There is one of Mrs. Blackburn's drawings, which throws a curious light on the situation at this junc-

---

[1] The scene is at St. Mary's Isle, and the younger boy is Lord Charles Scott, then four years old (September 1843).—Mr. A. Macmillan, the publisher, in particular, has a vivid recollection of Maxwell's ingenious ways of entertaining children, exhibiting his colour-top, showing them how to make paper boomerangs, etc.

ture.   Master James is in the duck-pond, in a wash-
tub, having ousted the ducks, to the amusement of the
young "vassals," Bobby and Johnny, and is paddling
himself (with some implement from the dairy, belike),
out of reach of the tutor, who has fetched a rake,
and is trying forcibly to bring him in.   Mr. Clerk
Maxwell has just arrived upon the scene, and is looking on complacently, though not without concern.
Cousin Jemima has been aiding and abetting, and is
holding the leaping-pole, which has probably served
as a boat-hook in this case.

Glenlair
1841

The achievement of sailing in the tub was one in
which James gloried scarcely less than Wordsworth's
Blind Highland boy in his tortoise shell.   It is
referred to in the following letter, written by the boy
of ten years old to his father, who had gone for a
short visit to St. Mary's Isle :—

Glenlair House,
[Friday], 29th October 1841.

DEAR PAPA—We are all well.   On Tuesday we [1] sailed in the tub, and the same yesterday, and we are improving, and I can make it go without spinning ; [2] but on Wednesday they were washing, and we could not sail, and we went to the potatoes.   Yesterday they took up the Prince Regents, and they were a good crop.   Mr. —— and I went to Maggy's, but she was away at Brooklands, and so I came back and sailed myself, for Nanny said Johnny was not to go in, and Bobby was away.   Fanny was there, and was frightened for me, because she thought I was drowning, and the ducks were very tame, and let me go quite close to them.   Maggy is coming to-day to see the tubbing.   I have got no more to say, but remain, your affectionate son,

JAMES CLERK MAXWELL.

The episode of the tutor was not a happy one.   I would omit the fact, as well as the name, were I not convinced that this first experience of harsh treatment had effects which long remained,—not in any bitterness, though to be smitten on the head with a ruler and have one's ears pulled till they bled might naturally have operated in that direction,—but in a certain hesitation of manner and obliquity of reply, which Maxwell was long in getting over, if, indeed, he ever quite got over them.

---

[1] From the context "we" seems to include "Bobby," one of the young "vassals."

[2] To enable him to "trim the vessel," he had put a block of wood in the centre.   Sitting on this, and tucking his legs on either side, he could paddle about steadily and securely.   Mrs. Blackburn tells me that years afterwards at Ruthven, in Forfarshire, being desirous of inspecting a water-hen's nest on a deep pond, where there was no boat, she adopted the same method, and made the voyage both ways alone without the slightest uneasiness.

The young Spartan himself made no sign that his relations with his tutor were otherwise than smooth. But something awakened the suspicions of his aunt, Miss Cay,[1] and the discovery was then made. It was this, I believe, which finally roused Mr. Clerk Maxwell to carry into effect what had been for some time in debate, viz. to take his son to Edinburgh and put him to school.

---

[1] Two letters from Miss Cay to James, one dated Sept. 1, the other October 21, 1841, were kept by Mr. Clerk Maxwell, and have been preserved. Although not significant enough to be inserted here, they show the confidential intercourse which had sprung up between "Aunt Jane" and her sister's son. She writes of theological and other matters which would not generally be thought interesting to a boy of ten, thanks him for his thoughtfulness in getting ferns for her, and says, "I was glad to hear you were happy, with all your experiments and adventures." There had been a visit to Edinburgh that summer, and she writes as if anticipating that it would be soon repeated. There is also a reference to an elaborate set of Berlin-wool work for the furniture of the drawing-room at Glenlair, which had been begun in Mrs. Maxwell's lifetime, and was afterwards completed by Miss Cay.

CHAPTER III.

BOYHOOD—EDINBURGH ACADEMY—" OLD 31 "—HOLIDAYS
AT GLENLAIR—1841 TO 1844—ÆT. 10-13.

THE first school-days are not always a time of progress. For one whose home life has been surrounded
with an atmosphere of genial ideas and liberal pursuits, to be thrown, in the intervals of " gerund-
grinding," amongst a throng of boys of average intelligence and more than average boisterousness, is not
directly improving at the outset. Not that Maxwell
ever retrograded—for his spirit was inherently active;
but where the outward environment was such as
awakened no response in him, he was like an engine
whose wheels do not bite—working incessantly, but
not advancing much. If the Scottish day-school
system had not still been dominated by a tyrannous
economy, and by that spirit of *laisser faire* which in
education is apt to result in the prevalence of the
worst, much that was in Maxwell would earlier have
found natural vent and growth. As it was, he was of
course storing up impressions, as under any circumstances he would have been ; but his activities were
apt for the time to take odd shapes, as in a healthy
plant under a sneaping wind. Or, to employ another
metaphor, the light in him was still aglow, but in

passing through an alien medium its rays were often refracted and disintegrated. The crowd of " aimless fancies," whose influence upon his life he so touchingly deprecated at a later time,[1] were now most importunate; and, bright and full of innocence as they were, they produced an effect of eccentricity on superficial observers which he afterwards felt to have been a hindrance to himself. His mother's influence, had she lived, would have been most valuable to him at this time.

The journey from Glenlair had been broken at Newton [2] and at Penicuik, where a halt of some days was made. It was the middle of November, and a

18ᵗ Nov: 1841

Edinburgh.

1841. Æt.
10. season of snow and frost. Soon after dusk on the 18th of November, the whole family party, including the

---

[1] In the poem written after his father's death in 1856.

[2] Above, p. 4.

faithful domestic, Lizzy Mackeand,[1] arrived at the door of No. 31 Heriot Row, Mrs. Wedderburn's house in Edinburgh. This (with occasional intervals, when he was with Miss Cay) was to be James Clerk Maxwell's domicile for eight or nine years to come.

The " White Horse " seen through a lunette above the doorway, the quaint figure of the butler (nicknamed " Hornie " from the way he dressed his hair [2]), and other noticeable features of this dwelling, appear and reappear in the boy's letters to his father, which now become more frequent. For, although not choosing to be much separated from James, Mr. Clerk Maxwell could not be long absent from Glenlair, and henceforward he lived a divided life between the two, spending most of the winter evenings by his sister's fireside in Edinburgh, and during most of the spring and summer attending personally to the improvement of his estate.

The Edinburgh Academy, which had been founded in 1824, was in high favour with the denizens of the New Town. Lord Cockburn was one of the directors. The Rector, Archdeacon Williams, was an Oxford first-classman, a College acquaintance of John Lockhart's,[3] and an admirable teacher. He had at one time been an assistant master at Winchester, and had subsequently, at Lockhart's recommendation, been tutor to Charles, Sir W. Scott's second son. The boys in the junior

---

[1] Now Mrs. MacGowan, Kirkpatrick-Durham.

[2] His real name was James Craigie.

[3] The inscription beneath his bust in Balliol College, Oxford, is a

classes, however, knew little of him except by report, for the assistant masters were jealous of their independence.

Various entries in his Diary testify to the father's deliberate care in placing his son at the Academy. Everything which seemed material to the boy's advantage had no doubt been carefully considered ; but there was one serious omission, arising from Mr. Clerk Maxwell's inveterate disregard of appearances. The boy was taken to school in the same garments in which we have seen him at Glenlair. No dress could be more sensible in itself. A tunic of hodden gray tweed is warmer than a round cloth jacket for winter wear, and the brazen clasps were a better fastening for the square-toed shoes than an adjustment of black tape, which is always coming undone. But round jackets were *de rigueur* amongst the young gentlemen ; while it must be admitted that they were

---

just tribute to the memory of one who, though he had his foibles, was a born educator, and no ordinary man :

JOHN WILLIAMS, M.A.,
Archdeacon of Cardigan—1835-1858 ;
Rector of the Edinburgh Academy—1824-1847 ;
Warden of Llandovery College—1848-1853 ;
Who, by the geniality of his Character
And the vigour of his Intellect,
Won the hearts of his Pupils,
And gave his life to the study of the Classics in Scotland :
A Celtic Scholar,
An ardent lover of Wales and of the Welsh People,
After a long absence,
He returned to his Native Land
And devoted his great talents
To the instruction of his Countrymen :
Born 1792 ; Died 1858.
He resided at this College
Between the years 1810 and 1814.

See also Lockhart's *Life of Scott*, small edition of 1871, pp. 484, 744, 781.

equally intolerant of dandyism. A frill for a round collar was of course unendurable, and the Gallovidian clasps—not to mention the square toes—were an unheard-of novelty. A new boy, coming in the second month of the second year, must in any case have had something to undergo; but here was evident provocation to "a parcel of boys in their teens."[1]

What happened in the interval after the first lesson (in the space behind the second classroom) is best indicated in the words of the Psalmist:—"They came about me like bees."

"Who made those shoes?" was the first question; but it was never easy to get a direct answer from Maxwell, least of all on compulsion. Brought thus to bay, he had recourse to his natural weapon—irony. His answer was soon ready, and his tormentors might make of it what they list. In the broadest tones of his Corsock *patois* he replied to one of them,

> "Div ye ken, 'twas a man,
> And he lived in a house,
> In whilk was a mouse."

He returned to Heriot Row that afternoon with

---

[1] "I do believe the veriest fien's
In the world is a parcel of boys in their teens."

Fo'c's'le Yarns.

The palinode should also be quoted:—

> "'Fiends' I called them, did I? Well,
> I shouldn then. It's hard to tell;
> And it's likely God has got a plan
> To put a spirit in a man
> That's more than you can stow away
> In the heart of a child. But he'll see the day
> When he'll not have a bit too much for the work
> He's got to do." *Ibid.*

E

his tunic in rags and "wanting" the skirt; his neat
frill rumpled and torn;—himself excessively amused
by his experiences, and showing not the smallest sign
of irritation. It may well be questioned, however,
whether something had not passed within him, of
which neither those at home nor his schoolfellows
ever knew.

The nickname of "Dafty" which they then gave
him clung to him while he remained at school, and he
took no pains to get rid of it. His "quips and
cranks" were taken for "cantrips," his quick, short,
elfin laughter (the only sign by which he betrayed his
sensitiveness), was construed into an eldritch noise.
Never was cygnet amongst goslings more misconstrued.
Within the class-rooms things were not much more
prosperous at first. Our master, Mr. A. N. Carmichael,
was a good and experienced teacher, and an excellent
scholar, in a dryish way. He was the author of the
Edinburgh Academy Greek Grammar and of an
Account of the Irregular Greek Verbs, which has now
been superseded, but was justly respected in its day.
He was a good disciplinarian; but those junior classes
of sixty and upwards were too large and miscellaneous
for real teaching. He had an eye for talent, too, where
it was shown. But his first business was to hear our
tasks, and to let us take places in the class in proportion to the accuracy and readiness with which we said
them. Maxwell did not at once enter into the spirit
of this contest, in which the chief requisites, next to
average talent and intelligence, were push and promp-
titude. His first initiation in Latin had not been

pleasant to him, and the repetition *ad nauseam* of "*di, do, dum,*" by his new acquaintances, varied with the sound of the tawse, did not make the subject more attractive. Like the boy Teufelsdröckh, he seemed to hear at school innumerable dead vocables, but no language. His hesitation got worse and worse, and as his place in the class was not amongst the "best boys," some of his neighbours willingly did their utmost to disconcert him. On one occasion we shall find him humorously retaliating. He was not in the least inwardly perturbed by all this, nor bore any one the slightest malice. It was a new scene of life, which he contemplated with amused curiosity. But it was natural that his chief interest should not lie there. He seldom took part in any games, though he was loyally proud of the success of his school in them, and characteristically took some interest in the spinning of "pearies" (pegtops), and the collision of "bools" (marbles); but, when he could, preferred wandering alone, sometimes imprisoning the humble-bees on the green slope at the back and letting them go again, sometimes doing queer gymnastics on the few trees that were left,—availing himself, in short, of the scanty inlets by which Nature visited that shingly ground.[1] For his heart was at Glenlair, even when he made sport for the young Philistines of the Academy "yards." It should be added that his attendance was a good deal interrupted by delicate health.

---

[1] Things are altered now. For years past there has been an ample recreation-ground provided for the boys of the Academy.

1842-43.
Æt. 10-12. His life during this period was really centred in "Old 31." He was presently allowed to have a room to himself, in which he could read and draw and write, besides preparing for school. His cousin Jemima was at this time learning the art of wood-cutting, and he was allowed sometimes to dig away with her tools. The result was a series of rude engravings, to which allusions occur in his letters to his father; and a woodcut of his, representing the head of an old woman, still remains, with the date 1843 engraved on it. In the previous year he produced more than one elaborate piece of knitting. One of these, a sort of sling for holding a work-basket, with its proper name, "Mrs. Wedderburn's Abigail," worked into it, has been preserved by Mrs. Blackburn. The library at his new home was more extensive than at Glenlair. He came to know Swift and Dryden, and after a while Hobbes, and Butler's *Hudibras*. Then if his father was in Edinburgh they walked together, especially on the Saturday half-holiday, and "viewed" Leith Fort, or the preparations for the Granton railway, or the stratification of Salisbury Crags; always learning something new, and winning ideas for imagin-

Æt. 10. ation to feed upon.[1] One Saturday, February 12, 1842, he had a special treat, being taken "to see electro-magnetic machines."

---

[1] Less frequently, he would rove about alone. Professor Fleeming Jenkin remembers hearing him say that when he first saw the twisted piles of candles with which grocers decorate their windows, he was struck by the curious and complex curves resulting from the combinations of these simple cylinders, and was resolved to understand all about that some day.

Mr. Clerk Maxwell was much more like an elder brother than a "governor" to James, and there was nothing the boy could not or did not tell him,—none of his whimsical vagaries in which the father did not take delight. And when "his papaship" was alone at Glenlair, James would strive to cheer him in his solitude by concocting the wildest absurdities, inventing a kind of cypher to communicate some airy nothing, illuminating his letters after the fashion of his school copy-books, and adding sketches of school-life (*e.g.* the class-room in the absence of the teacher), et cetera. This series of letters is so curious that it has been thought worth while to reproduce one of the least elaborate of them in facsimile. Some of those omitted are still more interesting, as showing his love of drawing complicated patterns and arranging colours, and as marking the early and spontaneous development of "the habit of constructing a mental representation of every problem,"[1] which was in some degree an hereditary proclivity. In order, however, to judge fairly of these *enfantillages*, the reader must take into account the boy's affectionate solicitude to amuse his father, who was accustomed to receive whimsical familiarities from his young relatives in "Old 31."[2]

1842-43.
Æt. 10-12.

---

[1] Professor Tait. In the letter of January 18, 1840 (above, p. 33), in his ninth year, when, in speaking of his amusement of seal-engraving he says, "I made a bird and a beast," the words "bird" and "beast" are each accompanied with a sort of hieroglyphic representing the figure he had made upon the seal.

[2] This is clearly proved by a set of delightful rhyming epistles addressed to him by his niece, Isabella Wedderburn, afterwards Mrs. Mackenzie, then a bright young girl, between the years 1825 and 1827.

In Edinburgh, as at Glenlair, he was allowed to participate in the amusements of his elders. It is just worth mentioning that his first play was *As You Like It*, with Mrs. Charles Kean as Rosalind ; and more important to observe that on December 18th, *Æt.* 12. 1843, his father took him to a meeting of the Edinburgh Royal Society.

But at school also he gradually made his way. He soon discovered that Latin was worth learning, and the Greek Delectus interested him, when we got so far.[1] And there were two subjects in which he at once took the foremost place, when he had a fair chance of doing so ; these were Scripture Biography and English. In arithmetic, as well as in Latin, his comparative want of readiness kept him down.

On the whole he attained a measure of success which helped to secure for him a certain respect, and, however strange he sometimes seemed to his companions, he had three qualities which they could not fail to understand—agile strength of limb, imperturbable courage, and profound good nature. Professor James Muirhead remembers him as " a friendly boy, though never quite amalgamating with the rest." And another old class-fellow, the Rev. W. Macfarlane of Lenzie, records the following as his impression :—" Clerk Maxwell, when he entered the Academy, was somewhat rustic and somewhat eccentric. Boys called him 'Dafty,' and used to try to make fun of him. On one occasion I remember he turned with

---

[1] The *Academy Greek Rudiments* was purchased before leaving Edinburgh for the holidays, July 28, 1842.

tremendous vigour, with a kind of demonic force, on his tormentors. I think he was let alone after that, and gradually won the respect even of the most thoughtless of his schoolfellows."[1]

It was on some such occasion as that to which Mr. Macfarlane here refers,—somewhere in 1843 or 1844,—that my own closer intimacy and lifelong friendship with James Clerk Maxwell began. I cannot recall the exact circumstances, only the place in the Academy yards, the warm rush of chivalrous emotion, and the look of affectionate recognition in Maxwell's eyes. However imperturbable he was, one might see that he was not thick-skinned.

Shortly after this we became near neighbours, my mother's new domicile being 27 Heriot Row, and we were continually together for about three years.

His letters now refer with more of interest to his progress at school, especially to exercises in verse, and to outdoor recreation with companions; above all to his delight in bathing and in learning to swim. In this, as in everything he did, he invented curious novelties, and was particularly fond of mimicking his old acquaintance, the frog.

On Sundays he generally went with his father to St. Andrew's Church (Mr. Crawford's) in the forenoon, and, by Miss Cay's desire, to St. John's Episcopal Chapel in the afternoon, where, also by her desire, he was for a time a member of Dean Ramsay's catecheti-

---

[1] " One hautboy will," etc. From an entry in his father's Diary of May 19, 1847, it appears that he was even then not free from annoyance. And I can bear witness to the fact.

cal class. Thus, having of course learned "his questions " as a child, he became equally acquainted with
the catechisms both of the Scotch and of the English
Church, and with good specimens of the Presbyterian
and Episcopalian styles of preaching. He also went
regularly " to the dancing " at Mr. MacArthur's, where
he was distinguished for the neatness of his reel-steps,
especially of those curious ones which some of us found
most difficult, such as the "lock-step."

But more delightful than bathing, and more
interesting even than writing English verse, was the
achievement of which he writes casually to his father
June 1844. very shortly after his thirteenth birthday.   After
Æt. 13. describing the Virginian Minstrels, and betwixt inquiries after various pets at Glenlair, he remarks, as if
it were an ordinary piece of news, " I have made a
tetrahedron, a dodecahedron, and two other hedrons,
whose names I don't know."   We had not yet begun
geometry, and he had certainly not at this time
learnt the definitions in Euclid ; yet he had not
merely realised the nature of the five regular solids
sufficiently to construct them out of pasteboard with
approximate accuracy, but had further contrived
other symmetrical polyhedra derived from them,[1]
specimens of which (as improved in 1848) may be
still seen at the Cavendish Laboratory.

Who first called his attention to the pyramid,
cube, etc., I do not know. He may have seen an

---

[1] By producing the facets until their alternate planes intersected.
In the specimens still extant, the facets belonging to each plane of the
original polyhedron are distinguished by specific colouring.

account of them by chance in a book. But the fact
remains that at this early time his fancy, like that of
the old Greek geometers, was arrested by these types
of complete symmetry; and his imagination so
thoroughly mastered them, that he proceeded to
make them with his own hand. That he himself
attached more importance to this moment than the
letter indicates, is proved by the care with which he
has preserved these perishable things, so that they (or
those which replaced them in 1848) are still in existence after thirty-seven years.

LETTERS, 1842 to 1844.

[*April 1842.*]

MY DEAR PAPA—The day you went away Lizzy and I   *Æt.* 10.
went to the Zoological Gardens, and they have got an
elephant, and Lizzy was frightened for its ugly face.   One
gentleman had a boy that asked if the Indian cow was he.

Asky [1] thinks he is a scholar, and was for going with me
to the school, and came into the dancing to-day.

On Friday there was great fun with Hunt the Gowk; [2]
we could believe nothing, for the clocks were all "stopped,"
and everybody had a " hole in his jacket."   Does Margaret [3]
play on the trump [4] still ? and what are the great works ?
Does Bobby sail in the tub ?—I am, your obedient servant,
JAMES CLERK MAXWELL.

[*Illuminated letters at beginning, and border after.*]   *Æt.* 11.

MY DEAR MR. MAXWELL—I saw your son to-day, when
he told me that you could not make out his riddles.   Now,

---

[1] Pet name of a dog.   See illustration on p. 18.
[2] Scotch name for the license of April Fool's day.
[3] A daughter of his nurse.   [4] The Jew's harp.

if you mean the Greek jokes, I have another for you.   A
simpleton wishing to swim was nearly drowned.   As soon
as he got out he swore that he would never touch water
till he had learned to swim; but if you mean the curious
letters on the last page, they are at Glenlair.—Your aff.
Nephew,   JAMES CLERK MAXWELL.

I have cut a puggy[1] nut, and some of the oil came upon
my fingers, and it smelt like linseed Oil, but it did not hurt.
There was a boy that brought Sea fyke[2] to the school, and
put it down the boys' backs, for which he was condemned
to learn 12 lines for 3 days.   Talking about places, I am
14 to-day, but I hope to get up.   Ovid prophesies very
well when the thing is over, but lately he has prophesied a
victory which never came to pass.   I send you a Bag-
piper to astonish the natives with.[3]   I have got a jumping
paddock and a boortree gun.[4]   When are you coming?—
Your most obedient Sarvent,

                        JAS. ALEX. M'MERKWELL.[5]

Æt. 12.   *Envelope of 24th June 1843—*
                MR. JOHN CLERK MAXWELL,
                        Postyknowswere,
                            Kirkpatrick Durham,
                                Dumfries.

---

[1] *i.e.* Cashew nut.   This is characteristic.   Compare the description
of the boy-genius in *Fo'c's'le Yarns* :—

> "Bless ye ! the pisons was just like mate
> To Tommy, that liked to feel the strong
> They were, and rowlin' them on his tongue."

[2] A substance often found on the sea-shore.   It is of a honey-combed
structure, and consists, in fact, of the egg-capsules of the common
whelk (*Buccinum undatum*).   When dried and pulverised, it has an
irritating effect upon the skin.   Hence the local name :—"Fyke"=
fidget.   See Jamieson's *Dictionary*.

[3] This fantastic and elaborately-coloured illustration is certainly
sufficiently astonishing.   [4] *i.e.* A pop-gun of elder-wood.

[5] Anagram of James Clerk Maxwell.

My Dear Sir.

At the Indian lecture we heard Mr Catlin give a lecture as the bill says in his American dress in which he told us how these people killed the bufaloes with a bow

Mr Catlin killing the Buffalo

how his dress was like that of the Romans

At the theatre M Gouffe did wonders

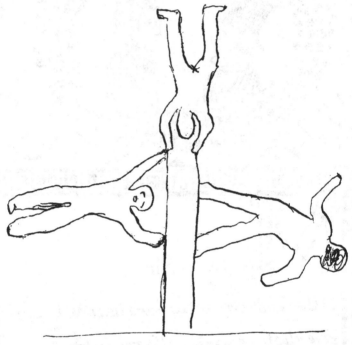

and Miss Ternan was very funny I have finished
the red bag Mrs Tis said you wrote Ill and Ungram
matically but I said you had been taught at the High
School when people did not know any better
The rest to be read backwards
through the paper

I am
Your most obedient servant

Guess who?

3ᵈ May 1843.

"Old 31," 28th March 1844.

MY DEAR FATHER—On *S*aturday last w*e* went to the Mar*I*ne Vill*A*;[1] it had a very strong M*a*rine scent, but I suppose it is all the better *f*or that.   I found out where shellfish breed; the*y* breed in sea fy*k*e; there were muscles, *c*ocles, and *o*ysters no b*i*gger than the*s*e O O O O f*a*stned to the fyke by filaments.   Nell and Frolic were immersed in the serene bosom of Ne*p*tune, from which with still quiverin*g* limbs the*Y* came *o*ut, but w*I*th very different feeli*N*G*s*; but Nell exited the comp*A*ss*IO*N of Me*d*dum,[2] and was carried by her.   I have fl*i*t*t*ed up to the little garret. W*h*at l*i*ke is the *n*ew ta*D*r*o*le ? a*n*d how is M*a*g*g*y getting on with *f* mmm;[3] how much m*o*r*e* is to be done conce*R*n*I*n*G* O fye, says the *p*ie.   J*o*hn's house is not finished ye*T*, I suppose.   There have been *l*etters from *U*ncle Rober*T*, dated G*i*bber Altar, but I have n*o*t *s*een or *h*eard what *i*s i*n* them fa*r*THE*r* than that he was to be at Suez on Monday last.   Lizzy says that when you come back it woul*D* not b*E* disp*L*easin*g* to her i*F* you *w*ould bri*n*g a bawl of gray wors*t*ed, w*h*Ich l*A*st word I suppose mea*N*s woolen thre*A*d. I have cast three sea*L*s of *L*ead from the life, or rather from the death; one of a cockle and two of m*U*scles, one of which is, or ra*T*her will be, on this letter ‡.   If you want to know more look along from the beginning of the letter to the mark ‡ for the red and blue letters[4] in order.

How are all the bodies and beasts,—Praecipue, Nanny, Maggy, Fanny, Bobby, Toby, and Marco.—Your obt. servt.,

JAS. ALEX. M'MERKWELL.

---

[1] Silverknowe, near Granton, which was being prepared as a residence for Mr. and Mrs. Mackenzie (Isabella Wedderburn).

[2] Mrs. Wedderburn.   [3] The sound of the " trump."

[4] Here represented by italics and small capitals respectively.   The italics spell—*Sea fyke is a good thing for polishing with.*   The small capitals spell—I AM COPYING AN OLD PRINT OUT OF THE DELFIAN (*i.e.* Delphin) SALLUST.   It need not be observed that the capricious spelling in these letters is merely a piece of " daftness."   The spelling of the letters of 1841 (*Æt.* 9-10) is faultless.

[*19th June 1844.*]   " *Old 31.*"

MY DEAR FATHER—On Wednesday I went to the
Virginian minstrels, in which some of the songs were sung,
the first line accompanied with clappers, the second on a
tamborine, the third on a banjo, like this, . . . played like a
guitar very quickly, and the fourth on the fiddle, and the
chorus by all together.   There were guesses [1] in abundance ;
and there was an imitation of a steam onion, and other things
which you will find in the bill.   On Saturday, having got
the play for verses on Laocoon, I went with Cha. H. Johnstone [2] so far, and then went to the murrain vile till Mrs.
M'Kenzie, Ninny, and $κυνη$ [3] went to visit Cramond, where
I played with the boies till high water ; and the minister's
young brother and the too boies and I doukit in C (big sea
as $κυνη$ calls it), and then dried ourselves after the manner
of Auncient Greeks ; we had also the luxury of a pail of
water to wash our feet in.

How is a' aboot the house now our Gudeman's at home ?
How are herbs, shrubs, and trees doing ?—cows, sheep, mares,
dogs, and folk ? and how did Nannie like bonny Carlisle?
Mrs. Robt. Cay was at the church on Sunday.[4]   I have
made a tetra hedron, a dodeca hedron, and 2 more hedrons that I don't know the wright names for.   How do
doos and Geraniuns come on.—Your most obt. servt.

JAS. ALEX. M^CMERKWELL.

1 2 5   12 7 4 13   3 6 11 8 9 10 14 15 16 17

---

[1] *i.e.* riddles or conundrums, of which the boy was fond.

[2] A son of Admiral Hope Johnstone, then living at Cramond, a
scion of the Johnstones, who in 15— had a feud with the Maxwells,
but in later times claimed kinship with them.

[3] "*i.e.* Coonie," viz. Mr. Colin Mackenzie, then a child of three.

[4] This helps to fix the date of this letter. Mrs. R. Cay joined her
husband in China early in the spring of 1845. Her son Alexander was
born May 7, and christened on Wednesday, June 26, 1844. Mr. C. Maxwell had left Edinburgh for Glenlair on June 7, taking with him six
pigeons in a basket, and some cuttings of pelargonium. His first entry
in the Diary after this, at Glenlair, is as follows :—" *Saturday, June 8.*
—Got home to dinner, and find all well. After dinner plant cuttings
of pelargonium from Killearn, and sort the doves in the new dove-cot."

10th July 1844.

DEAR FATHER—Excuse me on account of being so long
of writing, because of my being totally employed about preparations of verses, English and Latin.   I made four lines of
Latin one week, for which I got the play from ten ; but I am
not going to try for the prize, as when I lithp in numberth
it ith but a lithp, for the numberth do not come even with
the help of Gradus ; but I am making English ones on the
apparition of Creusa to Æneas in the end of the second
book.   Besides this, I am preparing the biography,[1] and
have been making a list of the kings of Israel and Judah.   I
have been going to Cramond and playing with the boys
every Saturday ; they went to Rayhills on the ninth.
Dooking[2] is grown fine and warm now.

> O father ! can it be that souls sublime
> return to visit our terrestrial clime ? [3]

Your obt. servt. and son to you,

JAMES CLERK MAXWELL.

I have been wavering about 14 for a good while in the
Latin.

HOLIDAYS AT GLENLAIR—1842 to 1844.

We can readily imagine the sense of enlargement
and release with which the boy went home to Glenlair
after his first long sojourn in Edinburgh.   The

Now, every letter received from James is recorded in the Diary, and
the only such entry between the limits of June 7-26 is on June 21st.
" THE THREE PAIRS OF DOVES ALL SITTING.   Recd. letters from Mrs.
Wed. and James."   It was a two-days' post.   The letters dated July
10 was received on July 12.   The " Verses on Laocoon " point to the
same date.   The translation from Virg. *Æn.* i. 159-169, which certainly preceded them, was given in on May 10, 1844.

[1] *i.e.* Scripture History.   [2] *i.e.* Bathing.

[3] From the poem on " Creüsa."

shadow which had fallen in 1839 was softened by
time, and society in the Happy Valley[1] resumed its

---

[1] This name was given to the Vale of Urr in the *Coterie-Sprache*,

aspect of harmless gaiety. Cousin Jemima was again there with her pencil. The "tubbing" was, of course, resumed, this time conjointly, and the scene of it was advanced from the duck-pond to the river, showing greatly increased confidence in navigation.[1]

There were nutting excursions, walks diversified with climbing, etc. etc. And in August and September 1843 there were again archery meetings at different houses in the valley, of one of which (the last) there has been preserved the following notice from one of the local newspapers of the time :—

Æt. 12.

ARCHERY IN " THE HAPPY VALLEY."

The Toxophilite Club of the Valley of Urr held their last meeting for the season on Mrs. Lawrie of Ernespie's lawn, on Tuesday the 12th curt. The club consists of from forty to fifty members.

Their meetings this summer have been quite charming. They ranged over the whole valley, on this fair lawn to-day, and on that the next ; and after their couple of hours of archery was over, a picnic took place on the spot. " God save the Queen" was invariably sung with the most graceful loyalty ; and the hospitable mansion adjoining gave them music and a hall for the evening quadrille, which wound up the delights of the day.

At every meeting some little prize was proposed to give zest to the sport ; Mr. Herries of Spottes, for instance, gave a case of ladies' arrows, which was shot for and gained by one of the lady competitors. Nor lacks the club its Laureate and its Painter to glorify the pastime. A scion of

---

and adopted even by the local newspapers in their notices of various social gatherings.

[1] Long afterwards, when asked by some one ignorant of Galloway, if there was boating on the Urr, he would answer by a grave reference to this incident.

the House of Middlebie has lent gallantry to the archers by his spirited songs; and a fair lady, a friend of the same house, has painted a couple of pieces, and presented them, the one to Mr. Lawrence of Largnean, the president of the club, and the other to Mrs. Bell of Hillowton, the lady patroness. The former picture represents William Tell aiming at the apple on his son's head; the latter, the chaste huntress Diana piercing a stag. Both are "beautiful exceedingly."

Thus well accomplished in every point and accessory of their beautiful pastime, loyal and happy are the Bowyers of Urr.

One of James's spirited songs, a parody of Scott, beginning "Toxophilite, the conflict's o'er," still exists in Cousin Jemima's handwriting, with a sketch for the picture of William Tell, in which the features of the House of Middlebie are idealised. The artist also proved the best shot on this occasion. The poem is not worth printing, though it has characteristic touches of grotesque ingenuity and humorous observation which are very curious in a boy of twelve.

Æt. 12. The summer of 1843 was also memorable for the completion of the New Offices at Glenlair. Whatever he may have intended before the death of his wife, Mr. Clerk Maxwell made no change in the dwelling-house during his lifetime. But these out-buildings had been designed by himself; he had drawn the working plans for the masons; he had acted as clerk of the works, rejecting unfit material, etc., and every detail had been executed under his own eye. So absorbed was he in the supervision, that he omitted his usual visit to Edinburgh in July. In one of Mrs.

Blackburn's drawings of the previous year, he is seen laying out the ground for the new offices, with James beside him intently contemplating his father's work. We may be sure that Mr. Maxwell had explained every step in the whole procedure, and equally sure that his son laid the lesson well to heart.

Soon after this he was provided with a new source of endless amusement in the " devil-on-two-sticks," which thenceforth became inseparable from the home life at Glenlair, and the companion of his holidays at Glasgow and elsewhere, even in the Cambridge time. In the family dialect it was humorously referred to *sotto-voce* as " the deil." There was nothing he could not do with that d——l. No performer on the slack or tight rope ever made such intricate evolutions and gyrations. His delight in it was like that which afterwards he used to take in the dynamical top.

The boy now came to know his own neighbourhood more widely. There were expeditions, visits, rides. The Covenanter's pool in the burn above Upper Corsock, New Abbey, Caerlaverock, and other places of traditional interest, were explored. And in the summer of 1844 there was a sort of driving excursion into the Cairnsmuir country, which is described in detail in the Diary.

1843-44.
Æt. 12-13.

It may be mentioned here in a general way, that the Christmas holidays were spent either at Penicuik (with skating, etc.) or Killearn, and afterwards sometimes at Glasgow with Professor and Mrs. Blackburn or Professor (now Sir William) Thomson.

F

CHAPTER IV.

THE commencement of the fifth year at the Academy was, for many of us boys, a time of cheerfulness and hope. The long period of mere drill and task-work was supposed to be over. We had learned the 800 irregular Greek Verbs, either by our own efforts, or by hearing others say them, and had acquired some moderate skill in Latin verse composition. On entering the rector's class-room, our less mechanical faculties were at once called into play. We found our lessons less burdensome when we had not merely to repeat them, but were continually learning something also in school. And the repetition of Virgil and Horace was a very different thing from the repetition of the rules of gender and quantity. Some foretaste of this more genial method had been afforded us in the previous year, when we had been encouraged to turn some bits of Virgil into English verse. But the change was, notwithstanding, considerable, and it was accompanied with another advance, which for Maxwell was at least equally important, for it was now that we began the serious study of geometry.

In October 1844 Mr. Clerk Maxwell and his sister, Mrs. Wedderburn, were both far from well, and James was received in Edinburgh by his aunt, Miss Cay. He writes to his father, October 14, 1844:—

I like P———[1] better than B———.[2] We have lots of jokes, and he speaks a great deal, and we have not so much monotonous parsing. In the English, Milton is better than history of Greece. . . . I was at Uncle John's,[3] and he showed me his new electrotype, with which he made a copper impression of the beetle. He can plate silver with it as well as copper, and he gave me a ⌷ thing with which it may be done. At night I have generally made vases.

This letter is sealed with the scarabæus referred to as "the beetle."

In the next letter we have a trace of his hesitation not being yet conquered:—

P——— says that a person † of education never puts in † hums and haws; he goes † on with his † sentence without senseless interjections.

N.B.—Every † means a dead pause.[4]

While thus privately retorting on his censor, he took a singular means for curing his own defect. He made a plan of the large window in the rector's room, and wrote the words of the lesson in the spaces of the frame-work. He conned his task in that setting, and,

---

[1] The boys' nickname for the Rector, Archdeacon Williams. Maxwell's first interview with him was as follows:—Rector: "What part of Galloway do you come from?" J. C. M.: "From the Vale of Urr, Ye spell it o, err, err, or oo, err, err."

[2] Ditto for Mr. Carmichael.

[3] Mr John Cay, Sheriff of Linlithgow. See above, p. 7.

[4] These pauses in the Rector's case were often filled, in less guarded moments, with "What you call," "Yes, yes," which he had a trick of interposing.

when saying it, looked steadily at the actual window, where, as he averred, the arrangement of the panes then helped to recall the order of the words. The only fear was that by changing his place in the class he might be obliged to stand sideways to the window.

Our mathematical teacher, Mr. Gloag, was a man who combined a real gift for teaching with certain humorous peculiarities of tone and manner. He was sometimes impatient, but had a kind heart, and we liked him all the better because we mimicked him.[1] Old academicians still delight in talking of him. He never allowed us to miss a step in any proof, and made us do many " deductions," which we puzzled out entirely without help. It must have been the companionship of Maxwell that made those hours so delightful to me. We always walked home together, and the talk was incessant, chiefly on Maxwell's side. Some new train of ideas would generally begin just when we reached my mother's door. He would stand there holding the door handle, half in, half out, while,

> " Much like a press of people at a door
> Thronged his inventions, which should go before,"

till voices from within complained of the cold draught, and warned us that we must part.

From some mathematical principle he would start off to a joke of Martinus Scriblerus, or to a quotation from Dryden, interspersing puns and other outrages on language of the wildest kind, " humming and haw-

---

[1] He once said to a nervous boy who had crossed his legs and was sitting uneasily, " Ha, booy ! are ye making baskets wi' your legs ?"

ing " in spite of P——; or in a quieter mood he would tell the story of Southey's *Thalaba*, or explain some new invention, which I often failed to understand. Our common ground in those days was simple geometry, and never, certainly, was emulation more at one with friendship. But whatever outward rivalry there might be, his companions felt no doubt as to his vast superiority from the first. He seemed to be in the heart of the subject when they were only at the boundary ; but the boyish game of contesting point by point with such a mind was a most wholesome stimulus, so that the mere exercise of faculty was a pure joy. With Maxwell, as we have already seen, the first lessons of geometry branched out at once into inquiries which soon became fruitful.

" Meantime, the rural ditties were not mute." Besides a serio-comic impromptu on the grievance of a holiday task, and other effusions concerning incidents of our school life, there was a romantic ballad written about Christmas 1844 or 5, and in July 1845 the prize for English verse was gained by the poem on the death of the Douglas, to which he refers in one of his letters to Miss Cay.[1]

1844-46.
Æt. 13-15.

But a prize of more consequence was the mathe-

1845.
Æt. 14.

---

[1] His turn for versifying may be traced back to his twelfth year, when, in one of his quaint letters to his father, there occur some lines (profusely illustrated) on the death of a goldfinch :—

> "Lo ! Ossian makes Comala fall and die,
> Why should not you for Richard Goldie cry," etc.

And in September 1843, as above mentioned, he wrote for the Archers in the " Happy Valley " a page and more of spirited verse.

matical medal, of which he writes to his aunt in a tone of undisguised though generous triumph. The following letters were written in June and July 1845:—

To Miss Cay.

June 1845.

I have drawn a picture of Diana,[1] and made an octohedron on a new principle, and found out a great many things in geometry. If you make two circles equal, and make three steps with the compasses (of any size), and cut them out in card, and also three equal strips, with holes at each end, and joint them with thread to the upper side of one circle and the lower side of the other; then if you put a pin through the centre of one and turn the other, the one will turn, and if you draw the same thing on both in the same position,— if you turn them *ever so,*—they will always be in the same position.

July 1845.

The subjects for prizes are as follows:—English Verses— The gude Schyr James Dowglas; Latin Hexameters and Pentameters—The Isles of Greece; Latin Sapphics—The Rhine. I have been getting information in many books for Douglas, but I found it so difficult not to Marmionise, that is, to speak in imitation of *Marmion,*—that I am making it in eight syllable lines. I have got Barbour's *Bruce,* Buke 20, which is a help in a different language, which is all fair; my motto is:—

> "Men . may . weill . wyte . thouch . nane . them . tell.
> How . angry . for . sorrow . and . how . fell .
> Is . to . tyne . sic . a . lord . as . he .
> To . them . that . war . of . hys . mengye . "[2]

Pa and I went to your house on Saturday and watered the plants. I have got the lend of the whole of Horne's

---

[1] *i.e.* A sketch from the antique.
[2] Barbour's *Bruce,* B. xx., ll. 507-10.

The Vampyre,
Compylt into Meeter, by
Jas Clerk Maxwell.

Thair is a Knichts rydis through the wood,
  And a douchty Knichts is hee,
And rure hee is on a message sent,
  Hee rydis sae hartilie.
Hee passit the aik, and hee passit the birk,
  And hee passit monie a tree,
Bot plesant to him was the sough sae slim,
  For beneath it hee did see

The boniest Ladye that ever hes saw,
    Scho wes sae schynand fair,
And there scho sat, beneath the sough,
    Kaiming hir gowden hair.
And then the Knichte: "O Ladye brichte,
"What chance hes broucht you here,
"Bot say the word, and ye schall gang
"Back to your kindred dear."
Then up and spok the Ladye fair—
    "I have nae friends or kin,
"Bot in a littel boat I live,
    "Amidst the waves loud din".
Then answered thus the douchty Knichte—
    "I'll follow you through all;
"For gin ye bee in a littel boat,
    "The world to it seemis small."
They gaed through the wood, and through the wood,
    To the end of the wood they came.
And when they came to the end of the wood,
    They saw the salt seafaem.
And then they saw the wee wee boat
    That daunced on the top of the wave,
And first got in the Ladye fair,
    And then the Knichte sae brave.

They got intothe wee wee boat,
    And rowed wi' a thair micht;
When the Knichts sae brave hee turnit about,
    And lookit at the Ladye bricht,

Hee lookit at hir bonie cheik,
    And hee lookit at hir twa bricht eyne,
Bot hir rosie cheik growe ghaistly pale,
    And cho seymit as scho did had bene.

The fause fause Knichte growe pale wi frichte,
    And his hair rose up on ende,
For ganeby days cam to his mynde,
    And his former luve hee kenned

Then spake the Ladye, "Thou fause Knichte
"Hast dene to mee much ill,
"Thou didst forsake me longago
    Bot I am constant still.

For though I ligg in this moole sae calde,
    At rest I canna bee,
Until I sucke the gude lyfe bludo
    Of the man that gart me dee.

Hee saw hir lippo were wet wi bludo,
    And hee saw hir lyfe lesse syne,
And loud hee cryd, "get frae my syde
    "Thou Vampyr corps uncleane."

Bot no, hee is in hir magic boat,
    And on the wyde wyde sea,
And the Vampyr suckis his guid lyfe bludë,
    Scho suckis hym till hee dee.
So now beware, whose're you are,
    That walkis in this lone woodë;
Beware of that deceitfull spright,
    The ghaist that suckis the bludë.

Introduction to the Knowledge of the Scriptures, and Prideaux's *Connection of the Old and New Testaments,* and Townshend's *Harmony,* which are of great use.[1]

July 1845.   *Æt.* 14.

I have got the 11th prize for Scholarship, the 1st for English, the prize for English verses, and the Mathematical Medal. I tried for Scripture Knowledge, and Hamilton in the 7th has got it. We tried for the Medal[2] on Thursday. I had done them[3] all, and got home at ½ past 2; but Campbell stayed till 4. I was rather tired with writing exercises from 9 till ½ past 2.[4]

Campbell and I went "once more unto the b(r)each" to-day at Portobello. I can swim a little now. Campbell has got 6 prizes. He got a letter written too soon congratulating him upon *my* medal; but there is no rivalry betwixt us, as B—— Carmichael says.

His aunt, Miss Cay, to whom these letters are addressed, had begun again to take more charge of him than in the preceding years. Mrs. Wedderburn's health was very uncertain. Cousin Jemima was grown-up and immersed in her own pursuits, and the companionship of his cousin, George Wedderburn, a young man about Edinburgh, and a humourist of a different order, was not in every way the most suit-

---

[1] Viz. for a competition in Scripture Knowledge, which was open to the 5th, 6th, and 7th classes. The prize was gained that year by one of the 7th.

[2] "Mathematical Prize" is added between the lines by Mr. Clerk Maxwell, who writes a P.S. to the letter.

[3] "The trial exercises," ditto, ditto.

[4] In these competitions we seem to have been allowed to stay till we had done all we knew. Witness the following extract from the Diary :—"1847, *July, Mon.* 19.—James at Academy trial for Prize for Scripture Knowledge. Worked from 9 to 5. Lewis Campbell and W. Tait worked till 6."

able for the growing boy. The Diary shows that he
was continually at his aunt's house, No. 6 Great Stuart
Street, and she is associated with some of my earliest
recollections of him. She sought to bring him out
amongst her friends, to soften his singularities, and to
make him more like other youths of his age. And he
would help her with patterns, arrangement of colours,
etc., as well as with her flowers. One of his earliest
applications of geometry was to set right the perspective of a view of the interior of Roslin Chapel on which
she was engaged.

Mr. Clerk Maxwell was a frequent visitor at the
Academy at this time. His broad, benevolent face
and paternal air, as of a gentler Dandie Dinmont,
beaming with kindness for the companions of his son,
is vividly remembered by those who were our schoolfellows in 1844-5.

The summer vacation of 1845 was spent almost
wholly at Glenlair. James passed a day now and
then at Upper Corsock with the Fletcher boys, [1] and
sometimes accompanied his father when he went out
shooting ; but he must have had abundance of time
for reading and for following his own devices. The
country gentlemen were particularly absorbed that year
in political excitement, and Mr. Clerk Maxwell was
often called away. The only event worth mentioning
was a "jaunt," evidently suggested by Miss Cay, to
Newcastle, Durham, and Carlisle, which gave Maxwell

---

[1] Sons of Colonel Fletcher of Upper Corsock.

his first direct impression of English Cathedral Architecture.

The taste thus formed was strengthened by a visit to Melrose in the following summer.

Saw the House of Abbotsford and antiquities in it, and go to Melrose. Got there about 2, and settle to remain all night. Spend the day and also the evening about the Abbey. Jane Cay and James drawing.—Diary, 1846, Sept. 10.

On returning to Edinburgh for the winter, Mr. Clerk Maxwell seems to have been roused by the expectation which his son's first school distinction had awakened amongst his kindred and acquaintance. He became more assiduous than ever in his attendance at meetings of the Edinburgh Society of Arts and Royal Society, and took James with him repeatedly to both. And so it happened that early in his fifteenth year the boy dipped his feet in the current of scientific inquiry, where he was to prove himself so strong a swimmer. In our walks round Arthur's Seat, etc., he had always something new to tell. For example, in February 1846, he called my attention to the glacier - markings on the rocks, and discoursed volubly on this subject, which was then quite recent, and known to comparatively few.

A prominent member of the Society of Arts at this time was Mr. D. R. Hay, the decorative painter, whose attempt to reduce beauty in form and colour to mathematical principles [1] had attracted considerable

1845.
Æt. 14.

---

[1] " First Principles of Symmetrical Beauty," by D. R. Hay, *Black-woods*, 1846.

attention amongst scientific men. Such ideas had a natural fascination for Clerk Maxwell, and he often discoursed on " egg-and-dart," " Greek pattern," " ogive," and what not, and on the forms of Etruscan urns. One of the problems in this department of applied science was how to draw a perfect oval; and

Æt. 14. Maxwell, who had by this time begun the (purely geometrical) study of Conic sections, became eager to find a true practical solution of this. How completely his father entered into his pursuit may best be shown by the following extracts from Mr. Clerk Maxwell's Diary :—

1846,
February.

W. 25.—Called on . . Mr. D. R. Hay at his house, Jordan Lane, and saw his diagrams and showed James's Ovals—Mr. Hay's are drawn with a loop on 3 pins, consequently formed of portions of ellipses.

Th. 26.—Call on Prof. Forbes at the College, and see about Jas. Ovals and 3-foci figures and plurality of foci. New to Prof. Forbes, and settle to give him the theory in writing to consider.[1]

March.

M. 2.—Wrote account of James's ovals for Prof. Forbes. Evening.—Royal Society with James, and gave the above to Mr. Forbes.

W. 4.—Went to the College at 12 and saw Prof. Forbes, about Jas. ovals. Prof. Forbes much pleased with them, investigating in books to see what has been done or known

---

[1] Part of the entry on the same day is :—" Parliament House.— Return with John Cay, called at Bryson's and suggested to Alexander Bryson my plan for pure iron by electro-precipitation from sulphate or other salt." It is interesting to observe this revival of his youthful ardour for science in the old companionship, following upon his sympathy with the efforts of his son.

in this subject.   To write to me when he has fully considered the matter.

Sa. 7.—Recd. note from Prof. Forbes :—

Edinburgh, 6th March 1846.

My DEAR SIR—I have looked over your son's paper carefully, and I think it very ingenious,—certainly very remarkable for his years ; and, I believe, substantially new.   On the latter point I have referred it to my friend, Professor Kelland, for his opinion.—I remain, dear Sir, yours sincerely,   JAMES D. FORBES.

W. 11.—Recd. note from Professor Forbes :—

3 Park Place, 11th March 1846.

My DEAR SIR—I am glad to find to-day, from Professor Kelland, that his opinion of your son's paper agrees with mine ; namely, that it is most ingenious, most creditable to him, and, we believe, a new way of considering higher curves with reference to foci.   Unfortunately these ovals appear to be curves of a very high and intractable order, so that possibly the elegant method of description may not lead to a corresponding simplicity in investigating their properties.   But that is not the present point.   If you wish it, I think that the simplicity and elegance of the method would entitle it to be brought before the Royal Society.—Believe me, my dear Sir, yours truly,

JAMES D. FORBES.

J. CLERK MAXWELL, Esq.

Th. 12.—Called for Prof. Forbes at the College and conversed about the ovals.

M. 16.—Went with James to Royal Society.

T. 17.—Jas. at Prof. Forbes's House, 3 Park Place, to Tea, and to discourse on the ovals.   Came home at 10.   A successful visit.

T. 24.—Cut out pasteboard trainers for Curves for James.

W. 25.—Call at Adie's,[1] to see about Report on D. R. Hay's paper on ovals.

Th. 26.—Recd. D. R. Hay's paper and machine for drawing ovals, etc.

M. 30.—Called on Prof. Forbes at College and saw Mr. Adie about report on Mr. Hay's paper.   Jas. ovals to be at next meeting of R.S.

---

[1] The Optician's.

M. 6.—Royal Society with Jas. Professor Forbes gave acct. of James's ovals. Met with very great attention and approbation generally.

The result of the attempt thus eagerly pursued, as communicated by Professor Forbes that evening, is embodied in the *Proceedings of the Edinburgh Royal Society*, vol. ii. pp. 89-93.

MONDAY, 6*th April* 1846.

SIR THOMAS M. BRISBANE, Bart., President, in the Chair.

The following communications were read :—

1. On the Description of Oval Curves, and those having a plurality of Foci. By Mr. CLERK MAXWELL, junior, with Remarks by Professor FORBES. Communicated by Professor FORBES.

Mr. Clerk Maxwell ingeniously suggests the extension of the common theory of the foci of conic sections to curves of a higher degree of complication, in the following manner :—

1. As in the ellipse and hyperbola, any point in the curve has the *sum* or *difference* of two lines drawn from two points or *foci* = a constant quantity, so the author infers that curves to a certain degree analogous may be described and determined by the condition that the simple distance from one focus, *plus* a multiple distance from the other, may be = a constant quantity; or more generally, *m* times the one distance + *n* times the other = constant.

2. The author devised a simple mechanical means, by the wrapping of a thread round pins, for producing these curves. See Figs. 1 and 2 (Plate 11). He then thought of extending the principle to other curves, whose property should be, that the sum of the simple or multiple distances of any point of the curve from three or more points or foci, should be = a constant quantity; and this, too, he has effected mechanically, by a very simple arrangement of a

string of given length passing round three or more fixed pins, and constraining a tracing point, P.   See Fig. 3. Further, the author regards curves of the first kind as constituting a particular class of curves of the second kind, two or more foci coinciding in one, a focus in which two strings meet being considered a double focus; when three strings meet a treble focus, etc.

Professor Forbes observed that the equation to curves of the first class are easily found, having the form—

$$\sqrt{x^2 + y^2} = a + b \ \sqrt{(x - c)^2 + y^2},$$

which is that of the curve known under the name of the First Oval of Descartes. Mr. Maxwell had already observed that, when one of the foci was at an infinite distance (or the thread moved parallel to itself, and was confined, in respect of length, by the edge of a board), a curve resembling an ellipse was traced ; from which property Professor Forbes was led first to infer the identity of the oval with the Cartesian oval, which is well known to have this property. But the simplest analogy of all is that derived from the method of description, r and r' being the radients to any point of the curve from the two foci.

$$mr + nr' = \text{constant},$$

which, in fact, at once expresses on the undulatory theory of light the optical character of the surface in question, namely, that light diverging from one focus F without the medium, shall be directly convergent at another point f within it ; and in this case the ratio $\frac{n}{m}$ expresses the index of refraction of the medium.

If we denote, by *the power of either focus*, the number of strings leading to it by Mr. Maxwell's construction, and if one of the foci be removed to an infinite distance,—if the powers of the two foci be *equal*, the curve is a parabola ; if the power of the nearer focus be greater than the other, the curve is an ellipse ; if the power of the infinitely distant focus be the greater, the curve is a hyperbola.   The first case evidently corresponds to the reflection of parallel rays to a focus, the velocity being unchanged after reflection ; the second, to the . refraction of parallel rays to a focus in a

Proceedings of the Edinburgh Royal Society, vol. ii.

PLATE XI.

Fig.1. Two Foci. Ratios 1:2.

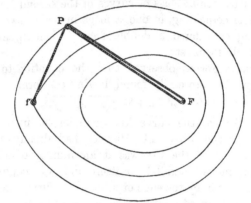

Fig.2. Two Foci. Ratios 2:3.

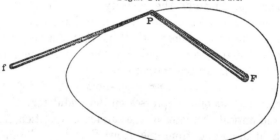

Fig.3. Three Foci. Ratios of Equality.

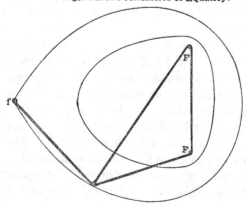

dense medium (in which light moves slower) ; the third case, to refraction into a rarer medium.

The Ovals of Descartes were described in his *Geometry*, where he has also given a mechanical method of describing one of them, but only in a particular case, and the method is less simple than Mr. Maxwell's. The demonstration of the optical properties was given by Newton in the *Principia*, Book i. Prop. 97, by the law of the sines, and by Huyghens in 1690, on the Theory of Undulations, in his *Traité de la Lumière*. It probably has not been suspected that so easy and elegant a method exists of describing these curves by the use of a thread and pins whenever the powers of the foci are commensurable. For instance, the curve, Fig. 2, drawn with powers 3 and 2 respectively, give the proper form for a refracting surface of glass, whose index of refraction is 1·50, in order that rays diverging from f may be refracted to F.

As to the higher classes of curves, with three or more focal points, we cannot at present invest them with equally clear and curious physical properties, but the method of drawing a curve by so simple a contrivance, which shall satisfy the condition,

$$mr + nr' + pr'' + \text{etc.,} = \text{constant,}$$

is in itself not a little interesting ; and if we regard, with Mr. Maxwell, the ovals above described, as the limiting case of the others by the coalescence of two or more foci, we have a further generalisation of the same kind as that so highly commended by Montucla, by which Descartes elucidated the conic sections as particular cases of his oval curves.

This was the beginning of the lifelong friendship between Clerk Maxwell and James D. Forbes. " I loved James Forbes " was his own emphatic statement to me in 1869. Maxwell's gratitude to all from whom he had received any help or stimulus was imperishable.

The curve-drawing, and the problems connected *Æt* 14-15.

with it, were by no means the only original investigations of this year.   Mr. John Scott, of Scott
Brothers, Greenock, remembers being in the attic of 31
Heriot Row, and seeing some preparations of jelly
with which James was experimenting there.   Mr.
Scott left Edinburgh in the summer of 1846.   What
was the exact object of these experiments and others
on *gutta percha* at this time is matter of conjecture.
There is little doubt that they prepared the way
for the investigation concerning the compression of
elastic solids.   But it seems probable that they
were immediately suggested by Forbes's *Theory of
Glaciers*, which had recently called attention to the
whole question of the difference between solid, liquid,
and " viscous " bodies, and the different effects of
gravitation and pressure as applied to them.[1]   Another
set of phenomena with which his mind was soon
afterwards engaged, viz. those of the refraction and
polarisation of light, were partly studied through
similar means.

The work of his cousin, who was now a rising
artist, still interested him.   An entry in the father's
Diary, December 5, 1845, has reference to this :—

Walk with Jas. and Jemima to Botanical Garden to
inspect palm-trees—for her sketching for a picture.

Either in this or the following year I remember
his raising the question, Whether it was not possible
to determine mathematically the curve of the waves

---

[1] See especially Forbes's paper on the " Viscous Theory of Glacier-
Motion " in the *Philosophical Transactions* for 1846, pp. 143-210.

on a particular shore, so as to represent them with
perfect truth in a picture.

After contributing to the *Proceedings* of the Edin- 1846-47,
burgh Royal Society, it might perhaps have been *Æt.* 15-16.
expected that Clerk Maxwell, although scarcely 15,
would at least have been taken from the Academy and
sent to the classes of Mathematics and Natural Philosophy at the University.   Instead of this, he simply
completed his course at school.   His inventions may
perhaps have interfered a little with his regular
studies—for he missed the Mathematical medal in
1846—but he was one of the few of the class which
he had joined in 1841 who continued at the Academy
until 1847.   And when he left, although still younger
than his competitors by about a twelvemonth, he was
not only first in mathematics and English, but came
very near to being first in Latin.   He had not yet
"specialised" or "bifurcated," although the bent of
his genius was manifest.   Nor have I ever heard him
wish that it had been otherwise.   On the contrary, he
has repeatedly said to me in later years that to make
out the meaning of an author with no help excepting
grammar and dictionary (which was our case) is one
of the best means for training the mind.   Some of his
school exercises in Latin prose and *verse* are still
extant, and, like everything which he did, are stamped
with his peculiar character.[1]   The first Greek play we
read (the *Alcestis* of Euripides) made an impression

---

[1] On his copy of Monk's *Alcestis*, which we read in the 6th class
1845-46), the owner's name is followed by an original distich :—
   " Si probrum metuis, nolito tangere librum,
      Nam magni domini nomina scripta vides."

on him to which he reverted in a conversation many
years afterwards.[1]   At the same time he had a quick
eye for the absurdities of pedantry.   One of the
teachers was apt to annoy our youthful taste by a
literal exactness in translating the Greek particles,
which would have pleased some more recent scholars.
Maxwell expressed our feelings on this subject in a few
lines, of which I can only recall the beginning :—

> " Assuredly, at least, indeed,
>   —*Decidedly* alsó . . ."

The frigid climax in " decidedly " is a good instance of
his roguish irony.

In September 1846 I made my first visit to Glenlair.   It was a time of perpetual gladness, but the
particulars are hardly worth recording.   James used
to sleep long and soundly, and seemed to be the whole
day at play, eagerly showing me his treasures and
accompanying each exhibition with lively talk,
sprinkled with innumerable puns.[2]   After such a
breakfast as became that land of milk and honey, there
was a long interval, while Mr. Maxwell was attending
to home business and deliberating what the " expedite"

---

[1] The following bit of Diary deserves to be quoted in this connection :—" *1845, Saturday, December 27.*—Jas. dined at Miss Cay's. I
went there to tea, taking Lewis and Robert Campbell.   Then we all
went to the play—'Antigone,' by Miss Faucit."

[2] I can only compare him to the fraternal spirit that William
Blake saw in vision " clapping its hands for joy " (Gilchrist's *Life of
William Blake*, vol. i. p. 59).   In the talk of that period a butterfly
was always a " flutterby ; " and idiotisms such as " used to could "
and " be-you-have yourself ! " were in common use.

should be. Miss Cay meanwhile was writing letters,
or finishing some drawing of Lincluden or New Abbey
(where they had lately been), and James would flit to
and fro between the little den, where his books and
various apparatus lived, and the drawing-room,[1] where
his father sat in the arm-chair, with Tobs on knee.
Ever and anon we boys would escape out of doors and
have a run in the field or the garden, or a bout with
the d——l. So the morning would pass till an early
luncheon, after which Tobin must do his various tricks;
then, if the men were busy, James would himself har-
ness Meg, the Galloway pony, for the drive of the
afternoon.   After dinner and Toby's second performance, and another turn at the deil, there would be
something more to see—Cousin Jemima's drawings,
recent diagrams or other inventions of his own, the
magic discs, etc. etc., the charm of the whole consisting in the flow of talk, incessant, but by no means
unbroken,

> " Changing, hiding,
>   Doubling upon itself, dividing,"

of which neither of us ever tired.   On Sunday there
was the drive to Corsock Church (where the absolute
gravity of his countenance was itself a study),[2] and the
walk home by the river, past the Kirk pool, renowned
for bathing, with conversations of a more earnest kind,

---

[1] The present dining-room.

[2] At church he always sat preternaturally still, with one hand
lightly resting on the other, not moving a muscle, however long the
sermon might be. Days afterwards he would show, by some remark,
that the whole service, whether good or bad, had been, as it were,
photographed upon his mind.

and a stroll on the estate in the afternoon ; or, if we stayed at home, he would show his favourite books and talk about them, till the evening closed with a chapter and a prayer, which the old man read to the assembled household.

Æt. 15. During the winter of 1846-47, James was unusually delicate. He was often absent from school, and seems not to have attended the meetings of the Societies. But of these his father was sure to give him a faithful report. He was certainly more than ever interested in science. The two subjects which most engaged his attention were magnetism and the polarisation of light. He was fond of showing " Newton's rings "—the chromatic effect produced by pressing lenses together—and of watching the changing hues on soap bubbles.

In the spring of 1847 (somewhere in April) his uncle, Mr. John Cay, whose scientific tastes have been mentioned more than once above, took James and myself (with whom he chose to share all such delights) to see Mr. Nicol, a friend of Sir David Brewster, and the inventor of the polarising prism.[1] Even before this James had been absorbed in " polarised light," working with Iceland spar, and twisting his head about to see " Haidinger's Brushes " in the blue sky with his naked eye. But this visit added a new and important stimulus to his interest in these phenomena, and the speculations to which they give rise.[2]

---

[1] He lived in Inverleith Terrace, Edinburgh.

[2] So far as I can recall the order in which his ideas on this subject were developed, the phenomena of complementary colours came

Shortly afterwards (May 25th) he went with his father to the cutler's to choose magnets suitable for experimenting.

And a little earlier in the same year (March 17), he was taken to hear a lecture[1] on another subject, which was also connected with his subsequent labours, and must have impressed him not a little at the time. This was the discovery by Adams and Leverrier simultaneously, through a striking combination of hypothesis and calculation, of the planet Neptune, which then first " swam into " human " ken."

The magnetic experiments were continued that autumn at Glenlair, as appears from two entries in the Diary :—

Sept. 3.—Walk round by smiddy ; gave steel to be made into bars for magnets for James.

Sept. 7.—James and Robert (Campbell) most of the time at the smiddy, and got the magnet bars.

My brother perfectly remembers the magnetising of these bars of steel.

Lastly, in 1847—unless my memory deceives me —James had commenced the study of chemistry, and had taken extra lessons in German.

There was an odd episode in our school life.   To keep our education " abreast of the requirements of

---

first, then the composition of white light, then the mixture of colours (not of pigments), then polarisation and the dark lines in the spectrum, and about the same time " the art of squinting," stereoscopic drawing, etc., then colour-blindness, the yellow spot on the retina, etc.

[1] The lecturer was Mr. Nichol, Professor of Astronomy at Glasgow, the father of the distinguished Professor of English Literature in the same University.

the day," etc., it was thought desirable that we should have lessons in "Physical Science." So one of the classical masters gave them out of a text-book. The sixth and seventh classes were taught together ; and the only thing I distinctly remember about these hours is that Maxwell and P. G. Tait seemed to know much more about the subject than our teacher did.

Maxwell and Tait were by this time acknowledged as the two best mathematicians of the school, and it was already prophesied that Tait, who was about fifteen, would some day be a Senior Wrangler. The two youths had many interchanges of ideas, and Professor Tait remembers that Maxwell had by this time proved, by purely geometrical methods, that the central tangential section of a " tore," or anchorring, is a pair of intersecting equal and similar curves, *probably circles*.

This is referred to in the following extract from Professor Tait's admirable summary :—

When I first made Clerk Maxwell's acquaintance about thirty-five years ago, at the Edinburgh Academy, he was a year before me, being in the fifth class while I was in the fourth.

At school he was at first regarded as shy and rather dull. He made no friendships, and he spent his occasional holidays in reading old ballads, drawing curious diagrams, and making rude mechanical models. This absorption in such pursuits, totally unintelligible to his schoolfellows (who were then quite innocent of mathematics), of course procured him a not very complimentary nickname, which I know is still remembered by many Fellows of this Society. About the middle of his school career, however, he surprised his companions by suddenly becoming one of the most

brilliant among them, gaining high, and sometimes the
highest, prizes for scholarship, mathematics, and English
verse composition.   From this time forward I became very
intimate with him, and we discussed together, with schoolboy enthusiasm, numerous curious problems, among which I
remember particularly the various plane sections of a ring
or tore, and the form of a cylindrical mirror which should
show one his own image unperverted.   I still possess some
of the MSS. we exchanged in 1846 and early in 1847. Those
by Maxwell are on " The Conical Pendulum," " Descartes'
Ovals," " Meloid and Apioid," and " Trifocal Curves."   All
are drawn up in strict geometrical form, and divided into
consecutive propositions.[1]   The three latter are connected
with his first published paper, communicated by Forbes to
this Society and printed in our *Proceedings*, vol. ii., under
the title " On the description of Oval Curves, and those
having a plurality of Foci " (1846).   At the time when
these papers were written he had received no instruction
in mathematics beyond a few books of Euclid and the merest
elements of Algebra.[2]

On the whole, he looked back to his schooldays
with strong affection ; and his only revenge on those
who had misunderstood him was that he understood
them.   To many of us, as we advance in life, the
remembrance of our early companions, except those
to whom we were specially drawn, becomes dim and
shadowy.   But Maxwell, by some vivid touch, has
often recalled to me the image of one and another of
our schoolfellows, whose existence I had all but forgotten.

---

[1] See note appended to this chapter (pp. 91-104) containing two
papers which must have been written about this time.

[2] *Proceedings of the Royal Society of Edinburgh*.   Session 1879-80.
p. 332.

The following letters to his father still belong to his schoolboy life :—

[*April*] 1847.

I have identified Descartes' ovals with mine. His first oval is an oval with one of its foci outside; the second is a meloid with a focus outside. It also comprehends the circle and apioid. The third is a meloid with both foci inside; and the fourth is an oval with both foci inside."

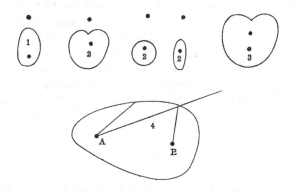

He says with regard to the last that rays from A are *reflected* to B, which I can disprove. . . .

[This is interesting as a proof of the complete independence and originality of the communication made to the Royal Society on April 6th, 1846. It is also important as an example of that love of comparing his own results with great authorities, which made Maxwell the most learned as well as the most original of scientific men. See above, p. 74, where Forbes promises to consult books on the subject, and the references to Descartes in the *Proceedings*. Forbes cannot, however, have made any very careful scrutiny into the history of the matter, else he would have mentioned that Descartes actually *figures* the cord by which his ovals are described, and his process is precisely that of Maxwell.]

[*1847.*]

I have made a map of the world, conical projection—

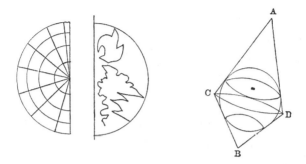

ACD is the one cone, BCD the other; the side AC = diameter of the base CD; therefore, when unrolled, they are each semicircles.

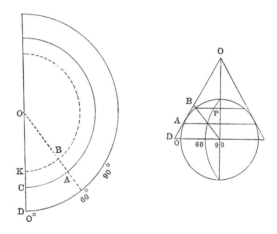

To make the map take OD = OD and OA = $\frac{3}{4}$ of OD, and describe a circle from O, with radius OA; this is the circle of contact. Let the longitude of a point P = 60°; take AOC in the map = 30°; let its latitude be 65°; subtract 30°, and take the tangent of 35, make AB = tangent. B is the point.

Mr. Clerk Maxwell made inquiries early in the summer of 1847, with a view to placing his son at College in November. In deciding not to continue his classical training, he appears to have been chiefly guided by some disparaging accounts of the condition of the Greek and Latin classes in comparison with those of Logic, Mathematics, and Natural Philosophy. The result was that in his seventeenth year Maxwell entered the second Mathematical Class, taught by Professor Kelland, the class of Natural Philosophy under J. D. Forbes, and the Logic Class of Sir William Hamilton. Classical reading, however, was not by any means relinquished, as the correspondence of 1847-50 clearly shows.

NOTE.

THE two following series of propositions concerning the Ovals belong to this period, and afford an illustration of Maxwell's mode of working in those early days. See above, p. 87. The only alterations made, with the exception of corrections of obvious slips of the pen, consist in a sparing insertion of stops. The MS., as will be seen by the *facsimile* at p. 104, has almost no punctuation.

I. OVAL.

Definition 1.—If a point move in such a manner that *m* times its distance from one point, together with *n* times its distance from another point, may be equal to a constant quantity, it will describe a curve called an Oval.

Definition 2.—The two points are called the foci, and the numbers signified by *m* and *n* are called the powers of the foci.

Definition 3.—The line joining the foci is called the axis.

PROPOSITION 1—PROBLEM.

To describe an oval with given foci, given multiples, and given constant quantity.

Let A and B be the given foci, 3 and 2 the multiples, and EF the constant quantity, it is required to describe an oval. At A and B erect two infinitely small cylinders. Take a perfectly flexible and inextensible thread, without breadth or thickness, equal to EF ; wind it round the focal cylinders and another movable cylinder C, so that the number of plies between A and C may be equal to *m*, that is 3, and the number between B and C equal to *n* or 2. Now move C in such a manner that the thread may be quite tight, and an oval will be described by the point.

F

For take any point C. There are m plies of thread between A and C, and n plies between B and C, which taken together make up the thread or the constant quantity, therefore m AC + n BC = EF.

The focus A, which has the greatest number of plies is called the greater focus, and B is called the less focus.

PROPOSITION 2—THEOREM.

The greater focus is always within the oval, but the less is within, on, or without the curve, according as the distance between the foci, multiplied by the power of the greater focus, is less equal or greater than, the constant quantity.

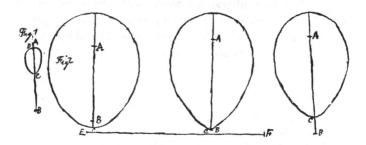

The greater focus is always within the Oval, for suppose it to be at A, Fig. 1, then m AC + n BC = constant quantity = m AD + m DC + n BC, and m AD + n DB = constant quantity = m AD + m DC + n BC = m AD + m DC + n BC, and n DC = m DC, and $m = n$, but $m > n$.

The less focus B is within the oval when m AB < EF the constant quantity.

For it is evident that BC (Fig. 2) $= \dfrac{EF - m AB}{m + n}$

It is in the curve when m AB = EF, this is evident.

It is without the curve when m AB > EF, for AC $= \dfrac{EF - n \, AB}{m - n}.$

PROPOSITION 3—THEOREM.

If a circle be described with a focus for a center, and the constant quantity divided by the power of that focus for a radius, the distance of any point of the oval from the other focus is to the distance from the circle as the power of the central focus is to the power of the other.

The circle EHP is described with the centre B and radius = constant quantity, divided by the power of B. At any point C, AC : CH :: power of B : power of A.

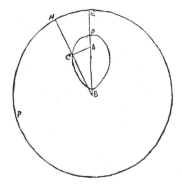

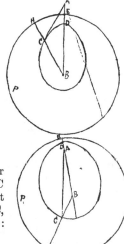

Let m = power of A, and n = power of B, n BH = constant quantity = n BC + n CH, and n BC + m AC = constant quantity = n BC + n CH, take away n BC, and n HC = m AC, therefore HC : CA : : m : n.   QED.

Cor. 1.—When the powers of the foci are equal the curve is an ellipse.

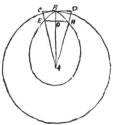

Cor. 2.—When the less focus is at an infinite distance the curve is an ellipse, for the circle becomes a straight line ; and

Cor. 3.—When the greater focus is at an infinite distance, the curve is an hyperbola for the same reason.

PROPOSITION 4—THEOREM.

When the less focus is in the curve, an angle will be formed equal to the vertical angle of an isosceles triangle, of which the side is to the perpendicular on the base, as the power of the greater focus to that of the less.

For let a circle be described, as in Prop. 3, it is evident that it will pass through B.   Take indefinitely small arcs CB = BD, join CA and DA, join EH.   EC : EB = power of B : power of A, and EC = BO, therefore EB : BO, power of A : power of B.

PROPOSITION 5—PROBLEM.

A point A, and a point B in the line BC being given, to find a point in the line as D, so that m AD + n BD may be a minimum.

Take a line HP, raise HX perpendicular, and from X as a centre describe a circle with a radius = $\dfrac{m\ \text{XH,}}{n}$ so that m : n : : XP : XH.

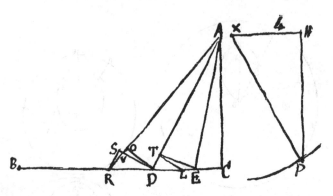

Draw AC perpendicular to BC, and make an angle CAD = HPX. D is the required point. If not, take any point E on the opposite side of D from B, make AT = AE.   Join TE, and make DTL = right angle, then DTE > right angle DTL, because ATE is less than a right angle (1.16), therefore L is within E.   And in the triangle DTL, DTL = PHX and TDL = HXP, therefore the triangles are similar, and TD : DL = (XH : XP) = $n : m$; and n DL = m DT, then nDE > mDT, add nBD and mTA or mEA, then nBD + nDE + mEA > nBD + mDT + mTA and nBD + mDA < nBE + mAE.

Now take a point R on the other side.   Join AR, and cut off AO = AD, make SRD = HXP = ADC. S will be beyond AR. Draw DS perpendicular to AD, then it will be perpendicular to RS, and below the line OD, then RSD similar to XHP and RS : RD = (XH : XP =) $n : m$, and m RS = n RD, but SR < RV < RO and n RD (= m RS) < m RO, add n BR and m OA or m DA, and n BR + n RD + m DA < n BR + m RO + m OA and n BD + m DA < n BR + m RA.   QED.

PROPOSITION 6—PROBLEM.

To draw a tangent to an oval from a focus without :—

Take m for the power of the greater focus, and n for that of the less, and find the angle ADB (Prop. 5). Upon AB describe a segment of a circle containing an equal angle. Join B and C, the point where it cuts the oval, B C is a tangent. For take any point O, join AO, m AO + n OB > m AC + n CB, therefore O is without the oval.

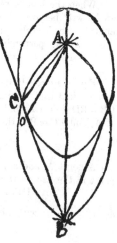

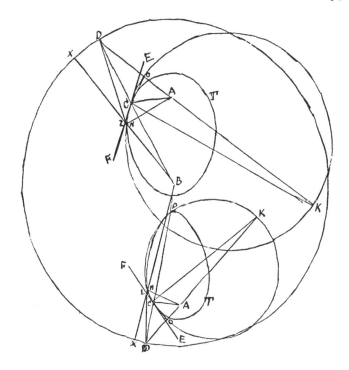

PROPOSITION 7—PROBLEM.

To draw a tangent to a given Oval, the foci and the ratio being given :—

It is required to draw a tangent at C to the oval CNT, ratio $m : n$. From B the less focus describe a circle as in (Prop. 3). Join CA, join CB, and produce to D, then $DC : CA :: m : n$. Join AD and produce it. Bisect the angle DCA by the line CO. $DO : OA ::$ $(DC : CA ::) m : n$. Draw CK perpendicular to CO, and describe the circle OCLK. Because OCK is a right angle, $OCK = DCO + KCB = OCA + ACK$. But $OCA = DCO$ ∴ $KCB = ACK$ and (6.3) $KA : KD :: CA : CD :: AO : OD$ ∴ $KA : KD :: AO : OD$, and at any point L in the circle, $AL : DL :: AO : OD$ (6 F), and $AL : DL :: n : m$; and the circle is wholly without the oval, and can only touch it in the two points C and P, where DB cuts the oval ; for suppose the circle coincided with the oval at L,

Join BL and produce to X, then $AL : DL :: n : m$ }
And by Prop. 3 . . $AL : XL :: n : m$ } ∴ $DL = XL$;
but $DL > XL$ (3.7), therefore the oval is within the circle. Therefore draw ECF a tangent to the circle, and as it is without the circle it is without the oval.

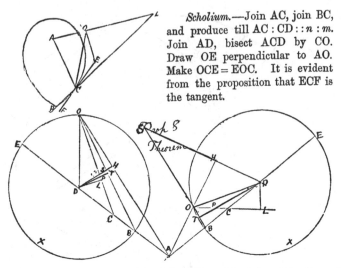

Scholium.—Join AC, join BC, and produce till AC : CD : : *n* : *m*. Join AD, bisect ACD by CO. Draw OE perpendicular to AO. Make OCE = EOC. It is evident from the proposition that ECF is the tangent.

If a line AE be cut in C and B, so that AB : BC : : AE : CE, and two semicircles BOE, BXE be described on BE, and AO, CO be drawn to O in the circumference, and perpendiculars DH, DL, be drawn from the centre, DH : DL : : AB : BC.

For AB : BC : : AE : CE ∴. AB : AE : : BC : CE ∴. (AE + AB) = (2 AB + 2 BD) = 2 AD : AB : : (CE + BC =) 2 BD : BC and AD : AB : : BD : BC ∴. (AD − AB =) BD : AB : : (BD − BC =) CD : BC ∴. BD : CD : : AB : BC.

Draw DT perpendicular to OB, then as it bisects the base it bisects the angle ODB and ODT = BDT, then in the triangles OTP, DPL, OTP = DLP and OPT = DPL ∴. PDL = POT. But POT = BOA (6 F Cor.) and BOA = SOH = PDL, and in the triangles SOH, STD, SHO = STD, and OSH = TSD ∴. SOH = SDT and SDT = PDL. But ODT = BDT ∴. CDL = ODH and DHO = DLC ∴. HOD, CDL are equiangular, and HD : DL : : (DO =) BD : CD, but BD : CD : : AB : BC ∴. HD : DL : : AB : BC.

Cor. Sine HOD : Sine DOL : : AB : BC.

Proposition 9—Theorem.

If lines be drawn from the foci to any point in the oval, the sines of the angles which they make with the perpendicular to the tangent are to one another as the powers of the foci.

Sine DCE : sine ACE : : power of A : power of B.

For describe the circle CXT as in Prop. 7, so that DT : TA : : DX : AX, then SC the tangent to the circle is a

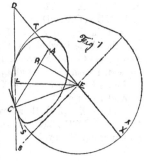

tangent to the oval, and CE the radius is perpendicular to it ; then by
Prop. 8 Corollary, sine DCE : sine ACE : : (DT : TA : :) power of A :
power of B.

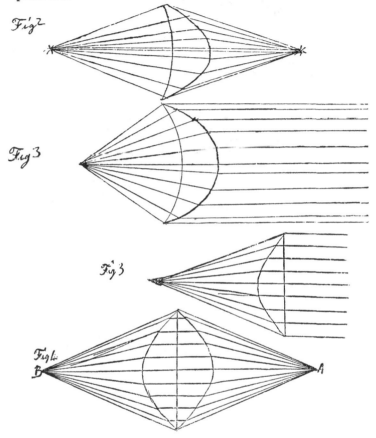

As the ratio of the sines of incidence and refraction is invariable
for the same medium, and as this ratio is as 2 to 3 in glass ; and as in
the oval at Fig. 1 the powers of the foci are as 2 to 3 ; if the oval
were made of glass, rays of light from B would be refracted to A.   If
now a circle be described from A, the case will not be altered, for the
rays are perpendicular to the circle as in fig. 2.

The ellipse and hyperbola, figs. 3 and 3 [*bis*], cause parallel rays to
converge, for (Prop. 3) they are ovals with one of the foci at an infinite
distance.

Fig. 4 is a combination of 2 hyperbolas, and rays from A are
refracted to B.

H

II. Meloid and Apioid.

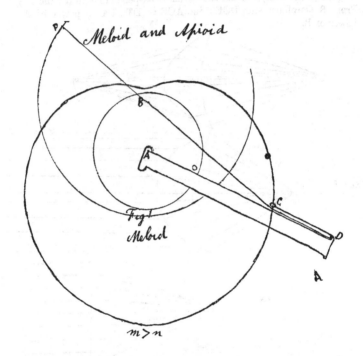

If a point C move so that $m\,CA \sim n\,CB = $ a constant quantity, it will describe a curve called a meloid when $m\,CA > n\,CB$, and an apioid when $n\,CB > m\,CA$. The points A and B are called the foci.

Proposition 1—Problem.

To describe a Meloid, Fig. 1, or Apioid, Fig. 2.

Take a rigid straight rod AD, centered at A. At D erect a small cylinder upon the rod, and also another at B on the paper. Then take a flexible and inextensible thread of such a length that $m\,AD \sim n$ times the thread = constant quantity; then wrap the thread round B and D and a movable cylinder C, so that there may be n plies between B and C, and m plies between C and D ; if now AD be turned round and the thread kept tight, the point C will describe a meloid or apioid.

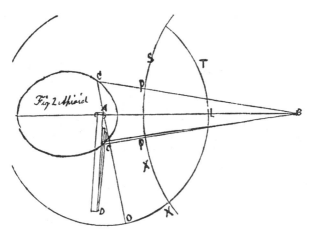

For take any point C, m DC $+ n$ CB = thread, and m DC $+ m$ AC $= m$ AD ; therefore take away m DC, and n CB $\sim m$ AC = thread $\sim m$ AD = constant quantity. As $m > n$, A is called the greater and B the less focus.

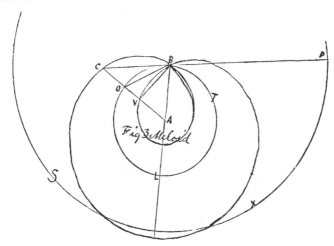

Proposition 2 is the same as proposition 2 of the Oval.

PROPOSITION 3—THEOREM.

If a circle be described with a focus for a center, and the constant difference divided by the power of that focus for a radius, the distance of any point in the curve from the other focus is to the distance from the circle as the power of the central focus is to the power of the other.

Let TLO, PXS be the circles: at any point C, BC : CO : : $m : n$ and PC : CA : : $m : n$.

For m CA $\sim n$ CB = constant difference, but m OA = constant difference $\therefore m$ CO = n CB and BC : CO : : $m : n$. QED.

And n BP = constant difference $\therefore m$ CA = n CP and PC : CA : : $m : n$. Cor. 1.—If the constant difference = 0 the curve is a circle. Cor. 2.— If an oval be described with a thread = constant difference, with the same foci and the same powers as the meloid, and any line be drawn from A, and CB, OB, VB be joined, the angle CBO = VBO.

For BC : CO : : $m : n$ and BV : VO : : $m : n$ $\therefore$ BC : CO : : BV : VO $\therefore$ BC : BV : : CO : VO and CBO = VBO (6.3).

PROPOSITION 4—THEOREM.

When the less focus is in the curve, an angle will be formed = that in the Oval (Prop. 4).

For take an indefinitely small arc DB in the circle, CBD = DBE (3. Cor. 2), and LBT = TBP $\therefore$ CBL = EBP.

Or it may be proved as in the oval. If the greater focus A is at an infinite distance the figure will appear thus :—

PROPOSITION 5—THEOREM.

If the distance between the greater focus and the point where the axis cuts the meloid, be to the distance between that point and the less focus, in a greater proportion than the power of the greater focus to that of the less, the curve is convex toward the greater focus at that point, but if the proportion is less, concave.

Let $AD : DB :: p : q.$

If $AD : DB > m : n$, CDE is convex towards A ; but if $AD : DB < m : n$, it is concave towards A. . . . . . Take C and E near D, and $CD = DE$. Join CE, CE cuts the axis in O. Draw the circle CLE from B, and CPE from A. Also draw HTK as in Prop. 3—

Then $BD : DT :: m : n :: BC : CH$, but $BC = BL$, and $CH = PT$ ∴ $BD : DT :: BL : PT$ ∴ $BD : BL :: DT : PT$ ∴ $BD : BL - BD :: DT : PT - DT$ ∴ $BD : DL :: DT : DP$ ∴ $BD : DT :: DL : DP$ ∴ $DL : DP :: m : n.$

As E and C are very near D, $AD : BD :: AC : BC$, but $PE = LE$ and $PCE : 2$ right angles $:: PE :$ circumference of CPE, and $LCE : 2 \llcorner :: LE :$ circumference of CLE, but circ. $CPE :$ circ. $CLE :: AC : BC :: AD : BD :: p : q$ and $PE = LE$ ∴ $PCE : 2 \llcorner :: q\ PE : q$ circ. CPE, and $LCE : 2 \llcorner :: p\ PE : (p$ circ. CLE or$)\ q$ circ. CPE ∴ $PCE : LCE :: q\ PE : p\ PE :: q : p$ ∴ $PCO : LCO :: q : p$ and $PO : OL :: q : p;$ and if $p : q > m : n$, $OL : OP > DL : DP$, and D is nearer to A than the line COE, and CDE is convex toward A ; but if $p : q < m : n$, D is on the opposite side, and it is concave.

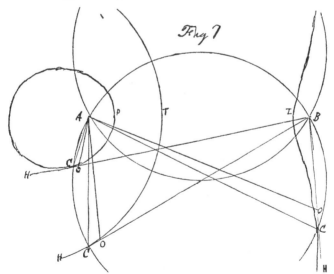

Fig 7

PROPOSITION 6—PROBLEM.

To draw a tangent to a meloid or apioid from a focus without :—

Take m for the power of the greater focus, and n for that of the less, and find the angle ADC (Prop. 5 of the Oval) upon AB describe a

segment ACCCB containing an equal angle. Join BC and produce to
H, BH is a tangent, for suppose H to be the end of the rod, and take
any point O, n CH + m CA < n OH + n OA, therefore O is without the
curve.

CP is an apioid, CT is a circle, and CL is a meloid, with A and B
as foci.

PROPOSITION 7—PROBLEM.

To draw a tangent to an apioid from any point in the same, the
foci and the ratio being given.

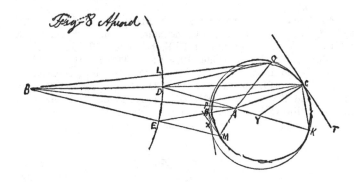

It is required to draw a tangent to the apioid at the point C.
Join BC, and draw a circle as in Prop. 3. Join AD, and produce it.
Make AP : PD : : n : m. Join CP, and draw CK at right angles to PC.
Describe a circle through P, C, K, and it was proved in Prop. 7 of the
Oval that if any point O be taken and DO, AO joined, DO : AO : : m : n.
Suppose O to be both in the circle and in the apioid, join BO, then
LO : AO : : m n, but DO : AO : m : n ∴ LO = DO, but LO < DO ∴ the
circle is without the apioid; therefore a tangent C T to the circle at C
is a tangent to the apioid.

AXIOM 1.

It is possible for a circle to be described touching any given curve
internally.

PROPOSITION 8.

To draw a tangent to a meloid at any point C.
Case 1.—Let the curve be concave towards B. Describe the circle
POCK as in Prop. 7 : it will be wholly within the meloid. At C
draw a tangent to the circle : it is also a tangent to the meloid. For

let RN be the tangent to the meloid, it must cut the circle (3.16), and therefore cuts the meloid.

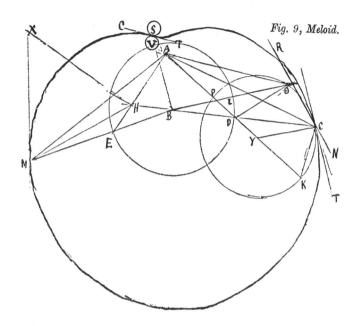

Fig. 9, Meloid.

Case 2.—When the curve is convex towards B, draw a circle V as before.   Draw CT a tangent to the circle ; it is also a tangent to the meloid.   For suppose a circle S drawn touching the curve internally, it must touch V and also CT, and any other line would cut S.   QED.

Scholium (see Figs. 8 and 9).—Let M be the point.   Join AM, BM, cut off ME, so that AM : ME : : power of B : power of A.   Join AE, bisect AME by MH, make XH perpendicular, make XMH = XHM, XM is a tangent.

PROPOSITION 9—THEOREM.

If lines be drawn from the foci to any point in a meloid or apioid, the sines of the angles which they make with the perpendicular to the tangent are to one another as the powers of

the foci See figs 8, 9
For CY is the perpendicular to the Tangent and (Prop 8 Oval Cor)
Sine ACY : Sine BCY :: (AP : PD) :: m : n

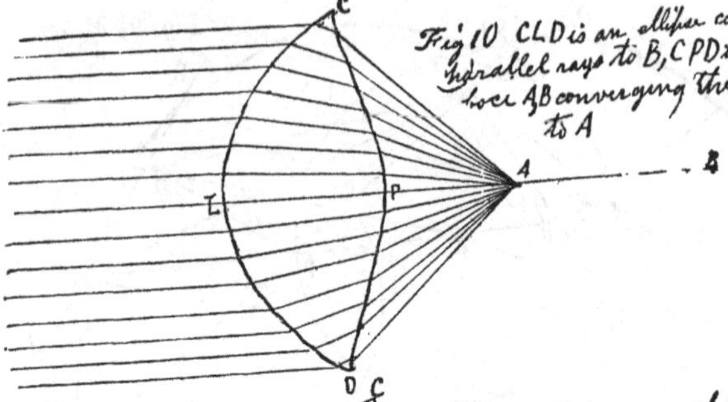

Fig 10 CLD is an ellipse converging
parallel rays to B, CPD is a meloid
foci AB converging the rays
to A

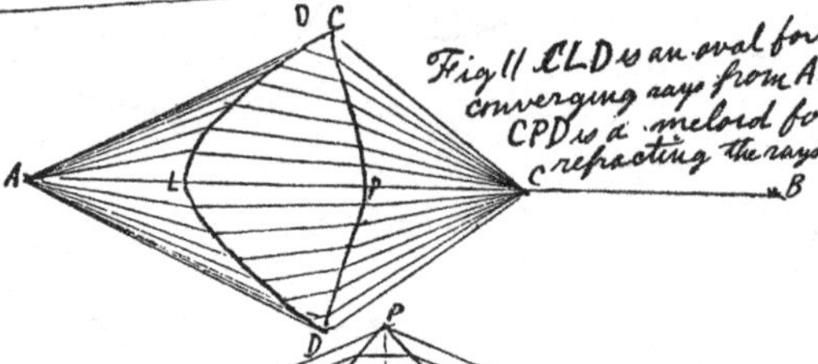

Fig 11 CLD is an oval foci A, B
converging rays from A to B
CPD is a meloid foci C, B
refracting the rays to C.

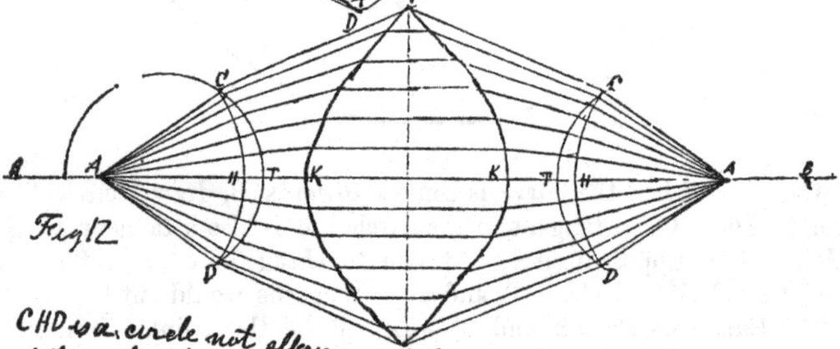

Fig 12

CHD is a circle not altering rays from A; CTD is a circle as in Prop 8
of this oval refracting the rays as if these had come from B
PKLR is a lens of hyperbolas refracting from B to B the whole
3 lenses refract from A to A

CHAPTER V.

OPENING MANHOOD—1847 TO 1850—ÆT. 16-19.

WHEN he entered the University of Edinburgh, James Clerk Maxwell still occasioned some concern to the more conventional amongst his friends by the originality and simplicity of his ways. His replies in ordinary conversation were indirect and enigmatical, often uttered with hesitation and in a monotonous key.[1] While extremely neat in his person, he had a rooted objection to the vanities of starch and gloves. He had a pious horror of destroying anything—even a scrap of writing paper. He preferred travelling by the third class in railway journeys, saying he liked a hard seat. When at table he often seemed abstracted from what was going on, being absorbed in observing the effects of refracted light in the finger-glasses, or in trying some experiment with his eyes—seeing round a corner, making invisible stereoscopes, and the like. Miss Cay used to call his attention by crying, "Jamsie, you're in a prop."[2] He never tasted wine ; and he spoke to gentle

[1] This entirely disappeared afterwards, except when ironically assumed. It was accompanied with a certain huskiness of voice, which was observed also in later years.

[2] " Prop." here and elsewhere is an abbreviation for " mathematical proposition."

and simple in exactly the same tone. On the other
hand, his teachers—Forbes above all—had formed the
highest opinion [1] of his intellectual originality and
force ; and a few experienced observers, in watching
his devotion to his father, began to have some inkling
of his heroic singleness of heart. To his college companions, whom he could now select at will, his quaint
humour was an endless delight. His chief associates,
after I went to the University of Glasgow, were my
brother, Robert Campbell (still at the Academy), P.
G. Tait, and Allan Stewart. [2] Tait went to Peterhouse, Cambridge, in 1848, after one session of the
University of Edinburgh ; Stewart to the same College
in 1849 ; Maxwell did not go up until 1850.

_Æt._ 16-19.   These three years—November 1847 to October
1850—were impartially divided between Edinburgh
and Glenlair. He was working under but slight
pressure, and his originality had the freest play. His
studies were multifarious, but the subjects on which
his thoughts were most concentrated during these
years were—1. Polarised light, the stereoscope, etc. ;
2. Galvanism ; 3. Rolling curves ; 4. Compression of
solids. That he early felt the necessity of imposing
a method on himself will appear from the letters.
His paper on Rolling Curves was read before the
Edinburgh Royal Society on February 19th, 1849, by
Professor Kelland (for it was not thought proper for

1 Forbes's certificate at the end of the second year goes beyond
the merely formal language of such documents :—" His proficiency
gave evidence of an original and penetrating mind."

2 Allan Stewart, Esquire of Innerhadden, Perthshire, C.E. _Æq._ 9th
Wrangler, 1853.

a boy in a round jacket to mount the rostrum there) ;
that on the Equilibrium of Elastic Solids in the
spring of 1850.

With regard to his class studies, it appears that
he attended Forbes for two sessions as a regular
student, and occasionally as an amateur student in
his third year ; Kelland for two sessions ; and Sir W.
Hamilton for two.   In his third session, while partially attending Forbes, he was a regular student in
the Classes of Chemistry (Professor Gregory), Practical Chemistry (Mr. Kemp), and Moral Philosophy
(Professor Wilson).

Mr. Macfarlan of Lenzie, in the letter already
quoted, says :—" He was in the Natural Philosophy
Class, Edinburgh University, the year before I was,
and was spoken of by Forbes and his fellow-students
as a discoverer in Natural Philosophy, and a very
original worker in Mathematics."

All scientific theories had an interest for him.   It
was at some time during these years that in a walk
towards Arthur's Seat he discoursed to me of Owen's
hypothesis of types of creation, not only with complete command of Owen's terminology, but with far-
reaching views of the questions to which the theory
led.   On the same occasion he made some characteristic remarks on the importance of cultivating the
senses, adding that he regarded dulness in that respect
as a bad sign of any man.

The lectures in Mental Philosophy, which were a
prominent element in the Scottish University curri-
culum, interested him greatly ; and from Sir William

Hamilton especially he received an impulse which never lost its effect. Though only sixteen when he entered the Logic Class, he worked hard for it, as his letters show; and from the Class of Metaphysics, which he attended in the following year, his mind gained many lasting impressions.[1] His boundless curiosity was fed by the Professor's inexhaustible learning; his geometrical imagination predisposed him to accept the doctrine of "natural realism;" while his mystical tendency was soothed by the distinction between Knowledge and Belief. The doctrine of a muscular sense gave promise of a rational analysis of the active powers. However strange it may appear that a born mathematician should have been thus influenced by the enemy of mathematics, the fact is indisputable that in his frequent excursions into the region of speculative thought, the ideas received from Sir William Hamilton were his habitual vantage-ground; the great difference being, that while Sir William remained for the most part within the sphere of Abstract Logic, Maxwell ever sought to bring each "concept" to the test of fact. Sir William in turn took a genial interest in his pupil, who was indeed the nephew of an old friend—Sheriff Cay having at one time been a constant companion and firm ally of Sir William's.[2] This is perhaps the most striking example

---

[1] A slight illustration of his devotion to Sir William's teaching, and also of his powers of endurance, is afforded by the following incident:—One day he sprained his ankle badly on the College stair-case; but, instead of going home, he attended Sir William Hamilton's class as usual, and sat through the hour as if nothing had happened.

[2] John Lockhart had made a third as the comrade of both.

of the effect produced by Sir William Hamilton on powerful young minds,—an effect which, unless the best metaphysicians of the subsequent age are mistaken, must have been out of all proportion to the independent value of his philosophy.

It was impossible that young Maxwell should listen to speculations about the first principles of things,—speculations, too, which, like all the Scottish philosophy, turned largely on the reality of the external world, — without eagerly working out each problem for himself. Besides various exercises done by him for the Logic Class, and, like all his youthful work, preserved by him with pious care, there is one which seems to have attracted special attention, and was found by Professor Baynes, when he came to assist Sir William, in the Professor's private drawer. This is so significant, and is so closely related to Maxwell's after studies, as to deserve insertion here.

ON THE PROPERTIES OF MATTER. *Æt.* 17.

These properties are all relative to the three abstract entities connected with matter, namely, space, time, and force.

1. Since matter must be in some part of space, and in one part only at a time, it possesses the property of locality or position.

2. But matter has not only position but magnitude; this property is called extension.

3. And since it is not infinite it must have bounds, and therefore it must possess figure.

These three properties belong both to matter and to imaginary geometrical figures, and may be called the geometric properties of matter. The following properties do not necessarily belong to geometric figures.

4. No part of space can contain at the same time more than one body, or no two bodies can coexist in the same space; this property is called impenetrability. It was thought by some that the converse of this was true, and that there was no part of space not filled with matter. If there be a vacuum, said they, that is empty space, it must be either a substance or an accident.

If a substance it must be created or uncreated.

If created it may be destroyed, while matter remains as it was, and thus length, breadth and thickness would be destroyed while the bodies remain at the same distance.

If uncreated, we are led into impiety.

If we say it is an accident, those who deny a vacuum challenge us to define it, and say that length, breadth and thickness belong exclusively to matter.

This is not true, for they belong also to geometric figures, which are forms of thought and not of matter; therefore the atomists maintain that empty space is an accident, and has not only a possible but a real existence, and that there is more space empty than full. This has been well stated by Lucretius.

5. Since there is a vacuum, motion is possible; therefore we have a fifth property of matter called mobility.

And the impossibility of a body changing its state of motion or rest without some external force is called *inertia*.

Of forces acting between two particles of matter there are several kinds.

The first kind is independent of the quality of the particles, and depends solely on their masses and their mutual distance. Of this kind is the attraction of gravitation and that repulsion which exists between the particles of matter which prevents any two from coming into contact.

The second kind depends on the quality of the particles; of this kind are the attractions of magnetism, electricity, and chemical affinity, which are all convertible into one another and affect all bodies.

The third kind acts between the particles of the same body, and tends to keep them at a certain distance from one another and in a certain configuration.

When this force is repulsive and inversely as the distance, the body is called gaseous.

When it does not follow this law there are two cases.

There may be a force tending to preserve the figure of the body or not.

When this force vanishes the body is a liquid.

When it exists the body is solid.

If it is small the body is soft; if great it is hard.

If it recovers its figure it is elastic; if not it is inelastic.

The forces in this third division depend almost entirely on heat.

The properties of bodies relative to heat and light are—
Transmission, Reflection, and Destruction,
and in the case of light these may be different for the three kinds of light, so that the properties of colour are—
Quality, Purity, and Integrity; or
Hue, Tint, and Shade.

We come next to consider what properties of bodies may be perceived by the senses.

Now the only thing which can be directly perceived by the senses is Force, to which may be reduced heat, light, electricity, sound, and all the things which can be perceived by any sense.

In the sense of sight we perceive at the same time two spheres covered with different colours and shades. The pictures on these two spheres have a general resemblance, but are not exactly the same; and from a comparison of the two spheres we learn, by a kind of intuitive geometry, the position of external objects in three dimensions.

Thus, the object of the sense of sight is the impression made on the different parts of the retina by three kinds of light. By this sense we obtain the greater part of our

practical knowledge of locality, extension, and figure as properties of bodies, and we actually perceive colour and angular
dimension.

And if we take time into account (as we must always
do, for no sense is instantaneous), we perceive relative
angular motion.

By the sense of hearing we perceive the intensity,
rapidity, and quality of the vibrations of the surrounding
medium.

By taste and smell we perceive the effects which liquids
and aeriform bodies have on the nerves.

By touch we become acquainted with many conditions
and qualities of bodies.

1. The actual dimensions of solid bodies in three dimensions, as compared with the dimensions of our own bodies.

2. The nature of the surface ; its roughness or smooth-
ness.

3. The state of the body with reference to heat.

To this is to be referred the sensation of wetness and
dryness, on account of the close contact which fluids have
with the skin.

By means of touch, combined with pressure and motion,
we perceive—

1. Hardness and softness, comprehending elasticity,
friability, tenacity, flexibility, rigidity, fluidity, etc.

2. Friction, vibration, weight, motion, and the like.

The sensations of hunger and thirst, fatigue, and many
others, have no relation to the properties of bodies.

LUCRETIUS ON EMPTY SPACE.

Nec tamen undique corporeâ stipata tenentur
Omnia naturâ, namque est in rebus Inane.
Quod tibi cognôsse in multis erit utile rebus
Nec sinet errantem dubitare et quærere semper
De summâ rerum ; et nostris diffidere dictis.
Quapropter locus est intactus Inane Vacansque
Quod si non esset, nullâ ratione moveri

Res possent; namque officium quod corporis extat
Officere atque obstare, id in omni tempore adesset
Omnibus. Haud igitur quicquam procedere posset,
Principium quoniam cedendi nulla daret res,
At nunc per maria ac terras sublimaque cæli
Multa modis multis variâ ratione moveri
Cernimus ante oculos, quæ, si non esset Inane,
Non tam sollicito motu privata carerent
Quam genita omninô nullâ ratione fuissent,
Undique materies quoniam stipata quiesset.[1]

In connection with his logical studies it should be
mentioned that Professor Boole's attempt, made about
this time, towards giving to logical forms a mathematical expression,[2] had naturally strong attractions for
Clerk Maxwell.

---

[1] Lucr. *de Rer Nat.*, i. 329-345.   The following Hamiltonian
notions will be found appearing from time to time in Maxwell's correspondence and occasional writings :—

1. Opposition of Natural Realism to " Cosmothetic Idealism " :—

2. Unconscious Mental Modifications :—

3. Distinction between Knowledge and Belief in relation to the
doctrine of Perception :—

4. The Infinite or Unconditioned.

The following passage is worth quoting here, although the experiment in question was probably well known to Maxwell before he went
to college :—

" . . . The experiment which Sir W. Hamilton quotes from Mr.
Mill, and which had been noticed before either of them by Hartley.

" It is known that the seven prismatic colours, combined in certain
proportions, produce the white light of the solar ray. Now, if the
seven colours are painted on spaces bearing the same proportion to one
another as in the solar spectrum, and the coloured surface so produced
is passed rapidly before the eyes, as by the turning of a wheel, the
whole is seen as white."—Mill *On Hamilton* (1st edition, 1865), p. 286.

[2] Mr. George Boole's first logical treatise, " The Mathematical
Analysis of Logic, being an Essay towards a Calculus of Deductive
Reasoning," was published in 1847 at Cambridge, by Macmillan,
Barclay, and Macmillan.

I

The metaphysical writers who had received most of his attention before going to Cambridge were Descartes and Leibnitz. He knew Hobbes well also, but chiefly on the ethical side.

When, in his address to Section A of the British Association at the Liverpool meeting in 1870, Maxwell spoke of the barren metaphysics of past ages, he knew the full force of his own words. And he certainly felt that his psychological studies had given him a distinct advantage in conceiving rightly the functions of the eye.

His grasp of Moral Philosophy at the age of nineteen,—when he had been stimulated to precise thought on the subject by listening to the vague harangues of Professor Wilson (Christopher North),—appears in some of his letters, and reveals an aspect of his genius of which too little is known ; and one which his subsequent career did not allow him to carry to perfection.

In the third year of his course, as above mentioned, besides attending Professor Wilson's lectures, he renewed his study of Chemistry under Professor Gregory, to whose laboratory he had unlimited access, and also to some extent continued his attendance on Professor Forbes.

It cannot be said that this period was unfruitful ; yet perhaps it is to be regretted that he did not go to Cambridge at least one year earlier. His truly sociable spirit would have been less isolated, he would have gained more command over his own genius, and his powers of expression would have more harmoniously developed. The routine of Cambridge would have

been not only more valuable but less irksome to him,
and he would have entered sooner and more fully upon
the study of mankind, for which he had such large
capacity, and opportunities hitherto so limited. He suffered less from isolation than most human beings, and
his spirit was deepening all the while ; yet the freedom
of working by himself during the summer months had
manifestly some drawbacks, and the tone of his correspondence shows that he felt the disadvantages of
solitude.

LETTERS, 1847 TO 1850—ÆT. 16-18.

To LEWIS CAMPBELL, Esq.

27 Heriot Row,[1]
Tuesday [16th Novr. 1847].

In Kelland we find the value of expressions in numbers   *Æt.* 16.
as fast as we can, the values of the letters being given ; light
work.   In Forbes we do Lever, which is all in Potter ; no
notes required, only read Pottery ware (light reading).
Logic needs long notes.   On Monday, Wednesday, Friday, I
read Newton's *Fluxions* in a sort of way, to know what I am
about in doing a prop.   There is no time of reading a book
better than when you need it, and when you are on the
point of finding it out yourself if you were able.

      .   .   .   .   .

    " Non usitata nec tenui ferar
      Pinnâ biformi per liquida æquora
      Piscis neque in terris morabor
      Longius "——but I will take to swimming
with a two formed oar with blades at right angles. . . .

            Yours, J. C. M., No. 2.

To THE SAME.

31 Heriot Row, Novr. 1847.

As you say, sir, I have no idle time.   I look over

―――――――――――――――――――――

[1] Above, p. 55.

notes and such like till 9.35, then I go to Coll., and I always go one way and cross streets at the same places; then at 10 comes Kelland. He is telling us about arithmetic and how the common rules are the best. At 11 there is Forbes, who has now finished introduction and properties of bodies, and is beginning Mechanics in earnest. Then at 12, if it is fine, I perambulate the Meadows; if not, I go to the Library and do references. At 1 go to Logic. Sir W. reads the first $\frac{1}{2}$ of his lecture, and commits the rest to his man, but reserves to himself the right of making remarks. To-day was examination day, and there was no lecture. At 2 I go home and receive interim aliment, and do the needful in the way of business. Then I extend notes, and read text-books, which are Kelland's *Algebra* and Potter's *Mechanics*. The latter is very trigonometrical, but not deep; and the Trig. is not needed. I intend to read a few Greek and Latin beside. What books are you doing? . . . In Logic we sit in seats lettered according to name, and Sir W. takes and puts his hand into a jam pig [1] full of metal letters (very classical), and pulls one out and examines the bench of the letter. The Logic lectures are far the most solid and take most notes.

Before I left home I found out a prop for Tait (P. G.); but he *will* not do it. It is " to find the algebraical equation to a curve which is to be placed with its axis vertical, and a heavy body is to be put on any part of the curve, as on an inclined plane, and the horizontal component of the force, by which it is actuated, is to vary as the nth power of the perpendicular upon the axis."

To the Same.

Glenlair, 26th April 1848.

. . . On Saturday, the natural philosophers ran up Arthur's Seat with the barometer. The Professor set it up at the top and let us pant at it till it ran down with drops. He did not set it straight, and made the hill grow fifty feet; but we got it down again.

---

[1] *i.e.* Jar.

We came here on Wednesday by Caledonian.   I intend to open my classes next week after the business is over.   I have been reading Xenophon's *Memorabilia* after breakfast; also a French collection book.   This from 9 to 11.   Then a game of the Devil, of whom there is a duality and a quaternity of sticks, so that I can play either conjunctly or severally.   I can jump over him and bring him round without leaving go the sticks.   I can also keep him up behind me.

Then I go in again to science, of which I have only just got the books by the carrier.   Hitherto I have done a prop on the slate on polarised light.   Of props I have done several.

1. Found the equation to a square.

2. The curve which Sir David Brewster sees when he squints at a wall.

3. A property of the parabola. . . .

4. The same of the Ellipse and Hyperbola. . . .

I can polarise light now by reflection or refraction in 4 ways, and get beautiful but evanescent figures in plate glass by heating its edge.   I have not yet unannealed any glass. . . .

I don't understand how you mug[1] straight on.   I suit my muggery to my temper that day.   When I am deep I read Xenophon's defence of $\Sigma\omega\kappa$; when not I read $\Sigma\omega\kappa$'s witty dialogues.   If I do not do this, I always find myself *reading Greek*, that is, reading the words with all their contractions, as a Jew reads Hebrew.   I get on very rapidly; but know nothing about the meaning, and do not even know but that I am really translating.

Please to write about your Prizes at College, and about coming here to mug?   You must learn the D——l.

Tatties is planting.—Yours, etc.

To the Same.

Glenlair, 5/6 July 1848.

I was much glad of your letter, and will be thankful for *Æt.* 17. a repetition.   I understand better about your not coming.

---

[1] *i.e.* Work.

I have regularly set up shop now above the wash-house at the gate, in a garret.[1]   I have an old door set on two barrels, and two chairs, of which one is safe, and a skylight above, which will slide up and down.

On the door (or table), there is a lot of bowls, jugs, plates, jam pigs,[2] etc., containing water, salt, soda, sulphuric acid, blue vitriol, plumbago ore; also broken glass, iron, and copper wire, copper and zinc plate, bees' wax, sealing wax, clay, rosin, charcoal, a lens, a Smee's Galvanic apparatus, and a countless variety of little beetles, spiders, and wood lice, which fall into the different liquids and poison themselves. I intend to get up some more galvanism in jam pigs; but I must first copper the interiors of the pigs, so I am experimenting on the best methods of electrotyping.   So I am making copper seals with the device of a beetle.   First, I thought a beetle was a good conductor, so I embedded one in wax (not at all cruel, because I slew him in boiling water in which he never kicked), leaving his back out; but he would not do.   Then I took a cast of him in sealing wax, and pressed wax into the hollow, and black-leaded it with a brush; but neither would that do.   So at last I took my fingers and rubbed it, which I find the best way to use the black lead.   Then it coppered famously.   I melt out the wax with the lens, that being the cleanest way of getting a strong heat, so I do most things with it that need heat. To-day I astonished the natives as follows.   I took a crystal of blue vitriol and put the lens to it, and so drove off the water, leaving a white powder.   Then I did the same to some washing soda, and mixed the two white powders together; and made a small native spit on them, which turned them green[3] by a mutual exchange, thus:—1. Sulphate of copper and carbonate of soda.   2. Sulphate of soda and carbonate of copper (blue or green).

With regard to electro-magnetism you may tell Bob that I have not begun the machine he speaks of, being occupied with better plans, one of which is rather down cast, however,

---

[1] In the "old house" (above, p. 26, l. 4); now fitted up as a gun-room.
[2] *i.e.* Jars.   [3] This is still remembered by "Lizzy," Mrs. MacGowan.

because the machine when tried went a bit and stuck; and
I did not find out the impediment till I had dreamt over it
properly, which I consider the best mode of resolving difficulties of a particular kind, which may be found out by
thought, or especially by the laws of association.   Thus, you
are going along the road with a key in your pocket.   You
hear a clink behind you, but do not look round, thinking it
is nothing particular; when you get home the key is gone;
so you dream it all over, and though you have forgotten
everything else, you remember the look of the place, but do
not remember the locality (that is, as thus, " Near a large
thistle on the left side of the road "—nowhere in particular,
but so that it can be found).   Next day comes a woman
from the peats who has found the key in a corresponding
place.   This is not " believing in dreams," for the dream
did not point out the place by the general locality, but by
the lie of the ground.

Please to write and tell how Academy matters go, if
they are coming to a head.   I am reading Herodotus, *Euterpe*,
having taken the turn; that is to say, that sometimes I can
do props, read diff. and Int. Calc., Poisson, Hamilton's dis-
sertations, etc.   Off, then I take back to experiments, history
of what you may call it, make up leeway in the newspapers,
read Herodotus, and draw the figures of the curves above.
O deary, 11 P.M.!   Hoping to see you *before* October. . . .
I defer till to-morrow.

July 6. To-day I have set on to the coppering of the
jam pig which I polished yesterday.

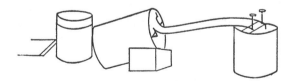

I have stuck in the wires better than ever, and it is
going on at a great rate, being a rainy day, and the skylight
shut and a smell of Hydrogen gas.   I have left it for an
hour to read Poisson, as I am pleased with him to-day.

Sometimes I do not like him, because he pretends to give information as to calculations of sorts, whereas he only tells how it might be done if you were allowed an infinite time to do it in, as well as patience. Of course he never stoops to give a particular example or even class of them. He tells lies about the way people make barometers, etc.

I bathe regularly every day when dry, and try aquatic experiments.

I first made a survey of the pool, and took soundings and marked rocky places well, as the water is so brown that one cannot see one's knees (pure peat, not mud). People are cutting peats now. So I have found a way of swimming round the pool without knocking knees. The lads[1] are afraid of melting, except one. No one here would touch water if they could help it, because there are two or three eels in the pool, which are thought near as bad as adders.

I took down the clay gun and made a centrifugal pump of it; also tried experiments on sound under water, which is very distinct, and I can understand how fishes can be stunned by knocking a stone.

We sometimes get a rope, which I take hold of at one end, and Bob Fraser the other, standing on the rock; and after a flood, when the water is up, there is sufficient current to keep me up like a kite without striking at all.

The thermometer ranged yesterday from 35° to 69°.

I have made regular figures of 14, 26, 32, 38, 62, and 102 sides of cardboard.

Latest intelligence—Electric Telegraph. This is going so as to make a compass spin very much. I must go to see my pig, as it is an hour and half since I left it; so, sir, am your afft. friend,   JAMES CLERK MAXWELL.

TO THE SAME.

Glenlair, 22d Sept. 1848.

. . . When I waken I do so either at 5.45 or 9.15, but I now prefer the early hour, as I take the most of my violent exercise at that time, and thus am *saddened down*, so that I

---

[1] See above, p. 93.

can do as much still work afterwards as is requisite, whereas if I was to sit still in the morning I would be yawning all day. So I get up and see what kind of day it is, and what field works are to be done ; then I catch the pony and bring up the water barrel.[1] This barrel used to be pulled by the men, but Pa caused the road to be gravelled, and so it became horse work to the men, so I proposed the pony ; but all the men except the pullers opposed the plan. So I and the children not working brought it up, and silenced vile insinuators. Then I take the dogs out, and then look round the garden for fruit and seeds, and paddle about till breakfast time ; after that take up Cicero and see if I can understand him. If so, I read till I stick ; if not, I set to Xen. or Herodt. Then I do props, chiefly on rolling curves, on which subject I have got a great problem divided into Orders, Genera, Species, Varieties, etc.

One curve rolls on another, and with a particular point traces out a third curve on the plane of the first, then the problem is :—Order I. Given any two of these curves, to find the third.

Order II. Given the equation of one and the identity of the other two, find their equation.

Order III. Given all three curves the same, find them. In this last Order I have proved that the equi-angular spiral possesses the property, and that no other curve does. This is the most reproductive curve of any. I think John Bernoulli had it on his tombstone, with the motto *Eadem mutata resurgo*. There are a great many curious properties of curves connected with rolling. Thus, for example,—

If the curve A when rolled on a straight line produces a curve C, and if the curve A when rolled up on itself produces the curve B, then the curve B when rolled upon the curve C will produce a straight line.

Thus, let the involute of the circle be represented by A,
the spiral of Archimedes by B,
and the parabola by C,
then the proposition is true.

---

[1] See above, p. 28, l. 6.

Thus the parabola rolled on a straight line traces a Catenary with its focus, an easy way to describe the Catenary. Professor Wallace just missed it in a paper in the Royal Society.

After props come optics, and principally polarised light.

Do you remember our visit to Mr. Nicol? I have got plenty of unannealed glass of different shapes, for I find window glass will do very well made up in bundles. I cut out triangles, squares, etc., with a diamond, about 8 or 9 of a kind, and take them to the kitchen, and put them on a piece of iron in the fire one by one. When the bit is red hot, I drop it into a plate of iron sparks to cool, and so on till all are done. I have got all figures up to nonagons, triangles of all kinds, and irregular chips. I have made a pattern for a tesselated window of unannealed glass in the proper colours, also a delineation of triangles at every principal inclination. We were at Castle-Douglas yesterday, and got crystals of salt Peter, which I have been cutting up into plates to-day, in hopes to see rings. There are very few crystals which are not hollow-hearted or filled up with irregular crystals. I have got a few cross cuts like ⬡ free of irregularities and long [wedge-shaped] cuts for polarising plates. One has to be very cautious in sawing and polishing them, for they are very brittle.

I have got a lucifer match box fitted up for polarising, thus. The rays suffer two reflections at the polar-

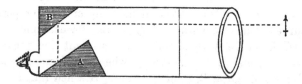

ising angle from glasses A and B. Without the lid it does for an analysing plate. In the lid there is set a plate of mica, and so one observes the blue sky, and turns the box round till a particular colour appears, and then a line on the lid of the box points to the sun wherever he is. Thus one can find out the time of day without the sun.

[*Here follow thirteen diagrams of patterns in triangles, squares, pentagons, and hexagons.*] These are a few of the figures one sees in unannealed glass.

Pray write soon and tell when, how, and where by, you intend to come, that you may neither on the other hand fall upon us at unawares, nor on the one hand break and not come at all. I suppose when you come I will have to give up all my things of my own devizing, and take Poisson, for the time is short, and I am very nearly unprepared in actual reading, though a great deal more able to read it.

I hope not to write any more letters till you come. I seal with an electrotype of the young of the ephemera. So, sir, I was, etc.

To the Same.

Bannavie, 6th July [*1849*].

This being a wet night, and I having exhausted my travelling props, set to to write to you about what I can *Æt.* 18. recollect of my past history. It is curious that though the remembrance of ploys remains longer than that of home doings, it is not so easily *imagined* after a short interval.

By *imagining* is here meant bringing up an accurate image of thoughts, words, and works, and not a mere geographical summary of voyages and travels.

But to the point. Perhaps you remember going with my Uncle John Cay (7th Class), to visit Mr. Nicol in Inverleith Terrace. There we saw polarised light in abundance. I purposed going this session but was prevented. Well, sir, I received from the aforesaid Mr. Cay a " Nicol's prism," which Nicol had made and sent him. It is made of calc-spar, so arranged as to separate the ordinary from the extraordinary ray. So I adapted it to a camera lucida, and made charts of the strains in unannealed glass.

I have set up the machine for showing the rings in crystals, which I planned during your visit last year. It answers very well. I also made some experiments on compressed jellies in illustration of my props on that subject The principal one was this :—The jelly is poured while hot into the annular space contained between a paper cylinder

and a cork: then, when cold, the cork is twisted round, and the jelly exposed to polarised light, when a transverse cross, ×, not +, appears with rings as the inverse square of the radius, all which is fully verified. Hip! etc. Q.E.D.

But to make an *abrupt transcision,*[1] as Forbes says, we set off to Glasgow on Monday 2d; to Inverary on 3d; to Oban by Loch Awe on 4th; round Mull, by Staffa and Iona, on (5th), and here on 6th. To-morrow we intend to get to Inverness and rest there. On Monday perhaps? to the land of Beulah, and afterwards back by Caledon. Canal to Crinan Canal, and so to Arran, thence to Ardrossan, and then home. It is possible that you may get a more full account of all these things (if agreeable), when I fall in with a pen that will spell; my present instrument partakes of the nature of skates, and I can hardly steer it.

There is a beautiful base here for measuring the height, etc., of Ben Nevis. It is a straight and level road through a moss for about a mile that leads from the inn right to the summit.

It is proposed to carry up stones and erect a cairn 3 feet high, and thus render it the highest mountain in Scotland.

During the session Prof. Forbes gave as an exercise to describe a cycloid from top of Ben Nevis to Fort William, and slide trees down it. We took an observation of the slide, but found nothing to slide but snow.

I think a body *deprived* of friction would go to Fort William in a cycloid in 49·6 seconds, and in 81 on an inclined plane. I believe I should have written the greater part of this letter to Allan Stewart, but I know not where he is, so you get it, and may read it or no as you like.

We will be at home between the 15th and 22d of this month, so you may write then, detailing your plans and specifying whether you intend to come north at all between this and November, for we would be glad if either you or Bob would disturb our solitude.

---

[1] See below, p. 138, l. 2. Forbes was extra-precise in articulation. Hence the spelling.

To the Same.

Glenlair, 19th October 1849.

Here is the way to dissolve any given historical event in a mythical solution, and then precipitate the seminal ideas in their primitive form. It is from Theodore Parker, an American, and treats of the declaration of American Independence. " The story of the Declaration of Independence is liable to many objections if we examine it *à la mode* Strauss. The Congress was held at a mythical town, whose very name is suspicious,—Philadelphia, brotherly love. The date is suspicious: it was the fourth day of the fourth month (reckoning from April, as it is probable that the Heraclidæ and Scandinavians, possible that the Americans, and certain that the Ebrews, did). Now 4 was a sacred number among Americans : the President was chosen for 4 years, 4 departments of affairs, 4 political powers, etc. The year also is suspicious. 1776 is but an *ingeni*[*ous*] ? combination of the sacred number, thus—

$$444$$
$$4$$
$$\overline{}$$
$$1776$$

Still further, the declaration is metaphysical and presupposes an acquaintance with the transcendental function on the part of the American people. Now the *Kritik of Pure Reason* was not yet published," etc.

To the Same.

October 1849.

Since last letter, I have made some pairs of diagrams representing solid figures and curves drawn in space ; of these pictures one is seen with each eye by means of mirrors, thus . . .

This is Wheatstone's Stereoscope, which Sir David Brewster has taken up of late with much violence at the Brightish Association. (The violence consists in making two lenses out of one by breaking it). (See Report). Last winter he exhibited at the Scottish Society of Arts Calotype pictures of the statue of Ariadne and the beast seen from two stations, which, when viewed properly, appeared very solid.

Since then I have been doing practical props on compression, and writing out the same that there may be no mistake. The nicest cases are those of spheres and cylinders. I have got an expression for the hardness of a cricket ball made of case and stuffing. I have also the equations for a spherical cavity in an infinite solid, and this prop: Given that the polarised colour of any part of a cylinder of unannealed glass is equal to the square of the distance from the centre (as determined by observation), to find—1st, the state of strain at each point; 2d, the temperature of each.

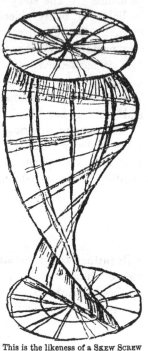

. . .

I have got an observation of the latitude just now with a saucer of treacle, but it is very windy.

Pray excuse this wickedly perplexed letter as an effect of the paucity of our communications. If you would sharpen me a little it would be acceptable, but when there is nobody to speak to one [*loses*] the gift of speech. . . .

This is the likeness of a SKEW SCREW SURFACE.

To the Same.

31 Heriot Row, Edinburgh, N.B.,
Monday, Nov. 5, 1849.

. . . . . .

I go to Gregory to Chemistry at 10, Morale Phil. at 12, and Pract. Chem. at 13, finishing at 14, unless perhaps I take an hour at Practical Mechanics at the School of Arts. I do not go to Sir W. H. logic, seeing I was there before.

Langhorne has got your Buchananic notes. Why do you think that I can endure nothing but Mathematics and Logic, the only things I have plenty of? and why do you presuppose my acquaintance with your preceptors, professors, tutors,[1] etc. ? . . .

I don't wonder at your failing to take interest in the exponential theorem, seeing I dislike it, although I know the use and meaning of it. But I never *would have*, unless Kelland had explained it. . . .

In your next letter you may give an abstract of Aristotle's *Rhet.*, for I do not attend Aytoun, and so I know not what Rhet. is. I know Logic only by Reid's account of it. I will tell you about Wilson's Moral Philosophy, provided always you want to know, and signify your desire.

To THE SAME.

[Edin.] 14th March 1850.

As I am otherwise engaged, I take this opportunity of provoking you to write a letter or two. I have begun to write Elastic Equilibrium, and I find that I must write you a letter in order *at least, indeed,*[2] to serve *on the one hand* as an excuse to myself for sticking up, and *on the other hand* as a sluice for all the nonsense which I would have written. I therefore propose to divide this letter as follows: —My say naturally breaks up into—1. Education; 2. Notions; 3. Hearsay.

1. Education—Public.

10-11.—Gregory is on Alloys of Metals just now. Last Saturday I was examined, and asked how I would do if the contents of a stomach were submitted to me to detect arsenic, and I had to go through the whole of the preparatory processes of chopping up the tripes, boiling with potash, filtering, boiling with $H C L$ and $K O C L O_5$, all which Kemp the *Practical* says are useless and detrimental processes, invented by chemists who want something to do. 11-12.—Prof. Forbes is on Sound and Light day about, as Bob well knows, and can tell you if he chooses. He (R. C.) has written an essay

---

[1] At Oxford.   [2] See above, p. 82.

on Probabilities, with very grand props in it; everything original, but no signs of reading, I guess. It was all written in a week. He has despaired of Optics.—12-1. Wilson, after having fully explained his own opinions, has proceeded to those of other great men: Plato, Aristotle, Stoics, Epicureans. He shows that Plato's *proof* of the immortality of the soul, from its immateriality, if it be a proof, proves its pre-existence, the immortality of beasts and vegetables, and why not transmigration? (Do you remember how Raphael tells Adam about meats and drinks in *Paradise Lost?*) (Greek Iambics, if you please.) He quarrels with Aristotle's doctrine of the Golden Mean,—" a virtue is the mean between two vices,"—not properly understanding the saying. He chooses to consider it as a pocket rule to find virtue, which it is not meant to be, but an apophthegm or maxim, or dark saying, signifying that as a hill falls away on both sides of the top, so a virtue at its maximum declines by excess or defect (not of virtue but) of some variable quantity at the disposal of the will. Thus, let it be a virtue to give alms with your own money, then it is a greater virtue to pay one's debts to the full. Now, a man has so much money: the more alms he gives up to a certain point, the more virtue. As soon as it becomes impossible to pay debts, the virtue of solvency decreases faster than that of almsgiving increases, so that the giving of money to the poor becomes a vice, so that the variable is the sum given away, by excess or defect of which virtue diminishes, say I; so that Wilson garbles Aristotle,—but I bamboozle myself. I say that some things are virtues, others are virtuous or generally lead to virtue. Substitute *goods* for virtues, and it will be more general: thus, Wisdom, Happiness, Virtue, are *goods*, and cannot be in excess; but Knowledge, Pleasure, and— what? (please tell me, Is it Propriety, Obedience, or what is it?) lead to the other three, and are not so much goods as tending to good; whereas particular knowledges, pleasures, and obediences may be in excess and lead to evils. I postpone the rest of my observations to my Collection of the Metaphysical principles of Moral Philosophy founded on the three laws of Liberty, Equality, Fraternity, thus expressed :—

1. That which can be done is that which has been done; that is, that the possibility (with respect to the agent) of an action (as simple) depends on the agent having had the sensation of having done it. [1]

2. That which ought to be done is that which (under the given conditions) produces, implies, or tends to the greatest amount of good (an excess or defect in the variables will lessen the good and make evil).

3. Moral actions can be judged of only by the principle of exchange; that is (1), our own actions must be judged by the laws we have made for others; (2), others must be judged by putting ourselves in their place.

22d March.

At Practical Mechanics I have been turning Devils of sorts. For private studies I have been reading Young's *Lectures,* Willis's *Principles of Mechanism,* Moseley's *Engineering and Mechanics,* Dixon on *Heat,* and Moigno's *Répertoire d'Optique.* This last is a very complete analysis of all that has been done in the optical way from Fresnel to the end of 1849, and there is another volume a-coming which will complete the work. There is in it besides common optics all about the other things which accompany light, as heat, chemical action, photographic rays, action on vegetables, etc.

My notions are rather few, as I do not *entertain* them just now. I have a notion for the torsion of wires and rods, not to be made till the vacation; of experiments on the action of compression on glass, jelly, etc., numerically done up; of papers for the Physico-Mathematical Society (which is to revive in earnest next session!); on the relations of optical and mechanical constants, their desirableness, etc., and suspension-bridges, and catenaries, and elastic curves. Alex. Campbell, Agnew, and I are appointed to read up the subject

---

[1] Maxwell often insisted on this in conversation, with especial reference to our command of the muscles depending on experience of the muscular sense.

of periodical shooting stars, and to prepare a list of the phenomena to be observed on the 9th August and 13th November.
The Society's barometer is to be taken up Arthur's Seat at
the end of the session, when Forbes goes up, and All students
are invited to attend, so that the existence of the Society
may be recognised.

I have notions of reading the whole of *Corpus Juris* and
Pandects in no time at all; but these are getting somewhat
dim, as the Cambridge scheme has been howked up from
its repose in the regions of abortions, and is as far forward
as an inspection of the Cambridge *Calendar* and a communication with Cantabs.

Mr. Bob is choosing his college. I rejected for him all
but Peter's, Caius, or Trinity Hall, the last being, though
legal, not in favour, or lazy, or something. Caius is popu-
lous, and is society to itself. Peter's is select, and knows
the University. Please give me some notions on these
things, both for Bob and me. I postpone my answer to
you about the Gorham business till another time, when also
I shall have read Waterland on *Regeneration*, which is with
Mrs. Morrieson, and some Pusey books I know. In the
meantime I admire the *Judgement* as a composition of great
art and ingenuity.

What cross influences had delayed his entrance at
Cambridge may be guessed at, but cannot be clearly
known. Mr. Clerk Maxwell was always slow in making up his mind, and the habit of *inertia* had grown
upon him. There had been a lingering expectation,
to which James alludes in the preceding letter, that
he would follow his father's profession, and become a
member of the Scottish Bar. And although he himself
felt, as he told me at the time, that it was "another
kind of laws" he was called upon to study, the
practical result of this conviction was slow in assert

ing itself. The fact that in going to Cambridge he decided against the profession which his friends had destined for him made the step a more serious one than it might have otherwise been. The close and constant intercourse between father and son made the parting more difficult. James's delicate health[1] would count heavily amongst the reasons *contra,* and certain floating prejudices about the " dangers of the English universities," Puseyism, infidelity, etc., had then considerable hold, especially on the Presbyterian mind. James himself was patient, and had hitherto decided nothing for himself. Only when Tait and Allan Stewart were already at Cambridge, my brother Robert destined for it, and myself at Oxford, his own voice was added to those which had long been urging the claims of Cambridge,[2] and then they prevailed. There had been searchings of heart on the subject as early as April 1849, when the following entries occur in the Diary :—

---

[1] On this subject the following entry from Mrs. Morrieson's Diary is of some interest :—" *5th Decr. 1846.*—L. and R. dined with Miss Cay. James Maxwell has been under her care during his father's absence and has been suffering very much from toothache and earache, in consequence of cutting his eye-teeth—an extraordinary thing at 15." To which may be added the following from his father's Diary :—" *1846, Sa., Dec. 12.*—Jas. still affected by the tooth. . . . Took a short walk, and came back by Mr. Nasmith (dentist), and went in, and on consultation, got the tooth drawn at once—it was nicely and quickly done, and Jas. never winced."

[2] Amongst these may be specially mentioned Mr. Hugh Blackburn and Dean Ramsay. Professor Forbes and Charles Mackenzie were also interested in the question. Professor Blackburn, in particular, insisted that the mathematical discipline of Cambridge would enable him to exercise his genius more effectively.

T. 17.—Called on Hugh Blackburn to talk about Cambridge.

W. 18.—Called on Capt. Wemyss to talk of Cambridge, and Prof. Forbes called on me and had a talk on James's studies, etc.

The only other document at my command which bears upon the point is a journal kept by my mother, then Mrs. Morrieson, in which she occasionally noted matters relating to her sons' friends. She was herself at this time (1850) making inquiries about Cambridge for my brother Robert. The following entries may be quoted :—

1850. *1st April.*—James Clerk Maxwell came in full of Forbes's recommendation of Trinity College above all others at Cambridge, and that Peterhouse was less expensive than Caius; that the latter is too full to admit of rooms, and freshmen are obliged to lodge out.

28th October.—I had a kind letter from Mr. C. M., from Glenlair, after placing his son James at Peterhouse. He has already distinguished himself at Edinburgh by papers on the compression of solids, and other scientific subjects, read for him at the Royal Society. His manners are very peculiar; but having good sense, sterling worth, and good humour, the intercourse with a College world will rub off his oddities. I doubt not of his becoming a distinguished man.

January 1851.—James Clerk Maxwell often comes in. He is full of genius. He went to Cambridge with Robert in October, R. to Caius, James to Peterhouse; but he is "migrating" to Trinity, and I have no doubt he will be a distinguished Philosopher some day.

In concluding the account of this period, I again beg leave to quote from Professor Tait's excellent paper :—

The winter of 1847 found us together in the classes of Forbes and Kelland, where he highly distinguished himself. With the former he was a particular favourite, being admitted to the free use of the class apparatus for original experiments. He lingered here behind most of his former associates, having spent three years at the University of Edinburgh, working (without any assistance or supervision) with physical and chemical apparatus, and devouring all sorts of scientific works in the library.[1]   During this period he wrote two valuable papers, which are published in our *Transactions*, on *The Theory of Rolling Curves* and on *The Equilibrium of Elastic Solids*.   Thus he brought to Cambridge, in the autumn of 1850, a mass of knowledge which was really immense for so young a man, but in a state of disorder appalling to his methodical private tutor.[2]   Though that tutor was William Hopkins, the pupil to a great extent took his own way, and it may safely be said that no high wrangler of recent years ever entered the Senate-house more imper-

---

[1] " From the University Library lists for this period it appears that Maxwell perused at home Fourier's *Theorie de la Chaleur*, Monge's *Géometrie Descriptive*, Newton's *Optics*, Willis's *Principles of Mechanism*, Cauchy's *Calcul Différentiel*, Taylor's *Scientific Memoirs*, and many other works of a high order.   Unfortunately no record is kept of books consulted in the reading-room."

[2] On the other hand it should be mentioned (though the statements are not contradictory) that Hopkins used to say he had never known Maxwell " make a mistake," *i.e.* he never misapprehended the conditions of any problem.   Of this fact, which was communicated to me by my brother, I have since received the following confirmation from Mr. W. N. Lawson, of the Equity Bar :—Mr. Lawson quotes from a diary kept by himself at the time—" *July* 15, 1853.—He (Hopkins) was talking to me this evening about Maxwell.   He says he is unquestionably the most extraordinary man he has met with in the whole range of his experience ; he says it appears impossible for Maxwell to think incorrectly on physical subjects ; that in his analysis, however, he is far more deficient ; he looks upon him as a great genius, with all its eccentricities, and prophesies that one day he will shine as a light in physical science, a prophecy in which all his fellow-students strenuously unite."

fectly trained to produce "paying" work than did Clerk
Maxwell. But by sheer strength of intellect, though with
the very minimum of knowledge how to use it to advantage
under the conditions of the Examination, he obtained the
position of Second Wrangler, and was bracketed equal with
the Senior Wrangler in the higher ordeal of the Smith's
Prizes. His name appears in the Cambridge *Calendar* as
Maxwell of Trinity, but he was originally entered at Peterhouse, and kept his first term there, in that small but most
ancient foundation which has of late furnished Scotland with
the majority of the professors of mathematics and natural
philosophy, in her four universities.

To the books mentioned by Professor Tait in his
footnote should be added Poisson's *Mechanics*, which
was taken out from the Advocates' Library 6th March
1848, and carried off into the country. A copy of
Fourier's *Theorie de la Chaleur* was ordered through
Maclachlan, 6th April 1849, for 25s.

Letters, April to September 1850.

To Lewis Campbell, Esq.

Glenlair, 26th April 1850.

As I ought to tell you of our departure from Edinburgh
and arrival here, so I ought to tell you of many other
things besides. Of things pertaining to myself there are
these:—The tutor of Peterhouse has booked me, and I am
booked for Peterhouse, but will need a little more booking
before I can write Algebra like a book.

I suppose I must go through Wrigley's problems and
Paley's Evidences in the same sort of way, and be able to
translate when required Eurip. *Iph. in Aulid.* In the
meantime I have my usual superfluity of plans.

1. Classics—Eurip. Ιφ. εν Αυλ. for Cambridge. (I hope
no Latin or Greek verses except for honours.) Greek

Testament, Epistles, for my own behoof, and perhaps some of Cicero *De Officiis* or something else for Latin.

2. Mathematics—Wrigley's Problems, and Trig. for Cambridge; properties of the Ellipsoid and other solids for practice with Spher. Trig. Nothing higher if I can help it.

3. Nat. Phil.—Simple mechanical problems to produce that knack of solving problems which Prof. Forbes has taught me to despise. Common Optics at length; and for experimental philosophy, twisting and bending certain glass and metal rods, making jellies, unannealed glass, and crystals, and dissecting eyes—and playing Devils.

4. Metaphysics — Kant's *Kritik of Pure Reason* in German, read with a determination to make it agree with Sir W. Hamilton.

5. Moral Philosophy—Metaphysical principles of moral philosophy. Hobbes' *Leviathan,* with his moral philosophy, to be read as the only man who has decided opinions and avows them in a distinct way. To examine the first part of the seventh chapter of Matthew in reference to the moral principles which it supposes, and compare with other passages.

But I question if I shall be able to overtake all these things, although those of different kinds may well be used as alternate studies.

I read in Edinburgh Wilson's Poems to see what he used to be like, and how much he had improved since then. Did you finish Festus?[1] I had only two days to read it, so that I skipped part of the long speech and a good deal of the jollification, which I think the dullest part of the book.

The opening makes one think that it is to be an imitation of the book of Job, but you soon see that you have to do with a dreamy mortal without a profession, but vain withal, and a hero among women, a jolly companion of some men, admired of students for talking of things which he knows not, nor can know, having a so-called philosophy, an intuitive science, and an underived religion, and with all these not perfect, but needing more expanded views of the

---

[1] Festus, a poem by Philip James Bailey, 1849.

folly of strict virtue and outward decency, of the magnificence, nay, of the duty of sinning, and of the identity of
virtue and vice, and of all opposites. He takes for his
friend one whom Wilson calls a very *poor devil*, who has
wonderful mechanical powers, but never attempts but once
the supposed object of his visit to earth, namely, tempta-
tion. He takes a more rational view of affairs than Festus
in general, but is so extremely refined from ordinary devils,
that the only passage sufficiently characteristic for ordinary
rapid readers to recognise is the sermon to the crowd, as the
speech in Hell is quite raw. He has not such an absolute
and intuitive sense of things as Festus, and does not change
so much according to his company. He seems a sincere,
good-natured, unselfish devil; while Festus is very change-
able, solemn when alone, jolly when with the jolly, drunk
with the drunk, open with Lucifer, reserved in good company, amorous with all women, talkative and serious. with
all angels and saints, stern towards the unfortunate, and in
all his affections altogether selfish.

The book is said to have a plan, but no plot. The plan
is an exposition of the state of a man's mind after having
gone through German metaphysics. It was one destitute of
notions, and has now been convinced that all these notions
are one and the same. It is neither meat, nor drink, nor
rank, nor money, nor any common thing he wants: "he is
sure it isn't," and he is sore troubled for want of some great
thing to do; and when L——r starts into proximity he is
the very being he wanted to speak to; "he knew who it
would be," and recognises him at once. An opportunity is
thus given for showing two ways of thinking about things,
and therein lies the matter of the book. This may be seen
in L——r's sermon and Festus' prayer. To turn and get
out of the confusion of this letter, pray let me hear your
opinion of the book. It may be considered thus:—

1. People read the book and wonder, why ?

It is not read for the sake of the story—that is plain;
nor for the clearness with which certain principles are
developed, nor for the consistency of the book, nor for the

variety of the characters ; there must therefore be something overpoweringly attractive to hold you to the book. Some say he has fine thoughts, sufficient to set up fifty poets; to which some may answer, Where are they ? Read it in a spirit of cold criticism, and they vanish. There is not one that is not either erroneous, absurd, German, common, or *Daft.* Where lies the beauty ? In the reader's mind. The author has evidently been thinking when he wrote it, and that not in words, but inwardly. The *benevolent reader* is compelled to think too, and it is so great a relief to the reader to get out of wordiness that he can put up with insanity, absurdity, profanity, and even inanity, if by so doing he can get into *rapport* with one who is so transcendental, and yet so easy to follow, as the poet. When Galileo set his [lamp] a-swinging by breathing on it, his power lay in the relation between the interval of his breaths and the time of vibration ; so in Festus the mind that begins to perceive that his train of thoughts is that of the poem is readily made to follow on. There are some passages where one breaks loose, especially the rhyming description of the subpœnaing of the planets, and the notion of the angel of the earth giving Festus a pair of bracelets, and the way in which F. improves his mind by travel.

. . . Beauty is attributed to an object when the subject anticipates pleasure in it. A true pleasure is a consciousness of the right action of the faculty or function or power. Happiness is the integral of pleasure, as wisdom is of knowledge. . . . Don't take all this about Festus for truth, as I don't believe much of it, and I'll maybe tell you a new story if you tell me one.

What of St. Peter, as compared with the Keys and with Bob ? [1]

FROM PROF. FORBES.

Edinburgh, 4th May 1850.

MY DEAR SIR—Professor Kelland, to whom your paper was referred by the Council R. S., reports favourably upon

---

[1] *i e.* Is Robert Campbell going to Peterhouse or Caius ?

it, but complains of the great obscurity of several parts, owing
to the *abrupt transitions* and want of distinction between
what is *assumed* and what is *proved* in various passages,
which he has marked in pencil, and which I trust that you
will use your utmost effort to make plain and intelligible.
It is perfectly evident that it must be useless to publish a
paper for the use of scientific readers generally, the steps of
which cannot, in many places, be followed by so expert an
algebraist as Prof. Kelland ;—if, indeed, they be *steps* at all
and not assumptions of theorems from other writers who are
not quoted.   You will please to pay particular attention to
clear up these passages, and return the MS. by post to Professor Kelland, West Cottage, Wardie, Edinburgh, so that he
may receive it by Saturday the 11th, as I shall then have
left town.—Believe me, yours sincerely,

JAMES D. FORBES.

To L. CAMPBELL, Esq.

[June ? 1850]

As there has been a long truce between us since I last
got a letter from you, and as I do not intend to despatch
this here till I receive Bob's answer with your address, I
have no questions to answer, and any news would turn old
by keeping, so I intend briefly to state my country occupations (otherwise preparation for Cambridge, if you please).
I find that after breakfast is the best time for reading Greek
and Latin, because if I read newspapers or any of those
things, then it is dissipation and ruin ; and if I begin with
props, experiments, or calculations, then I would be continually returning on them.   At first I had got pretty well
accustomed to regular study with a Dictionary, and did
about 120 lines of *Eurip.* a day, namely, 40 revised, 40 for
to-day, and 40 for to-morrow, with the looking up of to-
morrow's words.   As I am blest with Dunbar's *Lexicon*, it
is not very highly probable that I will find my word at all ;
if I do, it is used in a different sense from Dunbar's (so
much the better), and it has to be made out from the context (either of the author or the Dictionary).   So much for

regular study, which I have nearly forgot, for when I had got to the end of the first chorus I began to think of the rods and wires that I had in a box. They have entirely stopped *Eurip.*, for I found that if I spent the best part of the day on him, and took reasonable exercise, I could not much advance the making of the apparatus for tormenting these wires and rods. So the rods got the better of the Lexicon. The observations on the rods are good for little till they are finished; they are of three kinds, and are all distinguished for accuracy and agreement among themselves.

Thus a rod bent by a weight at the middle takes the form of a curve, which is calculated to be one of the fourth order. Let A C B be the rod bent by a weight at C. Mirrors fastened to it at A and B make known the changes of the inclination of the tangent to 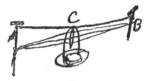 the rod there, and a lens at C projects an image of a copper scale of inches and parts from A to B, where it is observed, and so the deflection of the rod at C becomes known. Now the calculated value of the elasticity deduced from the deflection differs from that deduced from the observations on the mirrors by about $\frac{1}{140}$ of either, and as the deflection at C was about $\frac{1}{4}$ inch, the difference of the observed and calculated deflections is about $\frac{1}{280}$ of an inch, which is near enough for home-made philosophical instruments to go.

Thus you see I would run on about rods and wires, and weights, angles, and inches, and copper and iron, and silvered glass, and all sorts of practicalities. Where is now *Eurip.?*—Ay, where? On the top of the Lexicon, and behind bundles of observations and calculations. When will he come out? for he was a good soul after all, and wise (beg his pardon, *wiser*). For the rest I have been at Shakespeare and Cowper. I used to put Thomson and Cowper together (why?), and Thomson first; now they are reversed and far asunder.

As I suppose my occupations are not very like yours, I pray you send me an account of what Oxford notions you have got, either from Oxonians, books, or observation; and

as, if I was to question you, you could but answer my
questions, I leave you to question yourself and send me
some of the answers.

The only regular College science that I have thought
of lately is Moral Philosophy. Whether it is an Oxford
science I know not; but it must be, if not taught, at least
interesting; so I purpose to fill up this letter with unuttered
thoughts (or crude), which, as they are crammed into words,
may appear like men new waked from sleep, who leap in
confusion into one another's breeches, hardly fit to be seen
of decent men. Then think not my words mad if their
clothes fit them not, for they have not had an opportunity
of trying them on before.

There are some Moral Philosophers whose opinions are
remarkable for their general truth and good sense, but not
for their utility, fixity, or novelty.

They tell you that in all your actions you ought to be
virtuous, that benevolence is a virtue, that lawful rulers
ought to be obeyed, that a man should give ear to his conscience.

Others tell you of unalterable laws of right and wrong,
of Eternal truth and the Everlasting fitnesses of things.
Others of the duty of following nature, of every virtue
between two vices (Aristot.), and of the golden mean. That
a man should do what is best on the whole (1) for himself;
(2) for other men only, and *not* himself; (3) for the whole
universe, including himself, and so on. Now I think that
the answers to the following questions should be separate
parts of M. Ph.:—

1. What is man? This is the introduction, and is
called statical or proper Metaphysics.

2. What are the laws of human action? Action being
all that man does—thought, word, deed.

3. What are the motives of human actions?

4. What actions do men perform in preference to what
others, and why?

5. What is the principle by which men judge some
actions right, others wrong?

6. What do particular men think of this principle? What are their doctrines?

7. What is the best criticism of right and wrong, or what (to us) is absolute right?

8. What are the best motives of human actions?

9. How are these motives to be implanted without violating the laws of human action?

10. What might, or rather what *will*, mankind become after this has been effected?

Moral Philosophy differs from Nat. Phil. in this, that the more new things we hear of in Nat. Phil. the better; but in Mor. Phil. the old things are best, so that a common objection to Mor. Phil. is that everybody knows it all before. If a man tells you that tyranny and anarchy are bad things, and that a just and lawful government is a good thing, it sounds very fine, but only means that when men think the government bad from excess or defect they give it the name of tyranny and anarchy. The ancient virtue of Tyrannicide was a man's determination to kill the king whenever he displeased him. Thus it is easy to call a dog a bad name to beat him for. But there are other parts of Mor. Phil. in which there are differences of opinion, such as the nature of selfishness, self-love, appetites, desires, and affections, disinterestedness (what a word for a rush at!), which belong to the first three questions, and so on. I have told you something (p. 129) of three laws which I had been considering. In all parts of Mor. Phil. these three laws seem to meet one, and in each system of Morals they take a different form. Now, that I might not deceive myself in thinking that I was safe out of the hands of the philosophers who argue these matters, I have been looking into the books of Moralists the most opposed to one another, to see what it is that makes them differ, and wherein they agree. The three principles concerning the nature of man are continually changing their shape, so that it is not easy to catch them in their best shape. Nevertheless:

Lemma: Metaphysics.—A man thinks, feels, and wills, and therefore Metaphysicians give him the three faculties of cognition, feeling, and conation.

Cognition is what is called Understanding, and is most thought of generally.

Feelings are pleasures, pains, appetites, desires, aversions, approval and disapproval, love, hate, and all affections.

Conations are acts of will, whatever they be.

Now to move a man's will it is necessary to move his affections. (How? Wait!) For no convictions of the understanding will do, for a man does what he likes to do, not what he believes to be best for himself or others. The feelings can only be moved by notions coming through the understanding, for cognition is the only inlet of thoughts. Therefore, although it can be proved that self-love leads to all goodness, or, in other words, that goodness is happiness, and *self* loves happiness, yet it can also be proved that men are not able to act rightly from pure self-love; so that though self-love is a very fine theoretical principle, yet no man can keep it always in view, or act reasonably upon it. Now, most moralists take for granted that the end which men, good or bad, pursue is their own happiness, and that happiness, false or true, is the motive of every action, and that it is the only right motive. Others say that benevolence is the only virtue, and that any action not done expressly for the good of others is entitled to no praise.

Most of the ancients, and Hobbes among the moderns, are of the first opinion. Hutcheson and Brown (I think) are of the second, and call the first selfish Philosophers and the selfish school. A few consider benevolence to the whole universe as the proper motive of every action, but they all (says Macintosh) confound men's motives with the criterion of right and wrong, the reason why a thing is right, and that which actually causes a man to do it. In every book on Moral Philosophy some reference is made to that precept or maxim, which is declared to be the spirit of the law and the prophets (see Matt. vii. 12), and the application of it is a good mark of the uppermost thoughts or mode of thinking of the author.

Hobbes lays down as the first agreement of men to secure their safety, that a man should lay down so much of his natural liberty with respect to others, as he wishes that other men should to him. Hobbes having shown that men, in what the poets and moralists call a state of nature (that is, of equality and liberty, and without government), must be in a state of war, every man against every other, and therefore of danger to every man, deduces the obligation of obeying the powers that be from the necessity of Power to prevent universal war. Adam Smith's theory of Moral Sentiments (which is the most systematic next to Hobbes) is that men desire others to sympathise with them, and therefore do those things which may be sympathised with; that is, as Smith's opponents say, men ought to be guided by the desire of esteem and sympathy. Not so. Smith does not leave us there, but I suppose you have read him, as he is almost the only Scotch Moral Philosopher.

As it is Saturday night I will not write very much more. I was thinking to-day of the duties of [the] cognitive faculty. It is universally admitted that duties are voluntary, and that the will governs understanding by giving or withholding Attention. They say that Understanding ought to work by the rules of right reason. These rules are, or ought to be, contained in Logic; but the actual science of Logic is conversant at present only with things either certain, impossible, or *entirely* doubtful, none of which (fortunately) we have to reason on. Therefore the true Logic for this world is the Calculus of Probabilities, which takes account of the magnitude of the probability (which is, or which ought to be in a reasonable man's mind). This branch of Math., which is generally thought to favour gambling, dicing, and wagering, and therefore highly immoral, is the only " Mathematics for Practical Men," as we ought to be. Now, as human knowledge comes by the senses in such a way that the existence of things external is only inferred from the harmonious (not similar) testimony of the different senses, Understanding, acting by the laws of right reason, will assign to different truths (or facts, or testimonies, or what shall I call them) different degrees of probability. Now, as the senses give new testimonies continually, and as no man ever detected in them any real inconsistency, it follows that the probability and *credibility* of their testimony is increasing day by day, and the more a man uses them the more he believes them. He believes them. What is believing? When the probability (there is no better word found) in a man's mind of a certain proposition being true is greater than that of its being false, he believes it with a proportion of faith corresponding to the probability, and this probability may be increased or diminished by new facts. This is faith in general. When a man thinks he has enough of evidence for some notion of his he sometimes refuses to listen to any additional evidence *pro* or *con*, saying, "It is a settled question, *probatis probata*; it needs no evidence; it is certain." This is knowledge as distinguished from faith. He says, "I do not believe; I know." "If any man thinketh that he knoweth, he knoweth yet nothing as he ought to know." This knowledge is a shutting of one's ears to all arguments, and is the same as "Implicit faith" in one of its meanings. "Childlike faith," confounded with it, is not credulity, for children are not credulous, but find out sooner than some think that many men are liars. I must now to bed, so good night; only please to write when you get this, if convenient, and state the probability of your coming here. We perhaps will be in Edinburgh when the Wise men are there. Now you are invited in a corner of a letter by

JAMES CLERK MAXWELL.

Glenlair, 16th September 1850.

Professor W. Thomson has asked me [1] to make him some magne-crystallic preparations which I am now busy with. Now, in some of these bismuth is required, which is not to be found either in Castle-Douglas or Dumfries. I have, therefore, thought fit to request you, and do now request

---

1 This request had probably some connection with the meeting of the "wise men" in Edinburgh. Maxwell had been present, and had spoken in Section A. Professor Swan remembers the surprise felt by all but Forbes at seeing the beardless stripling rise to dispute some point in the colour theory with Sir David Brewster.

you, during your transit through Edinburgh on your way here, to go either to Mr. Kemp's establishment in Infirmary Street, beside the College, or to some other dealer in metals, and there purchase and obtain two ounces of metallic bismuth (called Regulus of Bismuth), either powder or lumpish—all one. Thus you may perceive that the end of this letter is in two ounces of Regulus of Bismuth, that is, the metal bismuth, which if you do bring it with you, will please me well. Not that I am turned chemist. By no means ; but common cook. My fingers are abominable with glue and chalk, gum and flour, wax and rosin, pitch and tallow, black oxide of iron, red ditto and vinegar. By combining these ingredients, I strive to please Prof. Thomson, who intends to submit them to Tyndall and Knoblauch, who, by means of them, are to discover the secrets of nature, and the origin of the magne-crystallic forces.

Now, if by coming here you could turn me from a cook to a grammarian by an irresistible influence you would do well ; but if you remember the way I used to translate at the Academy, distorting the Latin of Livy to mean what I had preconceived, you will understand that at first I had not only to find out what the author meant, but to become convinced that it could not be what I thought it was.

John Wilson's lectures on Moral Philosophy do not improve on reconsideration ; they become indistinct and are resolved into the excellence of happiness, the acquiredness of conscience, and general good-humour, philanthropy and φιλαγαθια. Here is an outline of Abstract Mor. Phil. :—

1. The principles of the growth of the mind (that is, the acquisition of opinions, propensities, and abilities).

2. The principles of government (the governor suits his actions to the laws of the thing governed).

3. The principle of sympathy (sauce for goose is sauce for gander).

Out of these heads may one make something ?

As it is bed-time, and I have to put the glue and oxide of iron into shape to-night, I must stop here, and remain in hope of seeing you soon (say when).

CHAPTER VI.

UNDERGRADUATE LIFE AT CAMBRIDGE—OCTOBER 1850 TO
JANUARY 1854—ÆT. 19-22.

BEFORE placing his son at Cambridge Mr. Clerk
Maxwell had, as was usual with him, consulted various
persons, including Professor James Forbes and Professor Kelland of Edinburgh, Professors Thomson and
Blackburn of Glasgow, and Charles Mackenzie, afterwards Bishop of Natal, then a Lecturer of Caius
College, Cambridge.[1] Forbes strongly advised Trinity,
and offered an introduction to Whewell; but after
various reasons urged for Trinity, Caius, and Peterhouse, the decision was in favour of Peterhouse.

Maxwell's first impression of college life, like that
of some other clever freshmen, was not one of un-
alloyed satisfaction. He was transplanted from the
rural solitudes of Galloway into the midst of a society
which was of curious interest to him, but did not
make him feel immediately at home. He found himself amongst the freshmen spelling out Euclid again,
and again "monotonously parsing" a Greek play.
He had brought with him his scraps of gelatine, gutta-

---

[1] He was the younger brother of Mr. Mackenzie of Silverknowe,
who had married Isabella Wedderburn.

percha, and unannealed glass, his bits of magnetised steel, and other objects, which were apt to appear to the uninitiated as "matter in the wrong place." And this in the home of science! Nor were his experiments facilitated by the casual "dropping-in" of the average undergraduate.

His boyish spirits and his social temper, together with the novelty of the scene, and a deep-rooted presentiment of the possibilities of Cambridge, no doubt made even his first term a happy one. But there was an undercurrent of restlessness and misgiving. And this made him lend a readier ear to the advice which was pressed upon him from various quarters, that he should migrate to Trinity.

The ground of this advice was simply that from the large proportion of high wranglers at Peterhouse, and the smallness of the foundation, the chances of a fellowship there for a mathematical man were less than at Trinity College. And this was the reason which, together with his son's evident wish, most weighed with Mr. Clerk Maxwell. He was also struck with the fact that "Porter senior," who about this time became Maxwell's private tutor, recommended the change, although he himself belonged to Peterhouse. But the friend whose counsel in this whole matter was most prized both by father and son was Charles Mackenzie, of whom one who was his colleague at Caius has said that he was "the best of all men whom he had known."

On Maxwell's own mind, it need hardly be said, the prospect of a fellowship had little or no influence,

except in so far as he desired to please his father.
His own prime motive was undoubtedly the hope, in
which he was not disappointed, that the larger College
would afford him ampler opportunities for self-improvement.

To LEWIS CAMPBELL, Esq.

St. Peter's College, 18th Oct. 1850.

Æt. 19.   You tell me to lay my account with being dull at first,
and to condole with Robert, whereas there is continual
merriment (stop it!), and Robert is not settled yet.   As for
secrets of nature, they are not for Freshmen even to think
of.   Now for personal journal, with observations on the
manners and customs, etc. . . .

We spent the night at Peterborough, and saw the
Cathedral in the morning.   Very grand outside.   West end
a fine subject for calotype seen right perpendicular, and the
point P for a picture made by hand, fine weeping
willows, etc.

Proceeded to Ely with some Gloucester people we met
in the Cathedral, and inspected Ely Cathedral like regular
Archbishops.   Went up the steeple to see land like sea.
Heard all the people talking of the enclosure of the Wash
to be called Victoria county, and to be worth 30s. per acre.
Got to Cambridge, and called on Mr. Fuller,[1] after getting
room for my father (as the Bull was full) in a lodgement,
Got rooms in College, sitting and bed, six paces from Chapel.
and good light.   Had Tait to tea.   Next day breakfast with
Tait and Steele (of Glasgow and Ireland, and a future
wrangler), and so on in detail.

M'Kenzie came up to-day.   He took us to most of the
colleges.   Saw Newton and Bacon in Trin. Chapel.   At
Hall there was a proclamation to this effect nearly : —
Whereas (on the — day of —, 1850), application was
made to the Syndicate (or Senate or something), by William
Cooke, for leave to erect his equestrian establishment; and

---

[1] Then Tutor of Peterhouse, now Professor Emeritus of Aberdeen.

whereas, considering the immoral nature of said establishment, it was unanimously resolved to refuse leave. And whereas, notwithstanding said refusal, said W. C. did publicly notify his intention of putting up said estab.: Be it known, etc., that it is resolved and enacted, that if any undergraduate or graduate *in statu pupillari*, or tutor, or fellow, or master, etc., be caught at said establishment, he will be punished with expulsion, rustication, Castigation, or such other punishment as the case may require.[1]

So there is to be a quarrel between the Town and University about this, and also about whether they are to pay poor rate, as the University is supposed to be extra-parochial.

Prelim. exam. to-morrow at 9. Peter can't afford to pluck at it. C. H. Robertson has past *his* at Trin. He is in Ling's lodgings. He wants to keep quiet and to read by himself, and have only old acquaintances.

To a Friend suffering from Depression.

Hope to write you more again; but to conclude and get to bed: You are always talking of your withering up, awful change, etc. Now I have not a sermon on this subject by me, neither will I deliver an *extempore* one; but though I do not pretend to have examined you in all the branches, yet I would take the liberty to say that with respect to intellect, as the Laird of Dumbiedykes said, " It'll be growing when ye're sleeping,"—that is to say, what you take for corruption and decay is only stratification.[2]

The letters written to his father by Maxwell while an undergraduate at Cambridge have unfortunately disappeared. But something of their tenor may be

---

[1] Contrast with this the following entry in the Diary:—" 1842, *Feb.* 19. *Saturday.*—Go with James to Cooke's Circus at 2 P.M. at York Hotel, Arena—being James's first time of seeing such entertainments."   [2] See above, p. 113, note.

gathered from Mr. Maxwell's letters to his son. The
following extracts are related to his first term at
college :—

FROM HIS FATHER.

Glenlair, 22d October 1850.

Did Prof. Thomson catch you, and view your " dirt ;" [1]
and if so, what thought he thereof ?

Glenlair, 30th October 1850.

Who is the lecturer in the Greek play ? Did I see
him while at Cambridge ? I am sorry to hear the Greek
class is a bad one, for you would have got more good of it
if [it] had required you to work to maintain a good position
in it; but you should study your part well, for it is not
comparative excellence, but absolute, that will be of use in
University competitions.

Glenlair, 8th Novr. 1850.

You say your lecturer in Greek is good, so I hope you
profit accordingly, altho' your classfellows are not great
scholars. . . .

. . . . It would be necessary to take care there are no
mouse-holes. A very hungry Chapel mouse might come
through. There had been an entrance that way to the
Chapel. It would be to the organ loft.

Have you called on Profs. Sedgwick at Trin., and
Stokes at Pembroke ? If not, you should do both. Stokes
will be most in your way if he takes you in hand at all.
Sedgwick is also a great Don in his line, and if you were
entered in Geology would be a most valuable acquaintance ;
and, besides, not going to him would be uncivil, both to him
and to the Alisons, after their having arranged the introduction. Provide yourself with cards.

It might be worth your while to stop at York to
view it.

---

[1] " Jamsie's dirt," the disrespectful name for the bits of unannealed
glass, etc. etc., in the *Coterie-Sprache*

Glenlair, 13th Novr. 1850.

I am glad you have communication with Stokes and Mackenzie.

Is all Cambridge up in arms against the Pope and Cardinal Wiseman ? I cannot enter into all the fuss about it. If there is any law to hinder people calling themselves Cardinals or Archbishops, let it be acted on; but if there is no such law, let the assumption of empty titles . . . be laughed at.

Men of genius are often represented by themselves and others as owing little or nothing to their education. This certainly was not true of Maxwell, whose receptivity was only less than his originality. He laid a strong retentive grasp on all that was given to him, and set his own stamp on it in return. Both what was good and what was defective in his early training had left a lasting impress on him, and it was by no means an indifferent circumstance that, in the maturity of his powers, he entered Trinity College, Cambridge, at an advanced period of the long mastership of Whewell.

I well remember my surprise, not unmixed with needless pangs of boyish jealousy, on finding in the following summer that Maxwell had all at once made a troop of friends. Their names were always on his lips, and he loved by some vivid trait to indicate the character of each. His acquaintance was multifarious, and he had many pithy anecdotes to relate of other than his own particular associates.

He was at first in lodgings in King's Parade, where he " chummed " with an old Edinburgh schoolfellow,

Charles Hope Robertson.[1] It appears that the college lectures in mathematics were still felt to be rather elementary, but that he worked harder problems (some were of his own invention) with Tait and with Porter of Peterhouse, who was still his private tutor. A certain amount of classical reading was required of freshmen, and he still took his classics seriously. There is an allusion to the Ajax in one of his letters, and he makes critical remarks upon Demosthenes. There was also a lecture on Tacitus by a " deep, half-sentence lecturer." Either now or in the following year—at some time before the little-go examination of March 1852—he translated the choral odes of the Ajax into rhymed English verse,[2] and made a rough caricature of Ajax slaughtering the oxen.

His chief outdoor amusements were walking, bathing, and sculling. He was upset in his "funny" in May 1851—a trifling accident to so expert a swimmer. " But," writes a contemporary Cantab, " he richly deserved it. For he tried to take off his jersey after 'shipping' his oars. The oscillations of the funny became rapidly more extensive, in spite of his violent efforts at equilibrium." For winter recreation he ordered a pair of basket-sticks to be made at home.

He tried some odd experiments in the arrangement of his hours of work and sleep. But his father disapproved of such vagaries, and they were not continued long—although not entirely abandoned even when he had rooms in college. The authority just quoted says,

---

1 Now Rector of Smeeth, near Ashford, in Kent.
2 I did not know of this until the present year, 1881.

"From 2 to 2.30 A.M. he took exercise by running along the upper corridor, *down* the stairs, along the lower corridor, then *up* the stairs, and so on, until the inhabitants of the rooms along his track got up and lay *perdus* behind their sporting-doors to have shots at him with boots, hair-brushes, etc., as he passed."

Intellectual interests of all kinds surrounded him, and he soon began to lend new life to all. There were also in the Cambridge of this period religious influences of a remarkable kind. Apart from the Simeonite tradition which still lingered, there was a class of younger men, who, while faithful to a pious evangelical upbringing, had open and inquiring minds. The names of Henry and Frank Mackenzie of Trinity, the one senior to Maxwell, the other junior to him,[1] may be mentioned in particular. To such men, and to many others, the preaching of Harvey Goodwin, now Bishop of Carlisle, but at that time chiefly known as a mathematical authority, was full of interest.

Two events of some importance to the scientific world were the introduction of Foucault's pendulum experiment for proving the rotation of the earth, and the Great Exhibition of 1851.

1851.

Maxwell saw the pendulum experiment in Mr. Thacker's (the tutor's) rooms at Trinity in April or

---

[1] In 1852 H. M. had left Cambridge. F. M. went as a freshman to Trinity in October 1852. See a book named *Early Death not Premature*, being a Memoir of Francis L. Mackenzie, late of Trinity College Cambridge. With Notices of Henry Mackenzie, B.A. By the Rev. Charles Popham Miles, M.A., M.D., F.L.S., Glasgow. Nisbet and Co. Fourth Edition. 1861.

May, and wrote an account of it to his father, which
has been lost.

The " viewing" of the Crystal Palace, although a
thing to be done, was made less exciting than it would
otherwise have been by the constant habit of visiting
all manufactures according to opportunity. Maxwell
disclaims all " fanaticism" on the subject, and his
father writes that while a fortnight would be required
to see it properly, a great deal of it must be already
familiar to them both.

Æt. 20.   In the October term he joined the " team" of
Hopkins, the great private tutor, as a fifteenth pupil.

He also became a regular attendant of Professor
Stokes's lectures, and commenced his lifelong friendship with one whose original investigations were so
closely akin to some of his own.

Professor Stokes's kindness was greatly valued even
in those old days, and much more afterwards. The
young man's happiness in all ways at Trinity is manifested by the reappearance of the poetic vein, about
the time of the junior sophs' examination, in the *Lay
of King Numa*, dated 13th December 1851.

To this sketch of his first year at college there
must be appended one more reminiscence of Glenlair.
We met there in the autumn of that year, and I
remember that either then or in the previous year,
he put into my hands Carlyle's translation of Goethe's
Wilhelm Meister, remarking that it was a book to be
read with discretion.

In the autumn of 1850 the neighbouring estate of

Upper Corsock had been let to a shooting-party,[1] one of whom remarked to me what a pity it was that young Mr. Clerk Maxwell was " so little suited for a country life." I clearly recollect his look of exulting mirth when this was repeated to him. His disinclination to field sports was certainly not due to any lack of activity, nor even to his shortsightedness, which for other purposes was easily overcome, but simply to his love for animals. The moral of Wordsworth's *Hart-Leap Well* was not so much a principle as an instinct with him.[2] I remember his once speaking to me on the subject of vivisection. He did not condemn its use, supposing the method could be shown to be fruitful, which at that time he doubted, but—"Couldn't do it, you know," he added, with a sensitive wistful look not easy to forget. This is all I ever heard him say on the subject.

LETTERS, 1851.

To LEWIS CAMPBELL, Esq.

11th March 1851.
Lings, King's Parade, Cambridge.

. . . Have you read Soph. *Ajax*, or would you like to do it then?[3]

1851.
Æt. 19.

I have been trying an experiment of sleeping after hall. Last Friday I went to bed from 5 to 9.30, and read very hard from 10 to 2, and slept again from 2.30 to 7.

I intend some time to try for a week together sleeping from 5 to 1, and reading the rest of the morning. This is practical scepticism with respect to early rising.

---

[1] Including the present Earl Cairns.

[2] His uncle, Sir G. Clerk, is said to have had the same peculiarity.

[3] Viz. at the time of my proposed visit to Cambridge.

An Oxford man is reported to have complained of the lateness of morning chapel; he could not sit up for it.   I will have the most of my reading over by that time. . . .

Demosthenes goes on.   I begin to see what may be written in prose, and how ill it may be translated.

It is a γραφη παρανομων, so there is less declamation and more demonstration; but the arguments, small at first, are added as they proceed, and never left behind, so by oft repetition they seem stronger than they are.

Last night I searched for difficult problems to puzzle Steele and Porter junior with.   Here are some much more mild, which we freshmen get. . . .

It is twelve o'clock, and I have to do Demosthenes to-morrow before breakfast.   This implies Chapel, therefore bed, therefore I shut up.

FROM HIS FATHER.

Glenlair, 6th March 1851.

Simpson rages at present in the Electro-biology.   Dr. Alison is very wroth about it.   He says he has known two cases of nervous people whose minds were quite disordered by it.   I hope it is not in fashion at Cambridge, and at any rate that you do not meddle with it.   If it does any-thing, it is more likely to be harm than good; and if harm ensue, the evil might be irreparable, so let me hear that you have dismissed it; you have plenty better things in hand where you are.[1]

Glenlair, 29th April 1851.

Explain the pendulum experiment to me.   You used often to speak of the retardation of the Rotation of the Earth by the friction of the tides.

What is the Phosphate of Lime theory of mental progress?

---

[1] In the preceding Christmas vacation I was at a private " séance " in Edinburgh, where Maxwell—whether he shammed or not who can tell ?—was selected by the operator (a man named Douglas, I think), who vainly tried to make him seem to forget his name.

Glenlair, 18th May 1851.

Do you like the Trig. lectures A ?[1]   Tacitus is not new to you.   His style must be congenial to a deep, half-sentence lecturer.

Were you carrying your watch when you were upset in your funny ?   and if so, how did it agree with the douking ?[2]

I shall be glad to hear about the Pendulum.   Who is Thacker, who asked you to his rooms to see it ?

To Lewis Campbell, Esq.

Cambridge, 9th June 1851.

I find I owe you one letter this term.   I intended to write three days ago, but I am now refreshed by classical papers, and disburdened of half the subjects of examination.

On Friday we had Euclid, on Saturday Greek,—cram on both subjects; to-day Ajax and Tacitus translations.   I did no composition, but did various readings, strongly preferring certain of them for obvious reasons.

I find that 4 hours Euclid is worse than 2 3-hour papers of cram, though I sent up much more cram than Euclid.   This of itself shows that disburdening cram is not like grinding out Mathematics.   M—— in the Plato cram, writing a comparison of Cynics and Platonists, said that Platonism was a real live thing, but Cynicism was sleepy, and that even in its greatest ornament, Diogenes, the view of the universe was contracted to a front look-out from a wash-tub, and the *summum bonum* reduced to sunning one's self with eyes shut and buttons open.   This was to let off his jaw on first setting down, but he let it in among his papers, and could not get it out again.

Excuse my square sentences.   I have spent my curves on Tacitus, and I must now proceed to Trig.   Write.— Yrs.

To the Same.

8 King's Parade, 9th Nov. 1851.

I began a letter last week, but stopped short for   _Æt._ 20.

---

[1] Mr. Mathison's lecture on trigonometry.

[2] *i.e. Anglicè,* " How did the dip agree with it ? "

want of matter. I will not send you the abortion. Facts
are very scarce here. There are little stories of great men
for minute philosophers. Sound intelligence from New-
market for those that put their trust in horses, and Calen-
dristic lore for the votaries of the Senate-house. Man
requires more. He finds x and y innutritious, Greek and
Latin indigestible, and undergrads. nauseous. He starves
while being crammed. He wants man's meat, not college
pudding. Is truth nowhere but in Mathematics ? Is
Beauty developed only in men's elegant words, or Right in
Whewell's *Morality* ? Must Nature as well as Revelation
be examined through canonical spectacles by the dark-
lantern of Tradition, and measured out by the learned to the
unlearned, all second-hand. I might go on thus. Now do
not rashly say that I am disgusted with Cambridge and
meditating a retreat. On the contrary, I am so engrossed
with shoppy things that I have no time to write to you. I
am also persuaded that the study of x and y is to men an
essential preparation for the intelligent study of the material
universe. That the idea of Beauty is propagated by communication, and that in order thereto human language must
be studied, and that Whewell's *Morality* is worth reading, if
only to see that there *may be* such a thing as a system of
Ethics.

That few will grind up these subjects without the help
of rules, the awe of authority, and a continued abstinence
from unripe realities, etc.

I believe, with the Westminster Divines and their predecessors *ad Infinitum* that " Man's chief end is to glorify
God and to enjoy him for ever."

That for this end to every man has been given a pro-
gressively increasing power of communication with other
creatures. That with his powers his susceptibilities increase.

That happiness is indissolubly connected with the full
exercise of these powers in their intended direction. That
Happiness and Misery must inevitably increase with increasing Power and Knowledge. That the translation from
the one course to the other is essentially *miraculous*, while

the progress is natural.   But the subject is too high.   I will not, however, stop short, but proceed to Intellectual Pursuits.

It is natural to suppose that the soul, if not clothed with a body, and so put in relation with the creatures, would run on in an unprogressive circle of barren meditation.   In order to advance, the soul must converse with things external to itself.

In every branch of knowledge the progress is proportional to the amount of facts on which to build, and therefore to the facility of obtaining data.   In the Mathematics this is easy.   Do you want a quantity?   Take x; there it is!—got without trouble, and as good a quantity as one would wish to have.   And so in other sciences,—the more abstract the subject, the better it is known.   Space, time, and force come first in certainty.   These are the subjects in Mechanics.

Then the active powers, Light, Heat, Electricity, etc. = Physics.

Then the differences and relations of Matter = Chemistry, and so on.

Here the order of advancement is just that of abstractedness and inapplicability to the actual.   What poor blind things we Maths. think ourselves!   But see the Chemists! Chemistry is a pack of cards, which the labour of hundreds is slowly arranging; and one or two tricks,—faint imitations of Nature,—have been played.   Yet Chemistry is far before all the Natural History sciences; all these before Medicine; Medicine before Metaphysics, Law, and Ethics; and these I think before Pneumatology and Theology.

Now each of these makes up in interest what it wants in advancement.

There is no doubt that of all earthly creatures Man is the most important to us, yet we know less of him than any other.   His history is more interesting than natural history; but nat. history, though obscure, is much more intelligible than man's history, which is a tale half told, and which, even when this world's course is run, and when, as some think, man may compare notes with other rational beings, will still be a great mystery, of which the beginning and the end are all that can be known to us while the intermediate parts are perpetually filled up.

So now pray excuse me if I think that the more grovelling and materialistic sciences of matter are not to be despised in comparison with the lofty studies of Minds and Spirits. Our own and our neighbours' minds are known but very imperfectly, and no new facts will be found till we come in contact with some minds other than human to elicit them by counterposition. But of this more anon.

FROM HIS FATHER.

Glenlair, 3d Novr. 1851.

How do you like and how do you profit by Hopkins' mode of driving? He should get one more pupil, and drive 16 in hand like Batty or Cook.

Glenlair, 18th Novr. 1851.

You seem to have great gaieties with College Parties with Scientific Dons. Do you take notes of Stokes' experiments on the bands of the Spectrum? Will they be suitable for repetition in the garret of the old House?[1]

The clearing away a bank of weeds was a sly trick of the Trinity 4-oar, and I think Peterhouse and Sidney justified in protesting.[2] What tribunal is there to settle such matters?

I copied from your letters plan and section and elevation of the Baskets for single stick, and committed the same to the Davie.[3]

In the spring of 1852 he got rooms in college

---

[1] See above, p. 118, l. 2.

[2] According to the " contemporary Cantab," Peterhouse and Sidney really won the race, in spite of the removal of the bank of weeds. The reason why Trinity got the cups was that the pistol of P. and S.'s umpire *missed fire.*   [3] David M'Vinnie, see above, p. 39.

(Letter G, Old Court, south Attics), succeeding to his friend Blakiston, and passed his Little-go. In April of the same year he gained his scholarship, and was now launched without further necessary distraction on the long pull of preparation for the Tripos under Hopkins. But his energies were by no means absorbed in this continuous "grind." He contributed various papers to the *Cambridge and Dublin Mathematical Journal*, did not confine himself to mathematics in the May examination, and in November, besides writing the comic *Vision of a University, of Pedantry, and of Philosophy*, made elaborate preparation for a College declamation, "the Scottish Covenanters" being the subject characteristically chosen by him.

In June he had paid a short visit to Oxford, and made a trip to Lowestoft with P. G. Tait, before settling at Cambridge (to read and bathe) for the vacation term.

Living in college and dining at the scholars' table, *Æt.* 21. he naturally became more intimate with the other scholars, and he appears especially to have sought contact with classical men. To the names of Cracroft, Whitt, and Blakiston, amongst his newer friends are now added in his correspondence those of " Droop the ingenuous," Gedge, Howard Elphinstone, Isaac Taylor, Maclennan,[1] and Vaughan Hawkins.

The idea of self-improvement in society had taken a firm hold of him, and he was conscious of the difficulty of guiding himself amongst so many cross

---

[1] The late J. F. Maclennan, author of *Primitive Marriage*, etc.

M

influences. He knew that he was involuntarily different with different men, and there are curious traces in his correspondence of the struggle to make the highest use of social circumstances. Those who saw him about this time after an interval were struck by a marked change in his countenance, which, as compared with the Edinburgh days, had very distinctly gained in manliness and gravity, and showed a certain massiveness in its proportions which they had not previously noticed. His dark brown eye seemed to have deepened, some parts of the iris being almost black. A slight contraction of the chest, and a stature which, although above the average, was not tall enough to carry off the weight of his brow, made him less handsome standing than sitting. But his presence had by this time fully acquired the unspeakable charm for all who knew him, which made him insensibly become the centre of any circle, whether large or small, consisting of his friends or kindred. His hair and incipient beard were raven black, with a crisp strength in each particular hair, that gave him more the look of a Nazarite than of a nineteenth century youth. His dress was plain and neat, only remarkable for the absence of everything adventitious (starch, loose collars, studs, etc.), and an " æsthetic " taste might have perceived in its sober hues the effect of his marvellous eye for harmony of colour.[1]

---

[1] He was the first who spoke to me (about 1865) of the principles of coloured glass, which have since become fashionable ; observing that it should be rich in sea-green, and not, " like the banners of the Assyrian, gleaming with purple and gold." In his critical studies of

The impression he made on older persons with whom he was less intimate may be gathered from the remarks of Dean Ramsay in a note to Miss Cay:—

I had great pleasure in seeing your nephew, young Clerk Maxwell. He is shrewd and cautious. He seems to like Cambridge, and I doubt not will distinguish himself. He is sparing in his words, but what he says is to the point.

The following contribution from the Rev. Charles Hope Robertson, the Rector of Smeeth (above, p. 152), throws light on more than one aspect of his life at this period :—

I was at Trinity College, Cambridge, in the same years as Clerk Maxwell, and for some time had lodgings in the same house. This, as well as knowing him before in Edinburgh, led to our frequently meeting.

He was of a very kindly disposition (under a blunt exterior), of which I can give an example. I had hurt my eyes a good deal with experiments on light, while working up for a course of Professor Forbes's lectures at Edinburgh College, and for a good part of my undergraduate course was able to use them very little. He used to find me sitting in my rooms with closed eyes, unable to prepare for next day's lectures, and often gave up an hour of his recreation time, to read out to me some of the book-work I wanted to get over. This infirmity prevented my reading for more than a moderate degree in mathematical honours ; but I should have been still worse off if he had not thus been " eyes to the blind" for me.

He had an innate reverence for sacred things, which I do not think was ever much disturbed by the scepticism

---

modern poets he used to note their fondness for particular colours,—*e.g.*, the uses of " white," " red," " black," " ruby," " emerald," " sapphire," in Tennyson and Browning.

fashionable among shallow scientific men. As my main
object in coming to Cambridge was to prepare for holy
orders, I had more interest in theological subjects than any
others. He knew this, and would refer to me points of
difficulty for our mutual consideration till we next met.
The result was useful to us both. On one occasion an
attack on the Mosaic history having perplexed him, he was
glad of an idea that occurred to me, that the account of
God's driving out the heathen by "sending the hornet"
before the Israelites, was not an idea likely to occur in a
book of human origin, where a leader would rather be apt
to magnify than to diminish his own and his people's
prowess. This exactly suited the sceptical state of some
friend he had been conversing with. On another occasion
he hit on a very beautiful mathematical illustration of St.
Paul's closing view of his career, in 2 Tim. iv. 6-8; "St.
Paul was looking backward, forward, and downwards—so
the *resultant was upwards.*"

As an original experimenter he was most ingenious in
contriving out of simple means apparatus for delicate experiments. I have to this day some crystals for showing
polarised light, which he gave me, cut and polished by most
simple rubbing, mounted with cardboard and sealing-wax.

His *Essay on the Rings of Saturn,*—showing how
mechanical principles required that these bodies were not
solid, but formed by multitudes of small bodies revolving
round Saturn, in bands of orbits,—has received abundant
confirmation from recent observation with large telescopes.
His own simple experiments with corks and rings suggested
the idea.

But while so ingenious himself, he had great difficulty
in imparting his ideas to others; consequently was not so
clear a lecturer or writer as might have been expected. It
was probably this that prevented his being senior wrangler
of his year.

I may mention that though not joining in the ordinary
games of young fellows, such as cricket and rowing, he was
very active; and I have seen him in bathing take a

running header from the bank, turning a complete somersault before touching the water.[1]

If shortly described, he might be said to combine a grand intellect with childlike simplicity of trust. He was too deep a thinker to be sceptical, but too well read not to feel for others' difficulties. All his experiments led him to greater reverence for the Great First Cause, heartily agreeing with Young's *Night Thoughts*, "An undevout astronomer is mad."

In the course of the winter he was elected a member of the Select Essay Club, the *crême de la crême* of Cambridge intellects, familiarly known (because limited to the number twelve) as "the Apostles." His contributions to this famous association still remain,[2] and present a curious reflection of the contemplative activities of his mind, which is far indeed from being engrossed with mere mathematics, but is rather, in the language of Plato, "taking a survey of the universe of things," πᾶσαν πάντων φύσιν ἐρευνωμένη τῶν ὄντων ἑκάστου ὅλου. Yet amidst this speculative ardour, and even wildness, we trace the persistence of certain root-ideas, and are often reminded of his intention (expressed with curious self-directed irony in 1850), "to read Kant's *Kritik* with a determination to make it agree with Sir William Hamilton."[3]

His coming of age, June 13, 1852, had been celebrated with a few quiet words in his father's

---

[1] Professor Tait says, "He used to go up on the pollard at the bathing-shed, throw himself *flat on his face* in the water, dive and cross, then ascend the pollard on the other side, project himself *flat on his back* in the water. He said it stimulated the circulation!"

[2] See chap. viii.   [3] P. 135.

letter of June 12, graced with the unusual addition
of a Scripture text—" I trust you will be as discreet
when Major as you have been while Minor (Prov.
x. 1[1])."

The six months from December 1852 to June 1853
were a time of great and varied mental activity. When
the Tripos work became most exacting, he seemed to
have the most spare energy.   No part of the rich
mental life of Trinity failed to touch and stimulate
him—from the Moral Philosophy of the Master, to
undergraduate discourses upon whist and chess. When
most burdened with analytical book-work, he yearned
the more deeply after comprehensive views of Nature
and Life, and found refreshment in metaphysical
discussion, and occasionally in theological controversy.   Even the " occult " sciences, in the contemporary shapes of electro-biology and table-turning, had
their share of ironical attention.

His relations with the dons, " scientific" and otherwise, whatever may have been their first impression of
him,[2] were, for the most part, smooth, but a humorous
passage of arms between him and the Senior Dean was
long remembered in Trinity.   The lines to J. A. Frere,
although somewhat personal, are too well known to be
omitted from his collected poems, and anything in
them which might give offence at the time is more
than redeemed by the large humanity of the conclud-

1 " A wise son maketh a glad father."

2 It is said that on the first occasion of his reading in chapel after
gaining his scholarship, his delivery of his first lesson (from the Book
of Job) in tones which, perhaps from nervousness, were unusually
broad, upset the gravity of every one, from the Master downwards.

ing stanza. The following letter, addressed (but perhaps not sent) to the same personage, throws an
amusing light on the circumstances under which the
parody of " John Anderson" was written :—

To THE REV. JOHN ALEXANDER FRERE.

Trin. Coll., 26th Feb. 1853.

DEAR SIR—Looking back on the past week I find I have
kept only seven chapels. I have no excuse to offer. The
reason, however, of the deficiency is this. Unaware that a
Saint's Day would occur in the course of the week I parted
with my surplice on Monday in order to have it washed.
I was thus prevented from appearing in chapel on the evenings of Wednesday and Thursday, as otherwise I would have
done. I might even after this have completed the requisite
number; but, unfortunately, reading till a late hour on Friday
night I found myself unable to attend chapel on Saturday
morning.

I can but hope that more forethought on my part may
prevent the recurrence of such accidents.

I have also to acknowledge the receipt of a small paper
from you relative to the observance of Sunday. I have
read it, and will keep it in mind.

Trusting that my past and future regularity may atone
for my present negligence, I remain, yours sincerely,

J. C. MAXWELL.

Rev. J. A. FRERE.

It was while staying up at Easter in the spring of *Æt.* 21.
1853, and working " at high pressure," that his longing
for the untrammelled and reverent investigation of
Nature's secrets found rhythmical expression in the
most serious of his poems, the " Student's Evening
Hymn," in which religious and philosophical aspirations are combined. Thus it was always with him ;
when most plunged in the minute investigation either

of phenomena or of abstract ideas, he was most eager
to rebound towards the contemplation of the whole of
things, and that which gives unity to the whole. Yet
no mind could be more averse from " viewiness," or
more determined to bring every statement to the test
of fact.

The brief remainder of that Easter vacation was
spent at Birmingham with his friend Johnson Gedge,
a scholar of Trinity, whose father was the headmaster
of King Edward's School. The exactions of Trinity
and Hopkins had only left him a few days of holiday,
and these were passed in the manner already mentioned,[1] in viewing the manufactures of Birmingham.
Mr. Clerk Maxwell's own delight in such things prevented him from realising how laborious a programme
he had suggested for the interval between two long
spells of severe head-work. Yet in the midst of it
Maxwell seems to have found time to contribute an
elaborate piece of humorous correspondence to the
King Edward's School Chronicle.

When working hardest he was never a recluse,
nor was he ever more sociable than in his third year
at college. He seems to have had some difficulty even
in avoiding supper parties, and one of his most
brilliant metaphysical *jeux d'esprit* purports to be a
mode of escaping from them. To the names already
mentioned as amongst his intimate friends, those of
Farrar [2] and Butler [3] are now added. Besides the

---

[1] Above, p. 7, note 1.
[2] The Rev. F. W. Farrar, D.D., Canon of Westminster.
[3] The Rev. H. Butler, D.D., Head-Master of Harrow.

metaphysical discussions, there were Shakespeare readings, of which he was an auditor sometimes, if not an actor in them.

Whilst speaking of these side-sparks from his anvil, it is right to keep in view the loyal and trustful spirit in which he did his regularly-appointed work. His words on this subject in one of the letters to Miss Cay, which will be given presently, are well worthy of separate quotation here :—

> " *Trin. Coll., 7th June 1853.*
>
> " If any one asks how I am getting on in Mathematics, say that I am busy arranging everything, so as to be able to express all distinctly, so that examiners may be satisfied now, and pupils edified hereafter. It is pleasant work, and very strengthening, but not nearly finished."

As became his kindliness, he was still mindful of the freshmen. Amongst these, Alexander Robertson, now Sheriff of Forfarshire, the brother of Charles, and Frank Mackenzie, were Scotsmen and compatriots. Another junior with whom he became still more intimate was "Freshman Tayler,"[1] so called, though now a junior Soph., in contradistinction to Isaac Taylor. Like Frank Mackenzie and others who have been mentioned, Freshman Tayler was of pious evangelical antecedents. Maxwell's own thoughts at this time, as has been seen, were taking a more decidedly religious colour, and this side also of his rich and deep nature received a fresh impulse in this critical year.

He had been persuaded to spend the short interval between the summer and vacation terms with the

1853.
Æt. 22.

---

[1] The Rev. G. W. H. Tayler, now Vicar of Trinity Church, Carlisle.

Rev. C. B. Tayler, rector of Otley, in Suffolk, who had been touched with Maxwell's kindness to his nephew. Here he found himself for the first time in the midst of a large and united English family, and in his half-speculative, half-emotional way, was contrasting what he saw with the experience of an only son, when he was suddenly taken ill. The long continuous strain of the past months had been too much for him, and indeed it appears that even in the early spring he had been physically below par.[1] The illness is described by Mr. Tayler as a sort of brain fever, and he was disabled by it for more than a month. The Taylers nursed him as they would have nursed a son of their own, and Maxwell, in whom the smallest kindnesses awakened lasting gratitude, was profoundly moved by this. He referred to it long afterwards as having given him a new perception of the Love of God. One of his strongest convictions thenceforward was that "Love abideth, though Knowledge vanish away." And this came to him at the very height of the intellectual struggle.[2]

---

[1] See the letter of 2d February 1853, in which his father refers to Miss Cay's advice that he should take wine.

[2] At the same time, it is not to be supposed that Maxwell was ever completely identified with any particular school of religious opinion. He was too much "the heir of all the ages," and, as he himself expressed it, "his faith was too deep to be in bondage to any set of opinions." Scottish Calvinism was the theological system which had most historical interest for him, and most claim on his hereditary piety. He was learned in the writings of Owen and Jonathan Edwards. But that which his latest pastor has called "his deep though simple faith," was not enclosed in any system. Even his youthful training (which in the case of one so loyal is not to be disregarded) was favour- able to a comprehensive view of Christianity. Beginning with the

The father was also much affected by this kindness shown to his son. Mr. Tayler had said nothing to make him anxious until the crisis of the illness was past. But when he knows all, there is something more than eloquence in his brief, inarticulate phrases of recognition:—

"With yours I have Mr. Tayler's letter. I do not write to him to-day. My only subject is thanks, and these are not to be measured in words—the strongest I can use; so at present give my respects and highest regards."

Though weakened by his illness, Maxwell was able to keep the vacation term, and profit by Hopkins' continued training, before going home for a few weeks of thorough refreshment. In the following term, with the Senate-House examination in immediate prospect, he was careful not to read inordinately hard.

In the autumn of this year the controversy which had been called forth by Professor Maurice's Theo-

---

Bible, which he knew by heart from a child, Presbyterian and Episcopalian influence had been blended as we have already seen. Hence, when he went to Cambridge, there was nothing strange to him in the service of the English Church. His mind was always moving (*elle planait*, as the French would say), above and " beyond these voices," yet they were not indifferent to him.

It is to be regretted that a letter on the Parties in the Church of England, written to his father in 1852, has not been found. His general approval of Hare's sermons, and his remarks on the Maurician controversy, indicate the direction which his thoughts on this subject must have taken. His interest in sermons, good and bad, was like Macaulay's interest in novels, or Charles Lamb's in old plays. Years after this, on board a friend's yacht one Sunday, he gave a sort of impromptu exposition of a chapter in the Book of Joshua, shouting up his remarks from below, which struck those who heard it as full of originality and wisdom.

logical Essays was brought to a crisis through his
deprivation of his office by Principal Jelf. Disputa-
tion on this theme was nowhere more rife than
amongst the scholars of Trinity, and Maxwell's remarks upon it will be read with interest even now.
He was from the first strongly attracted by Maurice's
combination of intense Christian earnestness with
universal sympathy, and although he sometimes felt
that the new teacher was apt to travesty the Popular
Theology in trying to delineate it, he had a deep
respect for what was positive in his doctrine. He was
still more drawn to him when he came to know him
personally,—no longer as a writer, but as a friend.
A mention of Thomas Erskine of Linlathen in Mr.
Maxwell's letter of 16th December 1853, was probably
occasioned by some question raised in connection with
Mr. Maurice.

The following letter, addressed to me by the Rev.
G. W. H. Tayler, will be read with interest in the
light of the preceding narrative :—

Holy Trinity Vicarage,
Carlisle, 4th March 1882.

MY DEAR SIR—You have asked me to send you some
account of James Maxwell, as I remember him during the
space of three years, 1852, 1853, and 1854. My first
acquaintance with him was about February 1852.   I
was soon attracted by the frankness of his manner and the
singular charm of his quaint and original remarks in conversation.

We undergraduates felt we had a very uncommon
personage amongst us; but we did not then appreciate his
rare powers. We had of course heard of the reputation
which he had at Edinburgh.

But this acute mathematician, so addicted even then to *original* research, was among his friends simply the most genial and amusing of companions, the propounder of many a strange theory, the composer of not a few poetic *jeux d'esprits.*

Grave and hard-reading students shook their heads at his discursive talk and reading, and hinted that this kind of pursuits would never *pay* in the long run in the Mathematical Tripos.

I have sometimes watched his countenance in the lecture-room. It was quite a study—there was the look of a bright intellect, an entire concentration on the subject, and sometimes a slight smile on the fine expressive mouth, as some point came out clearly before him, or some amusing fancy flitted across his imagination. He used to profess a dislike to reproducing speculations from books, or hearing opinions quoted taken bodily from books.

Yet he read a good deal in other lines of study than natural philosophy. Sir Thomas Brown's *Religio Medici* was one of his favourite books. Any such author, who propounded his speculations in a quaint, original manner was sure to be a favourite with him.

But I particularly remember his attraction to Sir Thomas Brown during the long vacation, when he was laid up with severe illness (a brain fever) in my dear uncle's house in June 1853. He came to stay at Otley, near Ipswich, of which my uncle was the rector. For a few days he was tolerably well, then suddenly fell ill, probably through overwork for his third year college examination. It was on his recovery from that illness that I seemed to know him better than ever. It was then that my uncle's conversation seemed to make such a deep impression on his mind. He had always been a regular attendant at the services of God's house, and a regular communicant in our College Chapel. Also he had thought and read much on religious subjects. But at this time (as it appears from his own account of the matter) his religious views were greatly deepened and strengthened.

I must add that I spent some little time in the long vacation of 1854 with Maxwell at Glenlair. His father was then living, and it was touching to witness the perfect affection and confidence which subsisted between father and son : the joy and satisfaction and exulting pride which the father evidently felt in his son's success and well-earned fame ; and, on the other hand, the tender, thoughtful care and watchfulness which James Maxwell manifested towards his father.

Maxwell has indeed left a very bright memory and example. We, his contemporaries at college, have seen in him high powers of mind and great capacity and original views, conjoined with deep humility before his God, reverent submission to His will, and hearty belief in the love and the atonement of that Divine Saviour who was his Portion and Comforter in trouble and sickness, and his exceeding great reward.—I remain, my dear sir, yours very truly,

G. W. H. TAYLER,
Vicar of Trinity Church, Carlisle.

Mr. Lawson of the Equity Bar, whose diary has preserved the remark quoted on p. 133, has also furnished me with the following vivid account of his impressions of Maxwell as an undergraduate :—

22 Old Square, Lincoln's Inn,
London, W.C., 6th January 1882.

I was in his year at Trinity, and knew him intimately, though our ways separated after I left Cambridge, and I scarcely ever saw him, except once or twice when he was Professor at King's College, and later on only at very long intervals, on an occasional visit to the University.

There must be many of his quaint verses about, if one could lay hands on them, for Maxwell was constantly producing something of the sort, and bringing it round to his friends, with a sly chuckle at the humour, which, though his own, no one enjoyed more than himself.

I remember Maxwell coming to me one morning with a

copy of verses beginning—"Gin a body meet a body Going through the air," in which he had twisted the well-known song into a description of the laws of impact of solid bodies.

There was also a description which Maxwell wrote of some University ceremony, I forget what, in which somebody "went before," and somebody "followed after," and "in the midst were the wranglers playing with the symbols."

These last words, however meant, were in fact a description of his own wonderful power. I remember one day in lecture, our lecturer had filled the black board three times with the investigation of some hard problem in Geometry of Three Dimensions, and was not at the end of it, when Maxwell came up with a question whether it would not come out geometrically, and showed how with a figure, and in a few lines, there was the solution at once.[1]

Maxwell was, I daresay you remember, very fond of a talk upon almost anything. He and I were pupils (at an enormous distance apart) of Hopkins, and I well recollect how, when I had been working the night before and all the morning at Hopkins's problems with little or no result, Maxwell would come in for a gossip, and talk on while I was wishing him far away, till at last, about half an hour or so before our meeting at Hopkins's, he would say—"Well, I must go to old Hop's problems;" and by the time we met there they were all done.

I remember Hopkins telling me, when speaking of Maxwell, either just before or just after his degree, "It is not possible for that man to think incorrectly on physical subjects;" and Hopkins, as you know, had had perhaps more experience of mathematical minds than any man of his time.

Of Maxwell's geniality and kindness of heart you will have had many instances. Every one who knew him

---

[1] Compare Plato, *Theæt.* 147, C D. A Cambridge friend who knew Maxwell at a later time, says of him, "One striking characteristic was remarked by his contemporaries at Hopkins's lectures. Whenever the subject admitted of it he had recourse to diagrams, though the rest might solve the question more easily by a train of analysis."

at Trinity can recall some kindness or some act of his which has left an ineffaceable impression of his goodness on the memory—for " good " Maxwell was, in the best sense of the word.

Mr. Lawson adds the following extract from his diary :—

Under date January 1. 1854 (Sunday evening), after saying I had been at tea at a friend's rooms, and naming the men who were there, of whom Maxwell was one, there is this note :—

" Maxwell, as usual, showing himself acquainted with every subject upon which the conversation turned. I never met a man like him. I do believe there is not a single subject on which he cannot talk, and talk well too, displaying always the most curious and out-of-the-way information."

1854.
Æt. 22.

The great contests of January came at last, with the result that Maxwell was Second Wrangler, Routh of Peterhouse being the Senior, and that Routh and Maxwell were declared equal as Smith's Prizemen.

A reminiscence of Professor T. S. Baynes which has reference to this time is interesting in connection with other signs of something exceptional in Maxwell's physical state during the previous year :—

He said that on entering the Senate-house for the first paper he felt his mind almost blank ; but by and by his mental vision became preternaturally clear. And, on going out again, he was dizzy and staggering, and was some time in coming to himself.

LETTERS, 1852-53.

To LEWIS CAMPBELL, Esq.

8 King's Parade, 10th Feb. 1852.

Æt. 20.   I was at Isaac Taylor's to-night. His father has come to

see him, a little, cold man, with a tremulous voice, who talks
about the weather as if he were upon oath, but who can
lift up his testimony against any unwarrantable statement.
Taylor junr. and Maclennan were talking about associations of workmen, Christian socialism, and so forth.   T.
junr. approved of the system where each workman has a
share in the firm.   M. liked one master better than many.

T. senr. described the inducement to hard work among
engineers at Manchester; the reward is not profit, but
situation.

There are advantages in subordination, besides good
direction, for it supplies an *end* to each man, external to
himself.   Activity requires Objectivity.—Do you ever read
books written by women about women?   I mean fictitious
tales, illustrating Moral Anatomy, by disclosing all thoughts,
motives, and secret sins, as if the authoress were a perjured
confessor.   There you find all the good thinking about
themselves, and plotting self-improvement from a sincere
regard to their own interest, while the bad are most dis-
interestedly plotting against or for others, as the case may
be; but all are caged in and compelled to criticise one
another till nothing is left, and you exclaim :—

> " Madam! if I know your sex,
> By the fashion of your bones."

No wonder people get hypochondriac if their souls are made
to go through manœuvres before a mirror.   Objectivity
alone is favourable to the free circulation of the soul.   But
let the Object be real and not an Image of the mind's own
creating, for Idolatry is Subjectivity with respect to gods.
Let a man feel that he is wide awake,—that he has something
to do, which he has authority, power, and will to do, and is
doing; but let him not cherish a consciousness of these
things as if he had them at his command, but receive them
thankfully and use them strenuously, and exchange them
freely for other objects.   He has then a happiness which
may be increased in degree, but cannot be altered in
kind.

N

To the Same.

8 King's Parade, 7th March 1852.

I have now nobody that I see too much of, though
I have got several new acquaintances, and improved several
old ones. I find nothing gives one greater *inertia* than
knowing a good many men at a time, who do not know
each other intimately. *N.B.*—*Inertia,* not = *laziness,* but
mass; i.e. if one knows a man, he forms an idea of your
character, and treats you accordingly. If one knows a
company of men, they are strong in union, and overawe
the individual. If one man only, we become mutual
tyrants. If several independently, every one plays the part
of Dr. Watt's celebrated "Busy Bee," and by mixing
according to every possible combination hit out the best
results. Now you see I am theorising again and preaching as of old; but the fact is, I am always laying plans and
preaching to myself till I seek for some one to whom I
may disgorge without fear of an immediate reply. Now, my
great plan, which was conceived of old, and quickens and
kicks periodically, and is continually making itself more
obtrusive, is a plan of *Search* and *Recovery,* or Revision and
Correction, or Inquisition and Execution, etc. The Rule of
the Plan is to let nothing be wilfully left unexamined.
Nothing is to be *holy ground* consecrated to Stationary Faith,
whether positive or negative. All fallow land is to be
ploughed up, and a regular system of rotation followed.
All creatures as agents or as patients are to be pressed into
the service, which is never to be willingly suspended till
nothing more remains to be done; *i.e.* till A.D.$+\infty$. The
part of the rule which respects self-improvement by means
of others is :—Never hide anything, be it weed or no, nor
seem to wish it hidden. So shall all men passing by pluck
up the weeds and brandish them in your face, or at least
display them for your inspection (especially if you make no
secret of your intention to do likewise). (I speak not here
literally of the case of those who revise each other's faults
every night, and quarrel before the month is out, but you
did not so misunderstand me.) Again I assert the Right of

Trespass on any plot of Holy Ground which any man has set apart (as the rustics did their Gude-man's Rig) to the power of Darkness.   Such places must be exorcised and desecrated till they become fruitful fields.   Again, if the holder of such property refuse admission to the exorcist, he *ipso facto* admits that it is consecrated, and that he fears the power of Darkness.   It may be that no such darkness really broods over the place, and that the man has got a habit of shutting his eyes in that field, which makes him think so.

Now I am convinced that no one but a Christian can actually purge his land of these holy spots.   Any one may profess that he has none, but something will sooner or later occur to every one to show him that part of his ground is not open to the public.   Intrusions on this are resented, and so its existence is demonstrated.   Now, I do not say that no Christians have enclosed places of this sort.   Many have a great deal, and every one has some.   No one can be sure of all being open till all has been examined by competent persons, which is the work, as I said before, of eternity. But there are extensive and important tracts in the territory of the Scoffer, the Pantheist, the Quietist, Formalist, Dogmatist, Sensualist, and the rest, which are openly and solemnly *Tabooed,* as the Polynesians say, and are not to be spoken of without sacrilege.

Christianity—that is, the religion of the Bible—is the only scheme or form of belief which disavows any possessions on such a tenure.   Here alone all is free.   You may fly to the ends of the world and find no God but the Author of Salvation.   You may search the Scriptures and not find a text to stop you in your explorations.

You may read all History and be compelled to wonder but not to doubt.

Compare the God of Abraham, Isaac, and Jacob with the God of the Prophets and the God of the Apostles, and however the Pantheist may contrast the God of Nature with the " Dark Hebrew God," you will find them much liker each other than either like his.

The Old Testament and the Mosaic Law and Judaism

are commonly supposed to be "Tabooed" by the orthodox.
Sceptics pretend to have read them, and have found certain
witty objections and composed several transcendental arguments against "Hebrew O' Clo'," which too many of the
orthodox unread admit, and shut up the subject as haunted.
But a Candle is coming to drive out all Ghosts and Bugbears.
Let us all follow the Light.

To MISS CAY.

8 King's Parade, 23d March 1852.

I received yours (of the 18th I suppose) on Saturday,
and began to muse on the difference of our modes of
life: your sickness—my health; your kind dealings with
neighbours—our utter independence of each other; you
visit without seeing people—we see without visiting each
other; you hear all about people's families and domestic
concerns—we do not, but we know exactly how everybody
is up in his different subjects, and what are his favourite
pursuits for the time.

The Little-go is now going on, so I am taking my
Easter vacation at this time. I do nothing but the papers
in the Senate-house, and then spend the day in walks and
company, reading books of a pleasant but not too light
kind, lest I should be disgusted with recreation.

I find myself quite at grass, and am sure that in 10
days I will be reading again as if I had been rusticated for
a year.

I never did such a feat as get up at 5 in the morning.
I get up at 6.30 for chapel in winter, and read in the day-
time, but I have now begun my summer practice of sleeping
in the mornings and reading at night, save when I get up
on a fine day to take a walk in the morning, which makes
me idle all day, and is sometimes agreeable.

I met old Isaac Taylor in his son's rooms some time
ago. He began by speaking of the weather in a serious
way, and went on to his Manchester concerns,—effective
motives to work, actual methods adopted, and so got into
the merits of socialism, joint-stock workmen's associations,

and so forth, appearing all the while to say nothing, but quietly feed on the wisdom of the undergrads., as they enounced their opinions.

FROM HIS FATHER.

Glenlair, 10th April 1852.

The Ordnance Surveyors are doing Contour Lines. The line 250 feet above sea-level passes just in front of the Bees.[1]

Glenlair, 25th April 1852.

I . . . congratulate you on your scholarship. You write of entering on the duties,—what are they? and what are the Privileges and Profits?

Glenlair, 12th May 1852.

Is M'Millan the Publisher of the *Cambridge and Dublin,* whereof William Thomson is editor? Have you sent him your prop. you were doing at Christmastime?

The gold-fever of Australian type prevails in these parts.

Glenlair, 19th May 1852.

[Prop. about resistance of sides and bottom of a meal-ark.]

. . . The meal, which may be called a fluid as much as a glacier.[2]

Query: Whether by putting bars beneath the bottom at points removed from the middle of the joist, the pressure would be more advantageously distributed?

Glenlair, 12th June 1852.

[Eve of James C. M.'s 21st birthday.]

I trust you will be as discreet when major as you have been while minor (Prov. x. 1).

Remember me to Tait. I am sorry I did not see him in Edinburgh to wish him joy of his honours.

---

[1] In the garden at Glenlair.   [2] See above, p. 80.

Glenlair, 29th June 1852.

Æt. 21.   Did you take to the Geology at all? I suppose that the cliffs at Lowestoft are Tertiary, with plenty of fossils.

Glenlair, 9th Novr. 1852.

Nativity of the Prince of Wales.

Received yours of St. Guy and William.

The Cambridge Commission as you report of it will not affect you in any way.

What sort of thing is "College Declamation"? You say you have chosen The Scottish Covenanters. Do you take the part of Advocate or Apologist for them? or do you try the impartial historian? It would be difficult to give the Scots Prelates their due without offending some of the Order. During the Persecutions both the civil and ecclesiastical government of Scotland was getting ripe for the Revolution.

Glenlair, 11th Novr. 1852.

It has been said, Had there been two other Leightons instead of Sharp of St. Andrews and —— —— of Galloway, Episcopacy would have been securely established.

Glenlair, 2d Feb. 1853.

Aunty Jane was saying—a glass of wine daily,—port for preference.

Glenlair, 12th Feb. 1853.

. . . *Mathematical Journal,* to which you send props. What props.? The one about the Pendulum?

To Lewis Campbell, Esq.

Trin. Coll., 20th Feb. 1853.

After Chapel I was at Litchfield's, where he, Farrar, Pomeroy, and Blakiston, discussed eternal punishment from 8 to 12. Men fall into absurdity as soon as they have settled for themselves the question of the origin of evil. A man whose mind is "made up" on that subject is contradictory on every other; one day he says that the man that can be happy in such a world is a brute, and the next day that if a man is not happy here he is a moping fool. At last they assert the Cretan dilemma, that if a man says that man is ignorant and foolish, it was ignorant and foolish to say so. Solomon, they say, was used up when he wrote Ecclesiastes, and said "all is vanity" in a relative sense, having himself been so. Solomon describes the search after Happiness for its own sake and for the sake of possession. It is as if a strong man should collect into his house all the beauty of the world, and be condemned to look out of the window and marvel that no good thing was to be seen. "No man can eat his cake and have it." I would add that what remains till to-morrow will stink.

As for evil being unripe good, I say nothing with respect to objective evil, except that it is a part of the universe which it may be the business of immortal man to search out for ever, and still see more beyond. We cannot understand it because it is relative, and relative to more than we know. But subjective evil is absolute; we are conscious of it as independent of external circumstances; its physical *power* is bounded by our finitude, bodily and mental, but within these its *intensity* is without measure. A bullet may be diverted from its course by the medium through which it passes, or it may take a wrong one owing to the unskilfulness of the shooter, or the intended victim may change his place; but all this depends, not on the will of the shooter, but on the ignorance of his mind, the weakness of his body, the resistance of inert matter, or the subsequent act of another agent; the bullet of the murderer may be turned aside to drive a nail, or what not, but his will is independent of all this, and may be judged at once without appeal.

> Yet still the lady shook her head,
> And swore by yea and nay,
> My whole was all that he had said,
> And all that he could say.

J. C. MAXWELL.

FROM HIS FATHER.

Edinburgh, 21st February 1853.

The Halo and accompaniment of the 15th had been very curious. I never saw the appearance of Mock Suns.

Lord Cockburn went in plain dress to the fancy ball. When the crowd hissed him, he said he was the minister that was to marry them all!!

To MISS CAY.

Trin. Coll., 11th March 1853.

I was so much among the year that is now departed that it makes a great difference in my mode of life. I have been seeking among the other years for some one to keep me in order. It is easier to find instructive men than influential ones. I left off here last night to go to a man's rooms where I met several others, who had gone a-prowling like me. . . . I have been reading Archdeacon Hare's sermons, which are good.

Trin. Coll., Feast of St. Charles II.

Pomeroy's mother and sister were up here lately. They used to be at Cheltenham. From them I learnt a good deal about the systematic and uncompromising mode of thinking and speaking which marks the great Irish Giant of Trinity. Bishop Selwyn of New Zealand prought[1] here yesterday about missions. He founded the Lady Margaret boat club at John's, and got the boat to the head of the river. He was 2d Classic in 1831, and still he is too energetic for his curates to keep up with him in his own visitations about the South Pole. He made a great impression on the men here by his plainness of speech and absence of all cant, whether he spoke of the doctrines of Christianity or the history of Pitcairn's Island. I have been reading various books, but few very entertaining. They are chiefly theories about things in general which take the fancies of men now-a-days. The only safe way to

---

[1] *i.e.* preached.

read them is to find out the facts first. With this pre-
caution they are tolerably transparent. I have been attending Sir James Stephen's lectures upon the causes of the first
French Revolution. They are now done, so I look in upon
Stokes' dealing with light.

FROM HIS FATHER.

Edinburgh, 13th March 1853.

Ask Gedge to get you instructions to Brummagem
workshops. View, if you can, armourers, gunmaking and
gunproving — swordmaking and proving — *Papier-maché*
and japanning — silver-plating by cementation and rolling
— ditto, electrotype — Elkington's Works — Brazier's works,
by founding and by striking up in dies — turning — spinning
teapot bodies in white metal, etc. — making buttons of sorts,
steel pens, needles, pins, and any sorts of small articles
which are curiously done by subdivision of labour and by
ingenious tools — glass of sorts is among the works of the
place, and all kinds of foundry work — engine-making —
tools and instruments (optical and philosophical) both coarse
and fine. If you have had enough of the town lots of
Birmingham, you could vary the recreation by viewing
Kenilworth, Warwick, Leamington, Stratford-on-Avon, or
suchlike.[1]

Glenlair, 29th April 1853.

You write (from King Edward's School, Birmingham)
about plans and visits, Freshman Tayler and two others
innominate.

Glenlair, 12th May 1853.

What do you know of Henry Mackenzie ? Do you find
Frank to be clever, good, agreeable, and wise, which you
state to be the desiderata for a friend ?

Here is a Prop. anent fuel. What would be the amount
of heat evolved in the combustion of a given weight of dry
wood compared with the same weight of coal ?

---

[1] This letter has been quoted above, p. 7, note.

Glenlair, The Day after the Wedding,[1]
1st June (1853).

I have yours of the day of the Restoration. . . . She
(Maria Clerk) also wrote about the new phase of animal
magnetism called table-Turning.   Do you know about that ?

Photography is also in the ascendant.   You will, no
doubt, be at Ipswich, I believe an ancient city, and hath
old kirks and sundries worthy of notice.   Is Otley towards
the sea ?   Douking, etc. ?

To Miss Cay.

Trin. Coll., 7th June 1853.

I have an engagement to go and visit a man in Suffolk, but
the spare bed is at present occupied by the " celebrated Dr.
Ting of America."   I only wait here for his departure.   I
spent to-day in a great sorting of papers and arranging of
the same.   Much is bequeathed to the bedmaker, and a
number of duplicate examination papers are laid up to give
to friends.

I intend to-morrow to get up early and make breakfast
for all the men who are going down, wakening them in good
time; then read Wordsworth's *Prelude* till sleepy; then
sally forth and see if all the colleges are shut up for the
season; and then go and stroll in the fields and fraternise
with the young frogs and old water-rats.   In the evening,
something not mathematical.   Perhaps write a biographical sketch of Dr. Ting of America, of whom you know as
much as I do.   To-morrow evening, or next day, our list
comes out.   You will hear of it from the Robertsons if in
town, or Mackenzie if not.   I have done better papers than
those of this examination; but if the examiners are not
satisfied with them it is not my fault, for they are better
than they have yet seen of mine.   If any one asks how I
am getting on in mathematics, say that I am busy arranging
everything so as to be able to express all distinctly, so that
examiners may be satisfied now and pupils edified hereafter.

---

[1] Viz. of Elizabeth M'Keand (see above, p. 47).

It is pleasant work and very strengthening, but not nearly
finished.

FROM HIS FATHER.

Glenlair, 24th June 1853.

I have just received your letter and Mr. Tayler's. You *Æt.* 22.
may be sure I am thankful to hear of your recovering,
although not previously made anxious about the illness. I
cannot but think of the fever fit you had in Edinburgh after
an Academy exam., when we had settled to go to Melrose—
that was in 1846.[1] Nothing can exceed the kindness of
Mr. and Mrs. Tayler, and I hope you will not need long
nursing. If you are well and not much hindered, you can
let me know more fully how you are getting on. Neither
you nor Mr. Tayler mention the day you were taken ill.
Mr. T.'s letter is dated 22d.

Glenlair, 28th June 1853.

I am most thankful and happy to hear of your con-
valescence through Mr. Tayler's most kind and daily
bulletins. I know not how sufficiently to thank Mr. and
Mrs. Tayler for their very great kindness. I think you may
be best to come home, when fit to travel, for further
recreation.

Glenlair, 1st July 1853.

Mr. Tayler says, both truly and kindly, "You must be
his guest till you are fit to travel." . . .

With yours I have Mr. Tayler's letter of 28th. I do
not write to him to-day. My only subject is thanks, and
these are not to be measured in words—the strongest that I
can use ; so at present give my respects and highest regards.

---

[1] From the Diary :—" *1846, July, W., 29.*—(Day of the prize-
giving at the Edinburgh Academy). Made all ready to start on journey
to-morrow morning. At night James complained of the light of the
candle hurting his eyes. *Th. 30.*—Bad, wet day. Jas. awoke at six ;
eyes weak and headache ; . . . seems to be a disorder from excitement
of school examinations."

To the Rev. C. B. Tayler.

Trin. Coll., 8th July 1853.
Evening Post.

My dear Friend—Your letter was handed to me by the postman as I was taking a walk after morning chapel. As I was engaged then, I thought I might wait till the evening. I breakfasted with Macmillan the publisher, who has a man called Alexander Smith with him, who published a volume of poems in the beginning of the year, which have been much read here, and, indeed, everywhere, for 3000 copies have been sold already. He is a designer of patterns for needlework, and he refuses to be made celebrated or to leave his trade. He speaks strong Glasgow, but without affectation, and is well-informed without the pretence of education, commonly so called. People would not expect from such a man a book in which the author seems to transfer all his own states of mind to the objects he sees. But he is young and may get wiser as he gets older. He sees and can tell of the beauty of things, but he connects them artificially. He may come to prefer the real and natural connection, and after that he may perhaps stir us all up by bringing before us real human objects of interest he has only dimly seen in the solitude of his youth.

I told you how I meant to go to Hopkins. He was not in. I had a talk with him on Sunday; he recommended light work for a while, and afterwards he would give me an opportunity of making up what I had lost by absence. Yesterday I did a paper of his on the Differential Calculus without fatigue, and as well as usual. Ask George how Mr. Hughes has arranged about Examinations. I will write to him soon, and send him a mass of papers in an open packet, to be taken twice a week, or not so often.

You dimly allude to the process of spoiling which has gone on during the last 2 years. I admit that people have been kind to me, and also that I have seen more variety than in other years; but I maintain that all the evil influences that I can trace have been internal and not external, you know what I mean—that I have the capacity of being more wicked than any example that man could set me, and that if I escape, it is only by God's grace helping me to get rid of myself, partially in science, more completely in society, —but not perfectly except by committing myself to God as the instrument of His will, not doubtfully, but in the certain hope that that Will will be plain enough at the proper time. Nevertheless, you see things from the outside directly, and I only by reflexion, so I hope that you will not tell me you have *little* fault to find with me, without finding that little and communicating it.

In the *Athenæum* of the 2d there is Faraday's account of his experiments on Table-turning, proving mechanically that the table is moved by the unconscious pressure of the fingers of the people wishing it to move, and proving besides that Table-turners may be honest. The consequence has been that letters are being written to Faraday boastfully demanding explanations of this, that, and the other thing, as if Faraday had made a proclamation of Omniscience. Such is the fate of men who make real experiments in the popular occult sciences,—a fate very easy to be borne in silence and confidence by those who do not depend on popular opinion, or learned opinion either, but on the observation of Facts in rational combination. Our anti-scientific men here triumph over Faraday.

I hope the Rectory has flourished during the absence of you and Mrs. Tayler. I had got into habits with you of expecting things to happen, and if I wake at night I think the gruel is coming.

Macmillan was talking to me to-day about elementary books of natural science, and he had found the deficiency, but had a good report of " Philosophy in Sport made Science in Earnest," which I spoke of with you. When I am settled I will put down some first principles and practicable experiments on Light for Charlie, who is to write to me and answer questions proposed ; but this in good time.—Your affectionate friend,   J. C. MAXWELL.

To LEWIS CAMPBELL, Esq.

Trin. Coll., 14th July 1853.

You wrote just in time for your letter to reach me as I reached Cambridge. After examination I went to visit the Rev. C. B. Tayler (uncle to a Tayler whom I think you have seen under the name of *Freshman*, etc., and author of many tracts and other didactic works). We had little expedites, and walks, and things parochial and educational, and domesticity. I intended to return on the 18th June, but on the 17th I felt unwell, and took measures accordingly to be well again—*i.e.* went to bed, and made up my mind to recover. But it lasted more than a fortnight, during which time I was taken care of beyond expectation (not that I did not expect much before). When I was perfectly useless, and could not sit up without fainting, Mr. Tayler did everything for me in such a way that I had no fear of giving trouble. So did Mrs. Tayler; and the two nephews did all they could. So they kept me in great happiness all the time, and detained me till I was able to walk about, and got back strength. I returned on the 4th July.

The consequence of all this is that I correspond with Mr. Tayler, and have entered into bonds with the nephews, of all of whom more hereafter. Since I came here I have been attending Hop., but with his approval did not begin full swing. I am getting on, though, and the work is not grinding on the prepared brain.

I have been reading *Villette* by Currer Bell *alias* Miss Brontë. I think the authoress of Jane Eyre has not ceased to think and acquire principles since that work left her hands.

It is autobiographic in form. The *ego* is a personage of great self-knowledge and self-restraint, strength of principle and courage when roused, otherwise preferring the station of an onlooker.

Then there is an excellent prying, upright, Jesuitical, and successful French school directress; a fiery, finical, physiognomic professor, priestridden, but taking his own way in benevolence as in other things, etc. etc.

Faraday's experiments on Table-turning, and the answers of provoked believers and the state of opinion generally, show what the state of the public mind is with respect to the *principles* of natural science. The law of gravitation and the wonderful effects of the electric fluid are things which you can ascertain by asking any man or woman not deprived by penury or exclusiveness of ordinary information. But they believe them just as they believe history, because it is in books and is not doubted. So that facts in natural science are believed on account of the number of witnesses, as they ought! I believe that tables are turned; yea! and by an unknown force called, if you please, the vital force, acting, as believers say, thro' the fingers. But how does it affect the table? By the *mechanical* action of the sideward pressure of the fingers in the direction the table ought to go, as Faraday has shown. At this last statement the Turners recoil.

To R. B. LITCHFIELD, Esq.

Coniston, 23d August 1853.

I came here with Campbell of Trin. Hall to meet his brother and another Oxford man called Christie.[1] We are all in a house[2] just above the lake, recreating ourselves and reading a little. Pomeroy is off to Ireland. I have seen a good deal of him, and we have read "at the same time successively" *Vestiges of Creation* and Maurice's *Theological Essays.* Both have excited thought and talk. . . . I was down after the Mug[3] with Tayler's uncle in Suffolk, and was taken in there. I was there made acquainted with the peculiar constitution of a well-regulated family, consisting entirely of nephews and nieces, and educated entirely by the uncle and aunt. There was plenty of willing obedience, but little diligence : much mutual trust, and little self-reliance. They did not strike out for themselves in different lines, according to age, sex, and disposition, but each so excessively sympathised (*bonâ fide*, of course) with the rest,

---

[1] W. Christie, Esq., advocate.   [2] Bank-ground.
[3] Trinity College Examination.

that one could not be surprised at hearing any one take part in criticising his own action.

In such a case some would recommend " a little wholesome neglect." I would suggest something like the scheme of self-emancipation for slaves. Let each member of the family be allowed some little province of thought, work, or study, which is not to be too much enquired into or sympathised with or encouraged by the rest, and let the limits of this be enlarged till he has a wide, free field of independent action, which increases the resources of the family so much the more as it is peculiarly his own.

I see daily more and more reason to believe that the study of the "dark sciences" is one which will repay investigation. I think that what is called the proneness to superstition in the present day is much more significant than some make it. The prevalence of a misdirected tendency proves the misdirection of a prevalent tendency. It is the nature and object of this tendency that calls for examination.

To Lewis Campbell, Esq.

Glenlair, 15th September 1853.

I see that Principal Jelf is going to "have up" Maurice for heresy published in his *Theological Essays.* The consequence will probably be that some others unconnected with Maurice will be set upon, and will perhaps join with him in self-defence, or at least be associated with him in popular opinion.

If the row becomes general it will be *the* controversy of the day. They have no firm and dogmatic statements to grapple with, but they will soon make them. All the ordinary disputes have been revivals of the *letter* of old contests. Here we have the very spirit of all reformations; an attempt must be made to find what is requisite to a Christian system, and whether the "variables" of such a system ought to remain constant, as they were at some arbitrary epoch (that of sect-founders, Fathers, General Councils, Reformers, etc.), and not rather to be trusted to the true and approved Christians of every age.

But he that is misty let him be misty still, and the same for him that is shallow ; but let him that is active not mar his activity by " tearing his neighbours in their slime," or by ascending into the thick mist and walking with " Death and Morning on the silver horns."

FROM HIS FATHER.

Glenlair, 10th October 1853.

I have set up the rain-gauge in the middle of the garden at the crossing of tho gooseberry bushes at the Camomile. I think it will do.

As to changing your rooms—I suppose from that, you have settled to continue for a time at Cambridge and to look out for a fellowship.

Glenlair, 28th October 1853.

Be sure to keep a long way within your powers of working, and then you may do well whatever you undertake.

Glenlair, 13th November 1853.

Your letter was chiefly a dissertation on the election of Examinators ; the names were all strange to me, except our old friend Charles Mackenzie.

TO MISS CAY.

Trin. Coll., 12th November 1853.

I am in a regular state of health though not a very regular state of reading, for I hold that it is a pernicious practice to read when one is not inclined for it. So I read occasionally for a week and then miss a few days, always remembering to do whatsoever College and Hopkins prescribe to be done, and avoiding anything more. Allan Stewart was up a week ago to be made a bye Fellow of Peter-house, so you may congratulate him when you see him. He is to be in Edinburgh this winter. Frank Mackenzie is up, and seems pretty well. He tells me that he does not sit up late ; but as I have not the management of his candles I do not know what that means with him. I have

o

not been up after twelve for a long time except on Saturdays when I am not reading. . . . You will have heard how the Council of King's College have sat upon Professor Maurice and intend to turn him out of the college. So there are pamphlets and replies on the meaning of the word "Eternal," and broadsides of the Record on the SIDE of the attack. I see that the Rev. Berkeley Addison is in trouble about the Scottish Reformation Society, for associating with non-episcopal clergymen.

To Lewis Campbell, Esq.

Trin. Coll., 3d Dec. 1853.

. . . We have the usual amount of discussion here on labour parliaments, multiplicity of votes, Eternity and Maurice and Jelf, or the contest between those who think that there is a real depth to which thought must go, though words cannot well follow it, and those who maintain that that which is not obvious to a man of sense, cannot be really connected with a religion which is not confined to deep thinkers, but professes to afford the highest principles to the simple. That is what most men discuss. Maurice has settled it for himself, believing that the things of which he treats do actually form the necessary thoughts of all men whether learned or no.

From his Father.

Glenlair, 16th December 1853.

I knew Thomas Erskine of Linlathen very well long ago. He and his mother and sisters lived in No. 30 Heriot Row. He came to the Bar in Edinburgh the year before me. He is related to George Dundas, and Stirlings, and Erskines, and many families we visited. For long he has lived at Linlathen, near Dundee, and is author of various religious books.

Your dissertation on the parties in the Church of England goes far beyond my knowledge. I would need an explanatory lecture first, and before I can follow the High, Broad, and Low, through their ramifications.

Penicuik, 30th December /53.

You will need to get muffetees for the Senate-Room. Take your plaid or rug to wrap round your feet and legs.

To Miss Cay.

Trin. Coll., 13th January 1854.

All my correspondents have been writing to me, which is kind, and have not been writing questions, which is kinder. So I answer you now, while I am slacking speed to get up steam, leaving Lewis and Stewart, etc., till next week, when I will give an account of the *five days*. There are a good many up here at present, and we get on very jolly on the whole, but some are not well, and some are going to be plucked or gulphed, as the case may be, and others are reading so hard that they are invisible. I go to-morrow to breakfast with shaky men, and after food 1 am to go and hear the list read out, and whether they are through, and bring them word. When the honour list comes out the poll-men act as messengers. Bob Campbell comes in occasionally of an evening now, to discuss matters and vary sports. During examination I have had men at night working with gutta-percha, magnets, etc. It is much better than reading novels or talking after $5\frac{1}{2}$ hours' hard writing.

Hunter is up here all the vacation. Do you know anything of him in Edinburgh? His father, who is dead, or his uncle, were known in Edinburgh, but I am not up in that subject. The present man is a freshman at Queen's, and is a thundering mathematician, is well informed on political, literary, and speculative subjects, and is withal a jolly sort of fellow with some human nature at the bottom, and lots of good humour all through. He does not talk much, and when he does it is broad Scotch and to the purpose. I hope to see more of him next term. Old Charlie Robertson is in better case I think than usual, and rejoices in the good opinion of several men whose opinion is most worth having. He has become better known and better estimated of late, especially since Sandy[1] came up.

---

[1] Alexander Robertson, Esq., now Sheriff of Forfarshire.

He did pretty well in the three days, and does not fret about
anything. The snow here is nearly gone, and it looks like
frost again. I have never missed a long tramp through the
slush day by day. When one is well soaked in a snow
wreath, cleaned and dried, and put beside a good fire, with
bread and butter and problems, one can eat and grind like a
miller. . . I have been reading a book of poems called
Benoni, by Arthur Munby of Trinity, which are above the
common run of such things (not *Lorenzo Benoni*, illustrated
by J. B., which I have seen but not read). Have you seen
the *Black Brothers*, a small book of Ruskin's, illustrated by
Doyle;—a good child's book, which big people ought to
read.

FROM HIS FATHER.

18 India St., 30th January 1854.

I heartily congratulate you on your place in the list.
I suppose it is higher than the speculators would have
guessed, and quite as high as Hopkins reckoned on   I wish
you success in the Smith's Prizes; be sure to write me the
result. I will see Mrs. Morrieson, and I think I will call
on Dr. Gloag to congratulate him. He has at least three
pupils gaining honours.

CHAPTER VII.

BACHELOR-SCHOLAR AND FELLOW OF TRINITY—
1854 TO 1856—ÆT. 22-24.

"In the Main of Light."

JAMES CLERK MAXWELL's position as Second Wrangler and equal Smith's Prizeman, gave deep satisfaction to his friends in Edinburgh. Any lurking wish that he had been Senior was silenced by the examples of William Thomson and Charles Mackenzie, as others have been since consoled with the examples of Maxwell and Clifford. His father was persuaded by Miss Cay to sit for his portrait to Sir John Watson Gordon, as a gift of lasting value to his son. James was not indifferent to these reflex aspects of his success; but the chief interest of the moment to him undoubtedly was that he was now free to prosecute his life-career, and to use his newly-whetted instruments in resuming his original investigations. His leisure was not absolute, for he took pupils as a matter of course, and the Trinity Fellowship was only to be gained by examination. But his freedom was as great as he himself desired, and it is a fact worthy of attention from "researchers," that Maxwell, with his heart fully set on physical inquiries, engaged of

1854.
Æt. 22.

his own accord in teaching, undertook the task of examining Cheltenham College, and submitted to the routine which belonged to his position at Cambridge. As a foretaste of delights in store, he had spent the evenings of the Senate-house days in (physico-) magnetic *séances* with his friends. But when actually emancipated he seems to have reverted principally at first to his beloved Optics. He makes inquiries about a microscope manufactory at Zurich; reads Berkeley's *Theory of Vision*—taking up Mill's *Logic* by the way, and finding there by no means the last word on the relation of sense to knowledge; looks up his stock of coloured papers furnished by D. R. Hay, and sets to work spinning and weaving the different rays; inquires him out colour-blind persons on all sides; and invents an instrument for inspecting the living retina, especially of dogs. By and by it is the Art of Squinting which again has charms for him, and he combines it with the teaching of solid and spherical geometry, by drawing wonderful stereoscopic diagrams. So far, his investigations oscillate between colour and form. But even the fascination of the Colour-Top [1] cannot

---

[1] The "colour-box," though perfected only in 1862, was in full operation in the study at Glenlair several years before this—I think as early as 1850. And even then he had begun spinning coloured discs, proportionately arranged, so as to ascertain the true "mixture of colours." He was fond of insisting, to his female cousins, aunts, etc., on the truth that blue and yellow do not make green. I remember his explaining to me the difference between pigments and colours, and showing me, through the "colour-box," that the central band in the spectrum was different from any of the hitherto so-called "primary

hold him long from searching into the more hidden things of Matter in Motion. Thus, in his letter to his father of May 15, 1855, after describing a successful exhibition of the Top and extemporary statement of his optical theories before the Cambridge Philosophical Society, he adds : — " I am reading Electricity and working at Fluid Motion." And on May 23—" I am getting on with my electrical calculations every now and then, and working out anything that seems to help the understanding thereof.' Some days earlier, May 5, he had written :—" I am working away at Electricity again, and have been working my way into the views of heavy German writers. It takes a long time to reduce to order all the notions one gets from these men, but I hope to see my way through the subject, and arrive at something intelligible in the way of a theory." These brief notices obviously refer to the studies which led up to his important paper on Faraday's Lines of Force, which was put into shape in the winter of 1855-6.

The " vassals " were not forgotten by him even when most occupied at Cambridge. The choice and

---

colours." His theory on this subject was gradually formed through an immense number of ingeniously arranged observations.

His interest in Faraday's investigations must have dated from a very early time, certainly before 1849. And now he sought to give to these speculations, at which his imaginative mind had long been working, precise mathematical expression. I wish I could recall the date (1857 ?) of a drive down the Vale of Orr, during which he described to me for the first time, with extraordinary volubility, the swift, invisible motions by which magnetic and galvanic phenomena were to be accounted for. It was like listening to a fairy-tale. For the substance of it see his papers in the *Philosophical Magazine* for 1861-2.

provision of suitable literature for the consumption of Sam Murdoch and Sandy Fraser is a frequent topic of correspondence between him and his father.

How earnestly he now set himself to make the most of life in a religious sense appears from a sort of aphorism on conduct which he wrote down originally for his own use, and afterwards communicated as a parting gift to his friend Farrar (now Canon of Westminster), who was about to become a master at Marlborough School. As a record of the spirit in which Maxwell entered at three-and-twenty on his independent career, this fragment[1] is of extraordinary value.

1854.   He that would enjoy life and act with freedom must
Æt. 23.   have the work of the day continually before his eyes.   Not yesterday's work, lest he fall into despair, nor to-morrow's, lest he become a visionary,—not that which ends with the day, which is a worldly work, nor yet that only which remains to eternity, for by it he cannot shape his actions.

Happy is the man who can recognise in the work of To-day a connected portion of the work of life, and an embodiment of the work of Eternity. The foundations of his confidence are unchangeable, for he has been made a partaker of Infinity. He strenuously works out his daily enterprises, because the present is given him for a possession.

Thus ought Man to be an impersonation of the divine process of nature, and to show forth the union of the infinite with the finite, not slighting his temporal existence, remembering that in it only is individual action possible, nor yet shutting out from his view that which is eternal, knowing that Time is a mystery which man cannot endure to contemplate until eternal Truth enlighten it.

---

[1] An autograph copy was found amongst his papers. Another copy, together with the interesting fact mentioned above, determining the date, has been supplied by Canon Farrar's kindness.

Meanwhile in his recreations he was as boyishly agile as ever, and his feats in bathing and gymnastics, though cautiously reported by him, somewhat alarmed his father, whose own health now showed signs of breaking.

His friendships went on multiplying. To the list already given must now be added in particular the names of Hort, V. Lushington, Pomeroy, and Cecil Monro. And his constant observation of character was gaining in fulness and precision. A letter to his father of 21st April 1855, besides referring to a pupil (Platt) who had gained a Trinity scholarship, contains a graphic delineation of several of his coevals who were going in for the first open competition for appointments in the Indian Civil Service. He confides to his father all his thoughts about them, in which speculative and personal interests are combined.

He had soon another outlet for this kind of sympathy. The children of his uncle, Mr. R. Dundas Cay, who had lately returned from Hong-Kong, were advancing in their education, and two of the boys, William and Charles, showed considerable promise in mathematics. He had spent some days with his cousins in the summer vacation of 1854 at Keswick, where there was also a Cambridge reading party under Mathison, the tutor of Trinity. After a joyous time with them he walked home to Galloway from Carlisle, and it was during this walk that he thought out certain improvements in his dynamical top. He took a continuous interest in the progress of his cousins, and prepared a special set of problems for the behoof of Willy, the eldest, who was by this time a student in Edinburgh.

Maxwell continued his contributions to the "Apostle" Essay Club, and on May 5, 1855, he read to them in his own rooms a paper on Morality, in which he summed up the principles and tendencies of the chief existing systems of moral philosophy, so resuming another thread of his earlier thought.[1]

This essay appears to have been promised early in the year, to judge from an allusion to the subject of it in a letter to C. J. Monro of February 7, which may be quoted here as showing also by what home cares his intellectual energies were interrupted or diversified. The reader of what precedes will not be misled by the light way in which he speaks of things which touched his heart so nearly.

I am at present superintending a course of treatment practised on my father, for the sake of relieving certain defluxions which take place in his bronchial tubes. These obstructions are now giving way, and the medico, who is a skilful bellowsmender, pronounces the passages nearly clear.

However, it will be a week or two before he is on his pins again, so would you have the goodness to tell Freeman to tell Mrs. Jones to tell those whom it may concern, that I cannot be up to time at all. . . . I may be up in time to keep the term, and so work off a streak of Mathematics, which I begin to yearn after. At present I confine myself to Lucky Nightingale's line of business, except that I have been writing descriptions of Platometers for measuring plane figures, and privately by letter confuting rash mechanics, who intrude into things they have not got up, and suppose that their devices will act when they can't. I hope that

---

[1] *Supra*, p. 114.

my absence will not delay the assumption of W. D. Maclagan.[1] He has been here all the vacation, if not there and everywhere. The foundation of Ethics, though it may have tickled my core, has not germinated at my vertex. Whether it will yet be laid bare, either as a paradox or a truism, is more than I can tell. Perhaps it may be a pun. . . .

I have now to do a little cooking and buttling, in the shape of toast and beef-tea and everfizzing draught. . . . Does Pomeroy flourish, and has he Crimean letters still?

It was only because his father insisted on his doing so that he returned to Cambridge at all at this time.

At the meeting of the British Association, held at *Æt.* 24. Glasgow in September 1855, Maxwell was present when Brewster made an attack on Whewell's optical theories, but he followed the example of the Master of Trinity, who was present, in saying nothing. He exhibited his Colour-Top, however, the same afternoon, by appointment, at Professor Ramsay's house, where Brewster had been expected, but did not appear. Maxwell at the same time renewed his intercourse with Dr. George Wilson, the Edinburgh Professor of Technology, whose likeness hung beside that of Forbes in his rooms at Trinity. Wilson brought out immediately afterwards a little book on Colour Blindness, in which the substance of his conversations with Maxwell is recorded.

In October he gained his fellowship at Trinity. 1855. It was his second trial, and his name appeared as one *Æt.* 24.

_____

[1] Now Bishop of Lichfield.   Maxwell had recommended him for co-optation by the "Apostles'" Club.

of three mathematicians who had been chosen from the bachelors of the second year.

He was at once appointed to lecture to the Upper Division of the third year in Hydrostatics and Optics, and, to reserve time for his own studies, he now desisted from taking private pupils. He had indeed enough to occupy him without burdening himself with them. Besides the lecture in hydrostatics and optics, for which he found it desirable to read beforehand "so as not to tell lies," he had a large share in " exercising the questionists," *i.e.* preparing pass-men for the final examination, by setting papers in arithmetic, algebra, etc., and looking them over with the writers individually.

He had also been asked to prepare a text-book[1] on optics, and made some plans for doing so, having previously resisted the solicitation of his friend Monro, who had urged upon him the task of " translating Newton." And it is a fact worthy of the attention of bachelor fellows, that young Clerk Maxwell thought it worth while to attend the lectures of the Professor of Mechanics [2] and to exchange ideas with him.

For Electricity and Magnetism he took out *Poisson* again, and presently began putting together more systematically his own ideas on Faraday's *Lines of Force*.

His interest in coevals and juniors, which even in his undergraduate days was often like that of an

---

[1] The MS. of a considerable part of this book is still extant. There appears to have been sometimes a contest in his mind between the claims of different subjects. " I will have nothing to do with optics," he was heard once to exclaim.   [2] Professor Willis.

elder brother, assumed a deeper and more authoritative cast. He read more widely than ever, and nothing, from the latest novel to the newest metaphysical system, escaped his penetrating mind. He never read without criticising, and his criticisms, often quaintly expressed, were always worth attending to. " I hope that analysis of Hegel has done the writer good," " Comte has good ideas about method, but no notion of what is meant by a person," " Some people keep water-tight compartments in their minds." Such were the sparks that flew about. Other examples, not less striking, will be found in the letters. His observation of social phenomena also took a new departure, and his remarks on life and manners were endlessly entertaining.

He was elected a member of the Ray Club, which he had attended as a visitor in the spring, and did not forsake the assembling of the " Apostles," as appears from at least two essays which can only be referred to this period. He also took an active interest in the scheme for the higher education of working men, which had been lately set on foot by Mr. Maurice.

Between whiles he found time for a full course of classical English reading. And as all that he read he read critically, and had it thereafter in perfect possession, his literary acquirements were by this time of no mean order. He was, at the same time, careful to maintain himself in proper physical condition, by a steady course of exercises at the new gymnasium, which proved a welcome refuge in the wet November of that year. It was an unhealthy season, and to all his other employments was now added that of helping to nurse his friend Pomeroy, who was struck down with bilious fever.

His many-sided nature was in full activity. It is most characteristic of him that at this important crisis of his intellectual life, the best hour of day after day was given ungrudgingly to the task of literally making a friend's bed in his sickness. A lighter trait of the same kind may be found in the fact that, in the midst of the fellowship examination, he had given his father detailed advice about the "vassals'" reading:—
"When Sam Murdoch has finished *Arabia*, there are the volumes of the Cabinet Library, called *Drake, Cavendish, and Dampier, and Circumnavigation of the Globe*, Humboldt's *Travels and Polar Regions*, but it would be better to change and try the third volume of *Household Words*."

His thoughts turned homewards the more often, because his father's health was now becoming a matter for grave anxiety. In going up to Trinity for the fellowship trials, he had been doubtful whether in any case it would be right for him to stay up for the rest of the term. And, although this question was decided in the affirmative, every letter home bears some trace of his unceasing solicitude.

Thus, at the close of a period of manifold brightness, there was some foreshadowing of darker days shortly to come, when Death would take his father from him, and make the first breach in the circle of his friends.

LETTERS, 1854 TO 1856.

FROM HIS FATHER.

India Street, 4th Feby. 1854.

I have got yours of the 1st inst., and to-night or on Monday I will expect to hear of the Smith's Prizes. I get congratulations on all hands, including Prof. Kelland and Sandy Fraser, and all others competent.

18 India St., 6th Feby. 1854.

George Wedderburn came into my room at 2 A.M. yesterday morning, having seen the Saturday *Times*, received by the express train, and I got your letter before breakfast yesterday. As you are equal to the Senior in the champion trial, you are but a very little behind him.

I am going to dine with John Cay, and with him proceed to the Royal Society. I may perhaps catch Prof. Gregory about the microscopist.

5th March 1854.

Mrs. Morrieson told me she had a poetical epistle from you on St. David's Day.[1]

Aunt Jane stirred me up to sit for my picture, as she said you wished for it and were entitled to ask for it, quâ wrangler. I have had four sittings to Sir John Watson Gordon, and it is now far advanced; I think it is very like. It is Kit-cat size, to be a companion to Dyce's picture of your mother and self, which Aunt Jane says she is to leave to you.

To R. B. LITCHFIELD, Esq.

Trin. Coll., 25th March 1854.

I am experiencing the effects of Mill, but I take him slowly. I do not think him the last of his kind. I think more is wanted to bring the connexion of sensation with Science to light, and show what it is not. I have been reading Berkeley on the *Theory of Vision,* and greatly admire it, as I do all his other non-mathematical works;

---

[1] Mrs. Morrieson's early home was in Montgomeryshire.

but I was disappointed to find that he had at last fallen into the snare of his own paradoxes, and thought that his discoveries with regard to the senses and their objects would show some fallacy in those branches of high mathematics which he disliked. It is curious to see how speculators are led by their neglect of exact sciences to put themselves in opposition to them where they have not the slightest point of contact with their systems. In the Minute Philosopher there is some very bad Political Economy and much very good thinking on more interesting subjects. Paradox is still sought for and exaggerated. We live in an age of wonder still.

To Miss Cay.

Trin. Coll., Whitsun. Eve, 1854.

I am in great luxury here, having but 2 pups., and able to read the rest of the day, so I have made a big hole in some subjects I wish to know. We have hot weather now, and I am just come from a meeting of subscribers to the Bathing Shed, which we organised into a Swimming Club so as to make it a more sociable affair, instead of mere "pay your money and use your key."

A nightingale has taken up his quarters just outside my window, and works away every night. He is at it very fierce now. At night the owls relieve him, softly sighing after their fashion.

I have made an instrument for seeing into the eye through the pupil. The difficulty is to throw the light in at that small hole and look in at the same time; but that difficulty is overcome, and I can see a large part of the back of the eye quite distinctly with the image of the candle on it. People find no inconvenience in being examined, and I have got dogs to sit quite still and keep their eyes steady. Dogs' eyes are very beautiful behind, a copper-coloured ground, with glorious bright patches and networks of blue, yellow, and green, with blood-vessels great and small.

Trin. Coll., 24th Novr. /54.

I have been very busy of late with various things, and

am just beginning to make papers for the examination at Cheltenham, which I have to conduct about the 11th of December. I have also to make papers to polish off my pups. with. I have been spinning colours a great deal, and have got most accurate results, proving that ordinary peoples' eyes are all made alike, though some are better than others, and that other people see two colours instead of three; but all those who do so agree amongst themselves. I have made a triangle of colours by which you may make out everything.

You see that W lies outside the triangle B, R, Y, so that White can't be made with Blue, Red, and Yellow; but if you mix blue and yellow you don't get green, but pink— 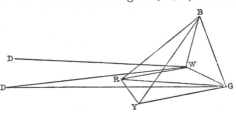 a colour between W and R. Those who see two colours only distinguish blue and yellow, but not red and green: for instance—

6 of blue and 94 of red make a red which looks to them like a gray made of 10 W and 90 Black.

40 of blue and 60 of green make 34 of W and 66 Black.

I should like you to find out if the Normans have got Bishop Percy's *Reliques of Ancient Ballad Poetry*, for if they have I would not send them a duplicate; if not I think the book would suit one-half of that family.

If you can find out any people in Edinburgh who do not see colours (I know the Dicksons don't), pray drop a hint that I would like to see them. I have put one here up to a dodge by which he distinguishes colours without fail. I have also constructed a pair of squinting spectacles, and am beginning operations on a squinting man.

To C. J. MONRO, Esq.

18 India Street, Edinburgh, 19th Feb. /55.

My steps will be no more by the reedy and crooked

P

till Easter term. My father's recovery is retarded by the
frosty weather, though we have got up an etherial mild-
ness here by means of a good fire and a towel hung up wet
at the other end of the room, together with an internal
exhibition of nitric ether.

I wrote to Mackenzie about putting a respectable man
in my rooms as a stopper to Cat's Hall men, Manns and
Boy Joneses, but I have not heard of his success. I have
no time at present for anything except looking through
novels, etc., and finding passages which will not offend my
father to read to him. He strongly objects to new-fangled
books, and knows the old by heart. But he likes the *Essays
in Intervals of Business,* cause why, they have not too many
words. The frost here has lasted long, and I am beginning
to make use of it. I get an uncle to take my place in the
afternoon, and I rush off to Lochend or Duddingston. I
have not yet succeeded in skating on one foot for an indefinite time and getting up speed by rising and sinking at the
bends of the path; but I attribute my failure to want of
faith, for I can get up speed for a single bend, only I always
slip at a certain critical turning. However, I have only
been 3 days, and I may do it yet. My plans are not
fixt, but I think it will be some while before my father is
on his pins again, and when he is I intend to look after him
still, but do a private streak of work, for I will soon be in a
too much bottled up condition of mathematics, from which
even mental collapse would be a relief. I have no intention
of doing a Newton or any elegant mathematics. I have a
few thoughts on top-spinning and sensation generally, and a
kind of dim outline of Cambridge palavers, tending to
shadow forth the influence of mathematical training on
opinion and speculation.

I suppose when my father can move I will see him out of
this eastern clime and safe located in Gallovidian westnesses,
and so be up in Cam. before the beginning of next term.

I should like to know how many kept baccalaurean
weeks go to each of these terms, and when they begin and
end. Overhaul the calendar, and when found make note of.

Is Pomeroy up, or where ?   This is the 2nd time of asking.

To his Father.

Trin. Coll., Saturday, 21st April 1855.
[Date in John C. M.'s hand.]

Lots of men are going in for the H.E.I.C.S. examination, —Pomeroy, B., C., D. (the best double degree for many years), E. (Senior Wrangler), etc., so I suppose the competition will be pretty active; but it is evident that these men will be totally different judges, etc., tho' they may be all good in examination subjects.

Pomeroy is a genial giant, generous and strong, but hasty in condemnation tho' slow to wrath.   B., intelligent and able to detect any humbug but his own ; but excitable, and impudent in the extreme to people he does not know. C. has strong feelings and affections, with a great amount of sympathy for all cases, but it is repressed for want of courage, and he is left with somewhat of a sneaking virtue of his own, always trying to put on the manners which suit those he is with.   D. is a good man of business, using up every scrap of his time most successfully, and honest, I believe.   E. is what I don't know, but I can conceive him reduced by circumstances to act the part of Sir Elijah Impey in India ; but I hope circumstances may be different, and then he may be a harmless mathematician or scientific referee, and leave a high reputation behind him.

Trin. Coll., Vesp. SS. Philipp. & S. Jac. 1855.

I have been working at the motion of fluids, and have got out some results.   I am going to show the colour trick at the Philosophical on Monday.   Routh has been writing a book about Newton in conjunction with Lord Brougham. Stokes is back again and lecturing as usual.

Saturday, 5th May 1855.

The Royal Society have been very considerate in sending me my paper on colours just when I wanted it for the

Philosophical here. I am to let them see the tricks on Monday evening, and I have been there preparing their experiments in the gas-light. There is to be a meeting in my rooms to-night to discuss Adam Smith's "Theory of Moral Sentiments," so I must clear up my litter presently. I am working away at electricity again, and have been working my way into the views of heavy German writers. It takes a long time to reduce to order all the notions one gets from these men, but I hope to see my way through the subject and arrive at something intelligible in the way of a theory.

Trin. Coll., 15th May 1855.

The colour trick came off on Monday, 7th. I had the proof sheets of my paper, and was going to read; but I changed my mind and talked instead, which was more to the purpose. There were sundry men who thought that Blue and Yellow make Green, so I had to undeceive them. I have got Hay's book of colours out of the Univ. Library, and am working through the specimens, matching them with the top. I have a new trick of stretching the string horizontally above the top, so as to touch the upper part of the axis. The motion of the axis sets the string a-vibrating in the same time with the revolutions of the top, and the colours are seen in the haze produced by the vibration. Thomson has been spinning the top, and he finds my diagram of colours agrees with his experiments, but he doubts about browns what is their composition. I have got colcothar brown, and can make white with it, and blue and green; also, by mixing red with a little blue and green and a great deal of black, I can match colcothar exactly.

I have been perfecting my instrument for looking into the eye. Ware has a little beast like old Ask,[1] which sits quite steady and seems to like being looked at, and I have got several men who have large pupils and do not wish to let me look in. I have seen the image of the candle distinctly in all the eyes I have tried, and the veins of the

---

[1] Above, p. 57.

retina were visible in some; but the dogs' eyes showed all the ramifications of veins, with glorious blue and green network, so that you might copy down everything. I have shown lots of men the image in my own eye by shutting off the light till the pupil dilated and then letting it on.

I am reading Electricity and working at Fluid Motion, and have got out the condition of a fluid being able to flow the same way for a length of time and not wriggle about.

Trin. Coll., Eve. of H. M. Nativity.

Wednesday last I went with Hort and Elphinstone to the Ray Club, which met at Kingsley of Sidney's rooms. Kingsley is great in photography and microscopes, and showed photographs of infusoria, very beautiful, also live plants and animals, with oxy-hydrogen microscope. . . . I am getting on with my electrical calculations now and then, and working out anything that seems to help the understanding thereof.

FROM HIS FATHER,

Glenlair, 21st May 1855.

Have you put a burn in fit condition to flow evenly, and not beat on its banks from side to side ? That would be the useful practical application.

FROM PROFESSOR J. D. FORBES.

Clifton, Bristol, 4th May 1855.

I left directions with Messrs. Neill & Co. to forward proofs of your paper, by inquiring at 18 India Street, and I understand that they were sent out on the 1st May.

I am informed that my note to you about some of my experiments on colour has been printed in the *Edinburgh Philosophical Journal*. This was by no means what I intended. . . . What I thought that you might do was to introduce into that part of your paper where you speak of what has been done or written on the subject, mention of the fact that as early as January 18— (I do not at the moment recol-

lect the year I stated to you) I had used the method of
rapid motion in blending colours; that I had endeavoured
to obtain an equation between certain mixed colours and
pure gray; and that I had pointed out before Helmholtz, or
I believe any one else, that a mixture of yellow and blue,
under these circumstances at least, does not produce green;
you yourself being a witness to what I then tried, though I
was prevented from resuming the subject by ill health and
some experimental occupations (conduction of heat) which I
considered more imperative.

I hope you will continue to prosecute your interesting
inquiries, and with an equal measure of success.

I address this to Cambridge, as I think you said you
should be there this month.

From the Same.

Clifton, Bristol, 16th May 1855.

I am much obliged by your note mentioning your
intention of referring to my experiments.

You inquire how I altered the proportions of the constituent colours. My plan was, in fact, the same as yours.
I had sectors much larger than I required of each colour,
making them overlap, and fixing them down by a screw at
the centre, pressing a disc of indiarubber on the discs.
When I got the anomalous result of blue and yellow, I
got Mr. Hay to make a disc of *many* alternating narrow
sectors merely to see whether it might be a physiological
effect from the imperfect blending of the colours.

I still think the experiment ought to be tried *without
motion*, by winding blue and yellow threads of silk or
worsted round a card and looking at it at a good distance,
or (as you proposed) by viewing it with a telescope out of
focus.

You will recollect that I had a whirling-machine (made
on purpose), in which a number of discs revolved simultaneously with equal velocities. I used black and white on
one of these; colours on another. Your teetotum, combining both, I consider preferable for experiments. By the

way, I did not get the teetotum you were to leave for
me.

P.S.—I hope you have got the proof of the plate as
well as of the paper.   If not, write to Messrs. Johnston,
engravers, 4 St. Andrew Square, Edinburgh.

To R. B. LITCHFIELD, Esq.

Trin. Coll., 6th June 1855.

It is hard work grinding out " appropriate ideas," as
Whewell calls them.   However, I think they are coming
out at last, and by dint of knocking them against all the
facts and $\frac{1}{2}$-digested theories afloat, I hope to bring them to
shape, after which I hope to understand something more
about inductive philosophy than I do at present.

I have a project of sifting the theory of light and making everything stand upon definite experiments and definite
assumptions, so that things may not be supposed to be
assumptions when they are either definitions or experiments.

I have been looking into all the dogs' eyes here to see
the bright coating at the back of the eye, thro' an instrument I made to that end.   The spectacle is very fine.   I
remember the appearance of Mungo's eyes at Cheltenham.
He would be the dog to sit.   Human eyes are very dark
and brown as to their retina, but you can see the image of
a candle quite well on it, and sometimes the blood-vessels,
etc.

FROM WILLIAM DYCE CAY, Esq., TO JOHN C. M.

(Glasgow, at the Meeting of the British Association).

18th September 1855.

Sir David Brewster was upon the triple spectrum.   As
far as I can understand, he believes the spectrum to be
composed of three colours—red, blue, and yellow; and that
the intermediate colours are composed of mixtures of these,
as, for example, the green from a mixture of blue and
yellow, which, I think, is different from what James be-

lieves. James did not say anything in the controversy
which followed his speech, as he was to meet Sir D.
Brewster at the Ramsays' afterwards, where he would have
his top and other apparatus to show him.

To his Father.

(After the Meeting of the British Association at Glasgow).

Holbrooke, by Derby, 24th Sept. 1855.

We had a paper from Brewster on the theory of three
colours in the spectrum, in which he treated Whewell with
philosophic pity, commending him to the care of Prof.
Wartman of Geneva, who was considered the greatest
authority in cases of his kind, cases in fact of colour-blindness. Whewell was in the room, but went out, and avoided
the quarrel; and Stokes made a few remarks, stating the
case not only clearly but courteously. However, Brewster
did not seem to see that Stokes admitted his experiments to
be correct, and the newspapers represented Stokes as calling
in question the accuracy of the experiments.

I am getting my electrical mathematics into shape, and I
see through some parts which were rather hazy before; but I
do not find very much time for it at present, because I am
reading about heat and fluids, so as not to tell lies in my
lectures. I got a note from the Society of Arts about the
platometer, awarding thanks, and offering to defray the ex-
penses to the extent of £10, on the machine being produced
in working order. When I have arranged it in my head, I
intend to write to James Bryson about it.

I got a long letter from Thomson about colours and
electricity. He is beginning to believe in my theory about
all colours being capable of reference to three standard ones,
and he is very glad that I should poach on his electrical
preserves.

Trin. Coll., 27th Sept. 1855.

. . . It is difficult to keep up one's interest in intellectual matters when friends of the intellectual kind are

scarce.[1]   However, there are plenty friends not intellectual, who serve to bring out the active and practical habits of mind, which overly-intellectual people seldom do.   Wherefore, if I am to be up this term, I intend to addict myself rather to the working men who are getting up classes, than to pups., who are in the main a vexation.   Meanwhile there is the examination [2] to consider.

Trin. Coll., 5th October 1855.

You say Dr. Wilson has sent his book.   I will write and thank him.   I suppose it is about colour-blindness.   I intend to begin Poisson's papers on electricity and magnetism to-morrow.   I have got them out of the library ; my reading hitherto has been of novels,—*Shirley* and *The Newcomes*, and now *Westward Ho.*

Trin. Coll., 10th October 1855.

Macmillan proposes to get up a book of optics, with my assistance, and I feel inclined for the job.   There is great bother in making a mathematical book, especially on a subject with which you are familiar, for in correcting it you do as you would to pups.—look if the principle and result is right, and forget to look out for small errors in the course of the work.   However, I expect the work will be salutary, as involving hard work, and in the end much abuse from coaches and students, and certainly no vain fame, except in Macmillan's puffs.   But, if I have rightly conceived the plan of an educational book on optics, it will be very different in manner, though not in matter, from those now used.

FROM HIS FATHER.

Glenlair, 10th October 1855.

The book sent by Dr. Wilson is the full edition about colour-blindness, with notes and appendices, containing your letter to him and notices of your communications to him on the subject.

---

[1] This is said *à propos* of a recent visit to a college friend who was settled as a clergyman in the country.

[2] For the Trinity College Fellowship.

To Lewis Campbell, Esq.

Trin. Coll., 17th October 1855.

I expect to be grinding this term.   There are lectures on hydrostatics and optics, papers for questionists to be set and read over with the men, which is procrastinatious. Besides this I may have to lecture the working men, and what spare time I have I intend to use on various subjects, which will keep me in work for some time to come, so I do not require any pupils to keep my hand in this term.   I was looking for Jowett's book in the library, but, as usual, all the new theology had been carried off in a lump by the M.A.'s, who get in the first day.   I wanted Ellicott, but he was out too, so I took Carlyle on the French Revolution.   I have been reading the English language, comprising Chaucer, Sir Tristram, Bacon F., Pope, Berkeley, Goldsmith, Cowper, Burns' letters, Isaac Taylor's *Saturday Night,* Carlyle, Ruskin, Kingsley, Maurice, and combining the whole with Trench on *English Past and Present,* and with all this I derive pleasure and information, but not a single glimmer of a theory about Words.

And yet I have presently to state whether words mould thought or thought brews words.   Is not one theory as good as another ?   Faith and a dale better too, if it was not for the sake of laying them together by the ears, which is a difficult task when you have to catch both yourself.

I was staying at the Blackburns' when I was at Glasgow, but they were away, and the Ramsays fed and tended me.   I found your photograph there, together with a few other pleasant recollections.   I have been over to H. M. Butler, who is come up again.   We were talking about Maurice, etc.   Maurice is a man I am loath to say nay to, or to accuse of wilful perversion of facts; but in some matters I think he is in great error, especially in his estimate of respectable ordinary Christians, as far as regards their creed. He cannot go too far in enforcing practice and work on people who were bound to it before, and theoretically confess it, but he is too hard upon the theories, and totally mis-

represents them.   I would rather be taken for a Yezide than
for one of Maurice's popular religionists.

To his Father.

Trin. Coll., 17th October 1855.

The lectures were settled last Friday.   I am to do the
upper division of the third year in hydrostatics and optics,
and I have most of the exercising of the questionists.

From his Father.

Glenlair, 20th October 1855.

If you do a book for M'Millan on optics, do not let him
hurry it on.   Take full time to yourself to revise and
re-revise the MS., and let anything published be creditable.
Do nothing in a careless manner, and so get a bad name.
A first work especially should be very carefully got up.

When you are set to lecture on hydrostatics and optics,
have you any apparatus for illustration ?

To his Father.

Trin. Coll., 25th October 1855.

I have refused to take pupils this term, as I want to get
some time for reading and doing private mathematics, and
then I can bestow some time on the men who attend lectures.

I go in bad weather to an institution just opened for
sports of all sorts — jumping, vaulting, etc.   By a little
exercise of the arms every day, one comes to enjoy one's
breath, and to sleep much better than if one did nothing but
walk on level roads.

1st November 1855.

I have been lecturing two weeks now, and the class
seems improving, and they come up and ask questions, which
is a good sign.

I have been making curves to show the relations of
pressure and volume in gases, and they make the subject
easier.   I think I told you about the Ray Club.   I was

elected an associate last Wednesday. . . . We had a discussion and an essay by Pomeroy last Saturday about the
position of the British nation in India, and sought through
ancient and modern history for instances of such a relation
between two nations, but found none.   We seem to be in
the position of having undertaken the management of India
at the most critical period, when all the old institutions and
religions must break up, and yet it is by no means plain
how new civilisation and self-government among people so
different from us is to be introduced.   One thing is clear, that
if we neglect them, or turn them adrift again, or simply
make money of them, then we must look to Spain and the
Americans for our examples of wicked management and
consequent ruin.

FROM HIS FATHER.

11th November 1855.

The platometer will require much consideration, both by
you and by any one that undertakes the making.   You need
hardly expect the details all rightly planned at the first;
many defects will occur, and new devices contrived to conquer unforeseen difficulties in the execution.   I would suspect
£10 would not go far to get it into anything like good
working order.   If the instrument were made, to whom is it
to belong ?   And if it succeeds well, for whose profit is all
to be contrived ?   Does Bryson so understand it as to be
able to make it ?   Could he estimate the cost, or would he
contract to get an instrument up ?   Fixing on a suitable
size is very important.

TO HIS FATHER.

Trin. Coll., 12th November 1855.

I attended Willis on Mechanism to-day, and I think I
will attend his course, which is about the parts of machinery.
I was lecturing about the velocity of water escaping from a
hole this morning.   There was a great noise outside, and we
looked out at a magnificent jet from a pipe which had gone
wrong in the court.   So that I was saved the trouble of
making experiments.

I was talking to Willis about the platometer, and he thinks it will work. Instead of toothed wheels to keep the spheres in position always, I think watch-spring bands would be better.

Trin. Coll., 25th November 1855.

I think I told you that Pomeroy was ill. He has had rather a sharp attack of bilious fever. His mother has come up. He was getting round on Thursday, but he saw too many people, and was rather the worse of it. However, the doctor says that the recovery simply requires attention, and patience, and no hurrying.

I have been reading old books of optics, and find many things in them far better than what is new. The foreign mathematicians are discovering for themselves methods which were well known at Cambridge in 1720, but are now forgotten.

I have got a contrivance made for expounding instruments. It is a squared rod, one yard long, on which slide pieces, which will carry lenses. Each piece has a wedge which fixes it tight on the rod, and a saw-shaft, with holes through it, for fastening the pasteboard frame of the lens. By means of this I intend to set up all kinds of models of instruments.

To R. B. Litchfield, Esq.

Trin. Coll., 28th November 1855.

I am busy with questionists pretty regularly just now, slanging them one after another for the same things. As they have just set upon me for the evening, I must stop now and get out some optical things to show them.

To his Father.

Trin. Coll., 3d December 1855.

I had four questionist papers last week, as my subjects come thick there; so I am full of men looking over papers. I have also to get ready a paper on Faraday's Lines of Force for next Monday.

Pomeroy is still very ill, but to-day he feels easier, and his mouth is not quite so dry and sore. He gets food every two or three hours, and port wine every time. I go up in the morning and look after the getting up and bed-making department along with the nurse, after which Mrs. Pomeroy comes, and the nurse goes to bed.

Maurice was here from Friday to Monday, inspecting the working men's education. He was at Goodwin's on Friday night, where we met him and the teachers of the Cambridge affair. He talked of the history of the foundation of the old colleges, and how they were mostly intended to counteract the monastic system, and allow of work and study without retirement from the world.

Trin. Coll., 11th December.

Last night I lectured on Lines of Force at the Philosophical. I put off the second part of it to next term. I have been drawing a lot of lines of force by an easy dodge. I have got to draw them accurately without calculation.

Pomeroy has been improving slowly, but sometimes stopping. He is so big that it requires a great deal to get up his strength again. I saw Dr. Paget at the Philosophical to-day, and he seemed to think him in a fair way to recover.

CHAPTER VIII.

THE description of Maxwell's life at Cambridge would be incomplete without some notice of the Essays written by him from time to time for the "Apostles'" Club. These range from the spring of 1853 to the summer of 1856. Thrown off, as such things are, in irresponsible gaiety of heart, mere "gardens of Adonis," as Plato would call them, they contain real indications of the writer's speculative tendencies, and are most characteristic of the activity and fulness of his mind, of his ironical humour, and of his provoking discursiveness and indirectness of expression. He is not "upon his oath," and often throws out tentatively a whole train of arguments or ideas.

1. "*Decision.*"   Written in February 1853.

After a humorous sketch of the distraction arising from *Æt.* 21. the different associations of term and vacation time, the question is raised whether on the whole a learned education is unfavourable to decision of character and opinion. The answer pointed at, though not distinctly given, is that high education may often unsettle opinion, but ought to strengthen character. It must suffice here to quote a few of the most characteristic passages :—

" . . . In this charitable (holiday) frame of mind, we

resolved to try the effect of our learning upon a mixed company.

"Not to dazzle them too much at first, we merely ventured to quote to an elderly lady a passage from Griffin on Presbyopic Vision. She intended to get a new pair of spectacles, and hoped that optical advice, fresh from College, might assist her in her choice. She was surprised to learn that she must ascertain the distance behind the retina at which the image of a distant object is formed, and that she might then determine from the proper formula the focal lengths of the lenses she required.

"Shocked at the unhesitating way in which we proposed the most barbarous if not impossible operations, she replied that she would rather try several pairs, and take those that suited her best."

" . . . When this indecision" (of opinion) "cannot be traced to hypochondria, we generally find indications of a defective appreciation of quantity and a deceptive memory.

"Its victims measure reasons by their number and not by their weight. They do not say 'so much,' but 'so many things to be said on both sides.' To make the number equal on both sides they will split an argument or state it in several ways. These ingenious self-tormentors have invented a form of reasoning which ought to take its place beside the 'reasoning in a circle.' We may call it reasoning in a corner, or tergiversation. . . . It derives its name from the motion of the imprisoned monarch in a drawn game at draughts, and is resorted to when pressed by a disjunctive argument. . . . In this way these clacking metronomes endeavour to transfer their inquietude to their neighbours."

" . . . It is this consciousness of aim that gives to their experience the character of self-education. While other men are drifted hither and thither by conflicting influences, their sails seem to resolve every blast in a favourable direction. To them catastrophes are lessons and mysteries illustrations. Every thing and every person is estimated by its effect in accelerating personal advancement.

"The aims thus adopted may be different in kind and value. One may aim at effective deeds, another at completeness, a third at correctness, a fourth at dignity, while another class estimates its progress by the universality of its sentiments and the comprehensiveness of its sympathy with the varieties of the human mind. Some, in short, attend more to self-government, and some to mental expansion. When these tendencies can be combined and subordinated, there emerges the perfectly educated man, who, in the rigidity of his principles, acts with decision, and in the expansibility of his sympathy tolerates all opinions."

2. *"What is the Nature of Evidence of Design?"*   1853.

"Design! The very word . . . disturbs our quiet discussions about *how* things happen with restless questionings about the *why* of them all. We seem to have recklessly abandoned the railroad of phenomonology, and the black rocks of Ontology stiffen their serried brows and frown inevitable destruction.

". . . The belief in design is a necessary consequence of the Laws of Thought acting on the phenomena of perception.

". . . The essentials then for true evidence of design are —(1) A phenomenon having significance to us ; (2) Two ascertained chains of physical causes contingently connected, and both having the same apparent terminations, viz., the phenomenon itself and some presupposed personality. . . . If the discovery of a watch awakens my torpid intelligence I perceive a significant end which the watch subserves. It goes, and, considering its locality, it is going well. . . . My young and growing reason points out two sets of phenomena . . . (*a*) the elasticity of springs, etc. etc., and (*b*) the astronomical facts which render the mean solar day the unit of civil time combined with those social habits which require a cognisance of the time of day.

" . . . It is the business of science to investigate these causal chains. If they are found not to be independent but to meet in some ascertained point, we must transfer the

Q

evidence of design from the ultimate fact to the existence of the chain. Thus, suppose we ascertained that watches are now made by machinery . . . the machinery including the watch forms one more complicated and therefore more evident instance of design."

" . . . The only subordinate centres of causation which I have seen formally investigated are men and animals; the latter even are often overlooked. But every well-ascertained law points to some central cause, and at once constitutes that centre a *being* in the general sense of the word. Whether that being be *personal* is a question which may be determined by induction. The less difficult question whether the *being* be *intelligent* is more practicable, and should be kept in view in the investigation of organised beings.

" The search for such invisible potencies or wisdoms may appear novel and unsanctioned. . . . For my part I do not think that any speculations about the personality or intelligence of subordinate agents in creation could ever be perverted into witchcraft or demonolatry.

" Why should not the Original Creator have shared the pleasure of His work with His creatures and made the morning stars sing together, etc.?

" I suspect that such a hope has prompted many speculations of natural historians, who would be ashamed to put it into words. [1]

" . . . Three fallacies—(1) Putting the final cause in the place of a physical connection, as when Bernoulli saw the propriety of making the curves of isochronous oscillations and of shortest time of descent both cycloids; (2) The erroneous assertion of a physical relation, as when Bacon supplemented the statement of Socrates about the eyebrows

---

[1] This Neo-platonic fancy (of $\delta\eta\mu\iota\text{ουργοί}$), with which the reader may *contrast* the serio-comic lines on " Paradoxical Philosophy," is embodied in the alternative title of the paper—" Ought the Discovery of a Plurality of Intelligent Creators to weaken our Belief in an Ultimate First Cause ?" A third title has been added later in the author's hand—" Does the Existence of Causal Chains prove an Astral Entity or a Cosmothetic Idealism ?"

by saying that pilosity is incident to moist places; (3) (and worst) applying an argument from final causes to wrongly asserted phenomena :—' Because water is incompressible, it cannot transmit sound, and therefore fishes have no ears.' Every fact here stated is erroneous."

In the course of this paper—in which are discernible the traces of early impressions derived through the poetry of Milton—there occurs also incidentally a statement of the Hamiltonian doctrine of Perception,[1] with the following significant corollary :—

" Perception is the ultimate consciousness of self and thing together.

" If we admit, as we must, that this *ultimate* phenomenon is incapable of further analysis, and that subject and object alone are immediately concerned in it, it follows that the fact is strictly private and incommunicable.   One only can know it, therefore two cannot agree in a name for it.   And since the fact is simple it cannot be thought of by itself nor *compared alone* with any other *equally simple fact.*   We may therefore dismiss all questions about the absolute nature of perception, and all theories of their resemblances and differences.   We may next refuse to turn our attention to perception in general, as all perceptions are particular."

3. *Idiotic Imps.*   Summer Term, 1853.

Starting from Isaac Taylor's *Physical Theory of another Life*, which Maxwell at this time seems to have regarded as in itself an innocent and rather attractive piece of fancy,— " the perusal of it has a tendency rather to excite speculation than to satisfy curiosity, and the author obtains the approbation of the reader, while he fails to convince him of the soundness of his views,"—he takes occasion from it to

---

[1] This statement concludes as follows :—" The late superfluity of assertions might have been avoided by simply, unintelligibly and therefore unanswerably, proclaiming myself a natural dualist, uncontaminated with the heresy of unitarianism or the pollution of cosmothetic idealism."

characterise the "Dark Sciences" to which Taylor's book may unintentionally lend encouragement—a result to be deprecated.

"The first question I would ask concerning a spiritual theory would be, Is it favourable or adverse to the present developments of Dark Science? The Dark Sciences . . . while they profess to treat of laws which have never been investigated, afford the most conspicuous examples of the operation of the well-known laws of association . . . in imitating the phraseology of science, and in combining its facts with those which must naturally suggest themselves to a mind unnaturally disposed. In the misbegotten science thus produced we have speciously sounding laws of which our first impression is that they are truisms, and the second that they are absurd, and a bewildering mass of experimental proof, of which all the tendencies lie on the surface and all the data turn out when examined to be heaped together as confusedly as the stores of button-makers . . . and those undigested narratives which are said to form the nutriment of the minute philosopher. . . . The most orthodox system of metaphysics may be transformed into a dark science by its phraseology being popularised, while its principles are lost sight of."

Three phases of dark science are described :—
"(1.) At first they were or pretended to be physical sciences. Their language was imitated from popular physics, and their professed aim was to explain occult phenomena by means of new and still more occult material laws. Experiments in animal magnetism were always performed with the nose carefully turned towards the north. In electrobiology a scrupulous system of insulation was practised at first, and afterwards, when galvanism became more popular than statical electricity, circuits were formed of alternate elements, those of one sex being placed between those of the other. . . . The fluid which in former times circulated through the nerves under the form of animal spirits, is in our day expanded so as to fill the universe, and

is the invisible medium through which the communion of
the sensitive takes place.

" (2.) The next phase of the dark sciences is that in
which . . . the phraseology of physics is exchanged for
that of psychology. In this stage we hear much of the power
of the will. The verb to will acquires a new and popular
sense, so that every one now is able to will a thing without
bequeathing it. People can will not to be able to do a
thing, then try and not succeed ; while those of stronger
minds can will their victims out of their wits and back
again.

" (3.) The third or pneumatological phase begins by dis-
trusting, as it well may, the explanations prevalent during
the former stages of apparitions, distant intercourse, etc. It
suggests that different minds may have some communion,
though separated by space, through some spiritual medium.
Such a suggestion if discreetly followed up might lead to
important discoveries, and would certainly give rise to
entertaining meditations. But the cultivators of the dark
sciences have done as they have ever delighted to do.
Their spirits are not content with making themselves present

'Where all the nerves of sense are numb,
Spirit to spirit, ghost to ghost.'

but they become the familiar spirits of money-making
media, and rap out lies for hours together for the amusement
of a promiscuous 'circle.' . . . While the believers sit
round the table of the medium and form one loop of the
figure 8, the spiritual circle enclosing the celestial mahogany
forms the upper portion of the curve, the medium herself
constituting the double point. But who shall say of the
dark sciences that they have reached the maximum of darkness ? Men have listened to the toes of a medium as to
the voice of the departed. Let them now stand about her
table as about the table of devils. If one spirit can wrap
itself in petticoats, why may not another dance with three
legs ? A most searching question truly ! And accordingly
the powerful analysis of Godfrey has led him to the conclusion that a table of which the plane surface is touched

by believing fingers may be transformed into a diaboloid of
revolution. . . . Will there be an interminable series of
such expressions of belief, each more unnatural than its
predecessor, and gradually converging towards absolute
absurdity ? "

4. *Has everything beautiful in Art its original in Nature ?*
   Spring of 1854—shortly after the Tripos Examination.

"As the possibility of working out the question within
the time forms no part of our specification, we may glance
at heights which mock the attenuated triangle of the mathematician, and throw our pebble into depths which his cord
and plummet can never sound."

Maxwell here takes his revenge upon the Senate-house
by becoming more discursive than ever.[1]

He begins by deprecating precise definitions and pro-
posing an appeal to facts.

His conclusion is as follows :—" Nothing beautiful can be
produced by Man except by the laws of mind acting in him
as those of Nature do without him ; and therefore the kind
of beauty he can thus evolve must be limited by the very
small number of correlative sciences which he has mastered ;
but as the Theoretic and imaginative faculty is far in advance
of Reason, he can apprehend and artistically reproduce
natural beauty of a higher order than his science can attain
to ; and as his Moral powers are capable of a still wider
range, he may make his work the embodiment of a still
higher beauty, which expresses the glory of nature as the
instrument by which our spirits are exercised, delighted, and
taught.   If there is anything more I desire to say it is that
while I confess the vastness of nature and the narrowness of
our symbolical sciences, yet I fear not any effect which
either Science or Knowledge may have on the beauty of
that which is beautiful once and for ever."

The following observations occur in the course of the
essay :—

---

[1] See above, pp. 167, 8.

"All your analysis is cruelly anatomical, and your separated faculties have all the appearance of preparations. You may retain their names for distinctness, but forbear to tear them asunder for lecture-room demonstration. . . . They separate a faculty by saying it is not intellectual, and then, by reasoning blindfold, every philosopher goes up his own tree, finds a mare's nest and laughs at the eggs, which turn out to be pure intellectual abstractions in spite of every definition.

"With respect to beauty of things audible and visible, we have a firm conviction that the pleasure it affords to any being would be of the same kind by whatever organisation he became conscious of it. . . . Our enjoyment of music is accompanied by an intuitive perception of the relations of sounds, and the agreement of the human race would go far to establish the universality of these conditions of pleasure, though Science had not discovered their physical and numerical significance."

Beauty of Form.—"A mathematician might express his admiration of the Ellipse. Ruskin agrees with him. . . . It is a universal condition of the enjoyable that the mind must believe in the existence of a law, and yet have a mystery to move about in. . . . All things are full of ellipses—bicentral sources of lasting joy, as the wondrous Oken might have said." Beauty of form, then, is—1. Geometrical; 2. Organic; 3. "Rivers and mountains have not even an organic symmetry; the pleasure we derive from their forms is not that of comprehension, but of apprehension of their fitness as the forms of flowing and withstanding matter. When such objects are represented by Art, they acquire an additional beauty as the language of Nature understood by Man, the interpreter, although by no means the emendator, of her expressions.

" . . . The power of Making is man's highest power in connexion with Nature."

Beauty of Colour.—"The Science of Colour does, indeed, point out certain arrangements and gradations which follow as necessarily from first principles, as the curves of the second order from their equations. These results of science are, many of them, realised in natural phenomena taking place according to those physical laws of which our mathematical formulæ are symbols; but it is possible that combinations of colours may be imagined or calculated, which no optical phenomenon we are acquainted with could reproduce. Such a result would no more prove the impropriety of the arrangement than ignorance of the planetary orbits kept the Greeks from admiring the Ellipse."

5. *Envelopment : Can Ideas be developed without Reference to Things as their developing Authorities ?* Summer Term, 1854.

Early in 1854 Maxwell had read J. H. Newman's[1] Essay on the Development of Christian Doctrine. He appears to have felt an inconsistency between the tenor of that work and its title, which set him meditating on the difference between true Development—*i.e.* Education—and Envelopment or Self-Involution, as a tendency incident to certain habits of thought.

He traces the working of this tendency in various subjects ending with theology, and then proceeds as follows :—

"Envelopment is a process by which the human mind, possessed with a preternatural impatience of facts and fascinated by the apparent simplicity of some half-apprehended theory, seeks, by involving the chain of its speculations in hopeless confusion, to round it, as it were, to a separate whole.

"Thus Mr. Newman and his predecessors take up some single practice of Christians, and by means of analogies derived from the practices of Egyptian priests, Roman emperors, or Jewish rabbis, they determine, most precisely, the situation, extent, and exposure of the place of purging by fire, together with all the technicalities, observances, and etiquette of that mysterious region. The convolutions of the brain are very wonderful."

---

[1] Cardinal Newman.

As a further illustration he proceeds to trace the genesis of phreno-mesmerism.

Against the Theory of the Development of Doctrine he sets the fact of the Education of Mankind.

The Essay is highly ironical and full of caustic touches, but it is difficult to detach them for quotation.

" One great art in argument when you have the first move is to divide everything into that which is and that which is not in some assigned class. In this way you make it the business of the opponent to discover what other important things there may be which may be said of the subject in hand.

" . . . These subtle differences when further multiplied by the application of the seven tests of development, would require a seven years' apprenticeship with Thomas Aquinas before anything could be said of them except assent or contempt.

" . . . In every human pursuit there are two courses— one, that which in its lowest form is called the useful, and has for its ultimate object the extension of knowledge, the dominion over Nature, and the welfare of mankind. The objects of the second course are entirely self-contained. Theories are elaborated for theories' sake, difficulties are sought out and treasured as such, and no argument is to be considered perfect unless it lands the reasoner at the point from which he started.

" . . . Some years ago I encountered a gentleman whose main object was to discover the musical relations of the number eleven. I hear on good authority that the question is not only more perplexed but more interesting than ever.

" . . . I have unaccountably passed over that Logic by means of which many a powerful mind has persuaded itself that it was usefully engaged while devoting a life to the defence or attack of the fourth figure of the syllogism, and that Metaphysics which even now seeks to find arguments about the operations of the senses, while it rejects the aid of physiology or any other appeal to facts.

"I now proceed to envelop an argument from one of those dark sciences which seem to have been sent up from Dom Daniel for the special purpose of displaying reasonings of this kind.

"It is well known that the brain is the organ of intellectual activity. It is held by all that the intellect is made up of many distinct faculties. Therefore the brain must be composed of corresponding organs." . . .

The Essay concludes quite seriously—"The education of man is so well provided for in the world around him, and so hopeless in any of the worlds which he makes for himself, that it becomes of the utmost importance to distinguish natural truth from artificial system, the development of a science from the envelopment of a craft."

6. *Morality.* May 1855. (See above, p. 202), *Is Ethical Truth obtainable from an Individual Point of View?*

An inquiry concerning the first principles of Moral Philosophy.

Of three criteria, fitness, pleasure, and freedom, the last is preferred, but is pronounced incomplete. Adam Smith's use of the principle of sympathy is then considered. "The repeated action of what Smith calls sympathy, calls forth various moral principles, which may be deduced, no doubt, from other theories, as necessary truths, but of which the actual presence is now first accounted for. . . . Instead of supposing the moral action of the mind to be a speculation on fitness, a calculation of happiness, or an effort towards freedom, he makes it depend on a recognition of our relation to others like ourselves." This method (that of self-projection) is pronounced the only true one, but is to be extended so as to embrace other relations than that of mere similarity.

Such is the bare outline of an essay which would fill at least a dozen pages. It touches on various themes, from the origin of law to the religious sanction of morals, and contains no little evidence of the writer's growing power of observing human life.

7. *Language and Speculation.*   Autumn of 1855.   (See
   above, p. 218.)   *Is the Modern Vocabulary of the English Language the Effect or the Cause of its Speculative
   State ?*

A series of observations on style, original but *very* discursive, chiefly aimed at certain literary affectations which
were then beginning to creep in.

   " . . . The new form of the old thought must be dressed
out with words, and must attract attention by bringing forward what should be kept in the background.   No wonder
the poor fellow thinks his head is turned, when he is trying
to see over the collar of his coat.
   " . . . By all means let us have technical terms belonging to every science and mystery practised by men, but let
us not have mere freemasonry or Ziph language by which
men of the same cult can secretly combine."

8. *Analogies.*   February 1856.   *Are there Real Analogies
   in Nature ?*

This essay contains a serious exposition of Maxwell's
deliberate views on philosophical questions, and is therefore
given here entire, not omitting the playful opening paragraph.
   " In the ancient and religious foundation of Peterhouse
there is observed this rule, that whoso makes a pun shall be
counted the author of it, but that whoso pretends to find it
out shall be counted the publisher of it, and that both shall
be fined.   Now, as in a pun two truths lie hid under one
expression, so in an analogy one truth is discovered under
two expressions.   Every question concerning analogies is
therefore the reciprocal of a question concerning puns, and
the solutions can be transposed by reciprocation.   But since
we are still in doubt as to the legitimacy of reasoning by
analogy, and as reasoning even by paradox has been pronounced less heinous than reasoning by puns, we must adopt

the direct method with respect to analogy, and then, if necessary, deduce by reciprocation the theory of puns.

" That analogies appear to exist is plain in the face of things, for all parables, fables, similes, metaphors, tropes, and figures of speech are analogies, natural or revealed, artificial or concealed. The question is entirely of their reality. Now, no question exists as to the possibility of an analogy without a mind to recognise it—that is rank nonsense. You might as well talk of a demonstration or refutation existing uncondi- tionally. Neither is there any question as to the occurrence of analogies to our minds. They are as plenty as reasons, not to say blackberries. For, not to mention all the things in external nature which men have seen as the projections of things in their own minds, the whole framework of science, up to the very pinnacle of philosophy, seems sometimes a dis- sected model of nature, and sometimes a natural growth on the inner surface of the mind. Now, if in examining the admitted truths in science and philosophy, we find certain general principles appearing throughout a vast range of subjects, and sometimes re-appearing in some quite distinct part of human knowledge ; and if, on turning to the con- stitution of the intellect itself, we think we can discern there the reason of this uniformity, in the form of a fundamental law of the right action of the intellect, are we to conclude that these various departments of nature in which analogous laws exist, have a real inter-dependence ; or that their rela- tion is only apparent and owing to the necessary condi- tions of human thought ?

" There is nothing more essential to the right under- standing of things than a perception of the relations of *number*. Now the very first notion of number implies a previous act of intelligence. Before we can count any number of things, we must pick them out of the universe, and give each of them a fictitious unity by definition. Until we have done this, the universe of sense is neither one nor many, but indefinite. But yet, do what we will, Nature seems to have a certain horror of partition. Perhaps the most natural thing to count " one " for is a man or human being, but yet it is very difficult to do so. Some count by heads, others by souls, others by noses; still there is a tendency either to run together into masses or to split up into limbs. The dimmed outlines of phenomenal things all merge into another unless we put on the focussing glass of theory and screw it up sometimes to one pitch of definition, and sometimes to another, so as to see down into different depths through the great millstone of the world.

"As for space and time, any man will tell you that ' it is now known and ascertained that they are merely modifications of our own minds.' And yet if we conceive of the mind as absolutely indivisible and capable of only one state at a time, we must admit that these states may be arranged in chronological order, and that this is the only real order of these states. For we have no reason to believe, on the ground of a given succession of simple sensations, that differences in position, as well as in order of occurrence, exist among the causes of these sensations. But yet we are convinced of the co-existence of different objects at the same time, and of the identity of the same object at different times. Now if we admit that we can think of difference independent of sequence, and of sequence without difference, we have admitted enough on which to found the possibility of the ideas of space and time.

"But if we come to look more closely into these ideas, as developed in human beings, we find that *their* space has triple extension, but is the same in all directions, without behind or before, whereas time extends only back and forward, and always goes forward.

"To inquire why these peculiarities of these fundamental ideas are so would require a most painful if not impossible act of self-excenteration; but to determine whether there is anything in Nature corresponding to them, or whether they are mere projections of our mental machinery on the surface of external things, is absolutely necessary to appease the cravings of intelligence. Now it appears to me that when we say that space has three dimensions, we not only express the impossibility of conceiving a fourth dimension, co-ordi-

nate with the three known ones, but assert the objective truth
that points may differ in position by the independent
variation of three variables. Here, therefore, we have a
real analogy between the constitution of the intellect and
that of the external world.

"With respect to time, it is sometimes assumed that the
consecution of ideas is a fact precisely the same kind as the
sequence of events in time. But it does not appear that
there is any closer connection between these than between
mental difference, and difference of position. No doubt it is
possible to assign the accurate date of every act of thought,
but I doubt whether a chronological table drawn up in this
way would coincide with the sequence of ideas of which we
are conscious. There is an analogy, but I think not an
identity, between these two orders of thoughts and things.
Again, if we know what is at any assigned point of space
at any assigned instant of time, we may be said to know all
the events in Nature. We cannot conceive any other thing
which it would be necessary to know; and, in fact, if any
other necessary element does exist, it never enters into any
phenomenon so as to make it differ from what it would be
on the supposition of space and time being the only necessary elements.

"We cannot, however, think any set of thoughts without
conceiving of them as depending on reasons. These reasons,
when spoken of with relation to objects, get the name of
causes, which are reasons, analogically referred to objects
instead of thoughts. When the objects are mechanical, or
are considered in a mechanical point of view, the causes are
still more strictly defined, and are called *forces*.

"Now if we are acquainted not only with the events,
but also with the forces, in Nature, we acquire the power of
predicting events not previously known.

"This conception of cause, we are informed, has been
ascertained to be a notion of invariable sequence. No doubt
invariable sequence, if observed, would suggest the notion of
cause, just as the end of a poker painted red suggests the
notion of heat, but although a cause without its invariable

effect is absurd, a cause by its apparent frustration only suggests the notion of an equal and opposite cause.

" Now the analogy between reasons, causes, forces, principles, and moral rules, is glaring, but dazzling.

" A reason or argument is a conductor by which the mind is led from a proposition to a necessary consequence of that proposition. In pure logic reasons must all tend in the same direction. There can be no conflict of reasons. We may lose sight of them or abandon them, but cannot pit them against one another. If our faculties were indefinitely intensified, so that we could see all the consequences of any admission, then all reasons would resolve themselves into one reason, and all demonstrative truth would be one proposition. There would be no room for plurality of reasons, still less for conflict. But when we come to causes of phenomena and not reasons of truths, the conflict of causes, or rather the mutual annihilation of effects, is manifest. Not but what there is a tendency in the human mind to lump up all causes, and give them an aggregate name, or to trace chains of causes up to their knots and asymptotes. Still we see, or seem to see, a plurality of causes at work, and there are some who are content with plurality.

" Those who are thus content with plurality delight in the use of the word force as applied to cause. Cause is a metaphysical word implying something unchangeable and always producing its effect. Force on the other hand is a scientific word, signifying something which always meets with opposition, and often with successful opposition, but yet never fails to do what it can in its own favour. Such are the physical forces with which science deals, and their maxim is that might is right, and they call themselves laws of nature. But there are other laws of nature which determine the form and action of organic structure. These are founded on the forces of nature, but they seem to do no work except that of direction. Ought they to be called forces ? A force does work in proportion to its strength. These *direct* forces to work after a model. They are *moulds*, not forces.

Now since we have here a standard from which deviation may take place, we have, besides the notion of *strength*, which belongs to force, that of *health*, which belongs to organic law. Organic beings are not conscious of organic laws, and it is not the conscious being that takes part in them, but another set of laws now appear in very close connexion with the conscious being. I mean the laws of thought. These may be interfered with by organic laws, or by physical disturbances, and no doubt every such interference is regulated by the laws of the brain and of the connexion between that medulla and the process of thought. But the thing to be observed is, that the laws which regulate the *right* process of the intellect are identical with the most abstract of all laws, those which are found among the relations of necessary truths, and that though these are mixed up with, and modified by, the most complex systems of phenomena in physiology and physics, they must be re-cognised as supreme among the other laws of thought. And this supremacy does not consist in superior strength, as in physical laws, nor yet, I think, in reproducing a type as in organic laws, but in being right and true; even when other causes have been for a season masters of the brain.

"When we consider voluntary actions in general, we think we see causes acting like forces on the willing being. Some of our motions arise from physical necessity, some from irritability or organic excitement, some are performed by our machinery without our knowledge, and some evidently are due to us and our volitions. Of these, again, some are merely a repetition of a customary act, some are due to the attractions of pleasure or the pressure of constrained activity, and a few show some indications of being the results of distinct acts of the will. Here again we have a continuation of the analogy of Cause. Some had supposed that in will they had found the only true cause, and that all physical causes are only apparent. I need not say that this doctrine is exploded.

"What we have to observe is, that new elements enter into the nature of these higher causes, for mere abstract reasons are simply absolute; forces are related by their strength; organic laws act towards resemblances to types; animal emotions tend to that which promotes the enjoyment of life; and will is in great measure actually subject to all these, although certain other laws of *right*, which are abstract and demonstrable, like those of reason, are *supreme* among the laws of will.

" Now the question of the reality of analogies in nature derives most of its interest from its application to the opinion, that all the phenomena of nature, being varieties of motion, can only differ in complexity, and therefore the only way of studying nature, is to master the fundamental laws of motion first, and then examine what kinds of complication of these laws must be studied in order to obtain true views of the universe. If this theory be true, we must look for indications of these fundamental laws throughout the whole range of science, and not least among those remarkable products of organic life, the results of cerebration (commonly called ' thinking '). In this case, of course, the resemblances between the laws of different classes of phenomena should hardly be called analogies, as they are only transformed identities.

" If, on the other hand, we start from the study of the laws of thought (the abstract, logical laws, not the *physiological*), then these apparent analogies become merely repetitions by reflexion of certain necessary modes of action to which our minds are subject. I do not see how, upon either hypothesis, we can account for the existence of one set of laws of which the supremacy is necessary, but to the operation contingent. But we find another set of laws of the same kind, and sometimes coinciding with physical laws, the operation of which is inflexible when once in action, but depends in its beginnings on some act of volition. The theory of the consequences of actions is greatly perplexed by the fact that each act sets in motion many trains of machinery, which react on other agents and come into regions of physical and metaphysical chaos from which it is difficult to disentangle them. But if we could place the telescope of theory in proper adjustment, to see not the physical events which form the subordinate foci of the disturbance propagated through the universe, but the moral foci where the true image of the original act is reproduced, then we shall recognise the fact, that when we clearly see any moral act, then there appears a moral necessity for the trains of consequences of that act, which are spreading through the world to be concentrated on some focus, so as to give a true and complete image of the act in its moral point of view. All that bystanders see, is the physical act, and some of its immediate physical consequences, but as a partial pencil of light, even when not adapted for distinct vision, may enable us to see an *object*, and not merely light, so the partial view we have of any act, though far from perfect, may enable us to see it morally as an act, and not merely physically as an event.

"If we think we see in the diverging trains of physical consequences not only a capability of forming a true image of the act, but also of reacting upon the agent, either directly or after a long circuit, then perhaps we have caught the idea of *necessary* retribution, as the legitimate consequence of all moral action.

"But as this idea of the *necessary* reaction of the consequences of action is derived only from a few instances, in which we have guessed at such a law among the necessary laws of the universe ; and we have a much more distinct idea of *justice*, derived from those laws which we necessarily recognise as supreme, we connect the idea of retribution much more with that of *justice* than with that of *cause and effect*. We therefore regard retribution as the result of *interference* with the mechanical order of things, and intended to vindicate the supremacy of the right order of things, but still we suspect that the two orders of things will eventually dissolve into one.

"I have been somewhat diffuse and confused on the subject of moral law, in order to show to what length analogy will carry the speculations of men. Whenever they see a relation between two things they know well, and think they see there must be a similar relation between things less known, they reason from the one to the other. This supposes that although pairs of things may differ widely from each other, the *relation* in the one pair may be the same as that in the other.   Now, as in a scientific point of view the *relation* is the most important thing to know, a knowledge of the one thing leads us a long way towards a knowledge of the other.   If all that we know is *relation*, and if all the relations of one pair of things correspond to those of another pair, it will be difficult to distinguish the one pair from the other, although not presenting a single point of resemblance, unless we have some difference of relation to something else, whereby to distinguish them. Such mistakes can hardly occur except in mathematical and physical analogies, but if we are going to study the consti- tution of the individual mental man, and draw all our arguments from the laws of society on the one hand, or those of the nervous tissue on the other, we may chance to convert useful helps into Wills-of-the-wisp.   Perhaps the ' book,' as it has been called, of nature is regularly paged ; if so, no doubt the introductory parts will explain those that follow, and the methods taught in the first chapters will be taken for granted and used as illustrations in the more advanced parts of the course ; but if it is not a ' book ' at all, but a *magazine*, nothing is more foolish to suppose that one part can throw light on another.

" Perhaps the next most remarkable analogy is between the principle, law, or plan according to which all things are made suitably to what they have to do, and the intention which a man has of making machines which will work.   The doctrine of final causes, although productive of barrenness in its exclusive form, has certainly been a great help to enquirers into nature ; and if we only maintain the existence of the analogy, and allow observation to determine its form, we cannot be led far from the truth.

" There is another analogy which seems to be supplant- ing the other on its own ground, which lies between the principle, law, or plan according to which the forms of things are made to have a certain community of type, and that which induces human artists to make a set of different things according to varieties of the same model. Here apparently the final cause is analogy or homogeneity, to the exclusion of usefulness.

"And last of all we have the secondary forms of crystals bursting in upon us, and sparkling in the rigidity of mathematical necessity and telling us, neither of harmony of design, usefulness or moral significance,—nothing but spherical trigonometry and Napier's analogies. It is because we have blindly excluded the lessons of these angular bodies from the domain of human knowledge that we are still in doubt about the great doctrine that the only laws of matter are those which our minds must fabricate, and the only laws of mind are fabricated for it by matter."

9. *Autobiography.* (Dated) 8th March 1856.
Is Autobiography possible?

Under the guise of an ironical paradox, that all biography is (1) impossible, (2) inevitable, Maxwell recommends the simple record of facts, and deprecates the method of introspection. Of many shrewd remarks occurring in the course of this essay, the following are the most noticeable :—

". . . When a man once begins to make a theory of himself, he generally succeeds in making himself into a theory.

". . . The truthfulness of the biography depends quite as much upon the relations which subsisted between the author and his subject, as upon his fidelity in collecting authentic accounts of his actions.

". . . It will be found that the motives under which the celebrated characters of history have acted, are, whether good or bad, pretty much of the same order of refinement, as long as we gather them from the same historian. It is when we pass from one historian to another that we discover a new order of motives, both in the good and the bad characters.

". . . People do not talk of you, or if they do, they

make blunders. But they *do* reflect you, and that more faithfully than your looking-glass.

". . . The stomach-pump of the confessional ought only to be used in cases of manifest poisoning. More gentle remedies are better for the constitution in ordinary cases.

". . . Every man has a right and is bound to become acquainted with himself; but he will find himself out better by intercourse with well-chosen reagents, than by putting on his own thumbscrews, or by sending round to his friends for their opinions. In the choice of reagents, the first thing to be avoided is incapability and insincerity, which generally go together.

". . . Suppose such a history or biography to exist, where actions are described without comment, but in a spirit faithful to the highest truth. It will be an indestructible picture of life, which cannot be distorted by future accidents, and which, by its clear arrangement and perfect simplicity, is sure to pass into our experience without that opposition which, by the constitution of man, accompanies the forcible administration of moral precept."

10. *Unnecessary Thought.* October (?), 1856. "*Is a horror of Unnecessary Thought natural or unnatural? Which does Nature abhor most, a superplenum or a vacuum of thought*"?

A great part of every life is necessarily unconscious or mechanical. "We have a natural and widespread aversion to the act of thinking, which exists more or less in all men." But, on the other hand, abstract thought needs to be continually checked through contact with reality, and it is more important that our thoughts should have a living root in experience, than that they should be perfectly self-consistent at any particular stage of their growth.

". . . They know the laws by heart, and do the calculations by fingers. . . . When will they begin to think? Then comes active life: What do they do that by? Precedent, wheel-tracks, and finger-posts.

". . . There is one part of the process at least to which

attention is unfavourable. I mean the very important and necessary operation of forgetting useless facts.

". . . Growth goes on in the mind as in the body by a process of appropriation and rejection; and the mental growth is rendered steady and real by its close connection with material and external things."

11 and 12.—Two unfinished Essays, on *Sensation,* and on *Reason and Faith,* may be probably assigned to the period of attendance on his father's illness in Edinburgh, in the spring of 1855. See his letter to C. J. Monro, of Feb. 19, on p. 210. " I have a few thoughts on . . . sensation generally, and a kind of dim outline of Cambridge palavers, tending to shadow forth the influence of mathematical training on opinion and speculation."

CHAPTER IX.

" And yet thy heart
The lowliest duties on herself did lay."

SOON after his return to Cambridge in February 1856 (after seeing his father comfortably established in Edinburgh), Maxwell heard from his old friend Professor Forbes that the Chair of Natural Philosophy at Marischal College, Aberdeen, was vacant, and he shortly afterwards became a candidate. He had never contemplated a life of entire leisure, but it may seem strange that Cambridge, where besides his lectureship he had various philanthropic interests, should not have afforded him a sufficient field for regular work. He foresaw that the Scotch appointment would please his father, and that the arrangement of session and vacation time would enable him to spend the whole summer uninterruptedly at Glenlair. Some expressions in his letters also seem to indicate that he rather shrank from the prospect of becoming a Cambridge "Don." He had observed the narrowing tendencies of college life, and preferred the rubs of the world.

His letters to his father and others at this time
sufficiently explain the course of his candidature, in
which the point most deserving notice is the generous
way in which he speaks of his rivals.   While treating
the whole matter with his usual grave irony, he seems
to have conducted his part of it with considerable
sagacity, and when he returned to Edinburgh about
the middle of March everything was well in train.
He had the pleasure of knowing that his father's
interest in the question was at least equal to his own,
and that the old man had been roused by it to some
return of his former vigour.   But the end was near.
After a few days spent in Edinburgh, the father and
son went home to Glenlair, as they had planned—a
matter of no small anxiety and difficulty.   The short
vacation had all but passed away, when, on Thursday
the 2d of April, just before his son was to have
returned to Cambridge, Mr. John Clerk Maxwell
suddenly expired.

The outward change was not very great.   Maxwell went up to Cambridge as usual.   Glenlair was
still his home.   His interest in his own subjects was
undiminished.   His candidature for Aberdeen continued.   But the personal loss to him was incalculable
and irreparable.   Their long daily companionship had
been followed by a correspondence which was all but
daily, by vacations spent together, and an uninterrupted interchange, whether present or absent, of
thoughts and social interests, both light and grave.
During the last six months it is true the old man had
been failing, and, to outward observers, was consider-

ably changed. But the change had only called out his son's affection into more active exercise, and had never checked the flow of communication by word or letter. What depth of feeling lay beneath Maxwell's quiet demeanour at this time may be inferred from the poem written at Cambridge during that summer term, and put into my hands when we met afterwards at Glenlair. Some lines of it may be appropriately inserted here:—

> " Yes, I know the forms that meet me are but phantoms of the brain,
> For they walk in mortal bodies, and they have not ceased from pain,
> Oh those signs of human weakness, left behind for ever now,
> Dearer far to me than glories round a fancied seraph's brow.
> Oh the old familiar voices ; oh the patient waiting eyes ;
> Let me live with them in dreamland while the world in slumber lies.
> For by bonds of sacred honour will they guard my soul in sleep
> From the spells of aimless fancies that around my senses creep.
> They will link the past and present into one continuous life ;
> While I feel their hope, their patience, nerve me for the daily strife.
> For it is not all a fancy that our lives and theirs are one,
> And we know that all we see is but an endless work begun.
> Part is left in nature's keeping, part has entered into rest ;
> Part remains to grow and ripen hidden in some living breast."

Such was James Clerk Maxwell during the " years of April blood."

LETTERS, 1856.

TO HIS FATHER.

Trin., 14th Feb. 1856.

Yesterday the Ray Club met at Hort's. I took my great top there and spun it with coloured discs attached to it. I have been planning a form of top, which will have more variety of motion, but I am working out the theory, so that I will wait till I know the necessary dimensions before I settle the plan.

I told Willie (Cay) how I had hung up a bullet by a combination of threads.

I have drawn from theory the curves which it ought to describe, and when I set the bullet a-going over the proper curve, it traces it out over and over again as if it were doing a pre-ordained dance, and kept a steady eye on the line on the paper. I have enlarged my stock of models for solid geometry, made of coloured thread, stretched between two pasteboard ends.

From Professor J. D. Forbes.

Edinburgh, 13th Feb. 1856.

You may not perhaps have heard that Mr. Gray, Professor of Natural Philosophy, Marischal College, Aberdeen, is dead. He was a pleasing and energetic person, in the prime of life and health, a few months ago, when I saw him last.

I have no idea whether the situation would be any object to you; but I thought I would mention it, as I think it would be a pity were it not filled by a Scotchman, and you are the person who occurs to me as best fitted for it.

Do not imagine from my writing that I have the smallest influence in the matter, or interest in it beyond the welfare of the Scottish Universities.

It is in the gift of the Crown. The Lord Advocate and Home Secretary are the parties to apply to. I am not acquainted with either.

In the Commissioners' Report of 1830 the emoluments are stated at about £350. But they are not always to be depended upon.

Another point. I think you ought certainly to be a Fellow of the Royal Society of Edinburgh. I shall be glad to propose you if you wish it.

To his Father.

Trin. Coll., 15th Feb. 1856.

Professor Forbes has written to me to say that the Professorship of Nat. Phil. at Marischal College, Aberdeen, is vacant by the death of Mr. Gray, and he inquires if I would apply for the situation, so I want to know what your notion or plan may be.   For my own part, I think the sooner I get into regular work the better, and that the best way of getting into such work is to profess one's readiness by applying for it.

The appointment lies with the Crown—that is, the Lord Advocate and Home Secretary.   I suppose the correct thing to do is to send certificates of merit, signed by swells, to one or other of these officers.

I am going to ask about the method of the thing here, and Thacker has promised to get me the Collego Testimonials.   If you see any one in Edinburgh that understands the sort of thing, could you pick up the outline of the process ?

In all ordinary affairs political distinctions are supposed to weigh a great deal in Scotland.   The English notion is that in pure and even in mixed mathematics politics are of little use, however much a knowledge of these sciences may promote the study of politics.   As to Theology, I am not aware that the mathematicians, as a body, are guilty of any heresies, however some of them may have erred.   But these are too mysterious subjects to furnish matter for calculation, so I may tell you that the reflecting stereoscope was finished yesterday, and looks well, and that I got a Devil made at the same time, which I play at the Gymnasium for relaxation and breathing time.

Forbes also suggests my joining the Royal Society.

Trin. Coll., 20th Feb. /56.

As far as writing (Testimonials) goes, there is a good deal, and if you believe the Testimonials you would think the Government had in their hands the triumph or downfall of education generally, according as they elected one or not.

However, wisdom is of many kinds, and I do not know which dwells with wise counsellors most, whether scientific, practical, political, or ecclesiastical.   I hear there are candidates of all kinds relying on the predominance of one or other of these kinds of wisdom in the constitution of the Government.

I had a letter from Dr. Swan of Edinburgh, who is a candidate, asking me for my good opinion, which I gave him, so far as I had one. His printed papers are good, and I hear he is so himself. Maclennan is also a candidate. He has the qualification of making himself understood.

The results of this term are chiefly solid Geometry Lectures, stereoscopic pictures, and optical theorems. My lectures are to be on Rigid Dynamics and Astronomy next term, so I do not expect to be out of work by reason of Aberdeen, and I have plenty to get through in those subjects.

I have been making more stereoscopic curves for my lectures. I intend to select some and draw them very neat the size of the ordinary stereoscopic pictures, and write a description of them, and publish them as mathematical illustrations. I am going to do one now to illustrate the theory of contour lines in maps, and to show how the rivers must run, and where the lines of watershed must be.

From his Father.

22d Feby. 1856.

. . . I believe there is some salary, but fees and pupils, I think, cannot be very plenty. But if the *postie* be gotten, and prove not good, it can be given up; at any rate it occupies but half the year.

To his Father.

Trin. Coll., 12th March.

I was at the Working College to-day, working at decimal fractions. We are getting up a preparatory school for biggish boys to get up their preliminaries. We are also agitating in favour of early closing of shops. We have got the whole of the ironmongers, and all the shoemakers but one. The booksellers have done it some time. The Pitt Press keeps late hours, and is to be petitioned to shut up.

I have just written out an abstract of the second part of my paper on Faraday's Lines of Force. I hope soon to write properly the paper of which it is an abstract. It is four weeks since I read it. I have done nothing in that way this term, but am just beginning to feel the electrical state come on again, and I hope to work it up well next term.

To Miss Cay.

Thursday Afternoon (3d April 1856.)

Dear Aunt—My father died to-day at twelve o'clock. He was sleepless and confused at night, but got up to breakfast. He saw Sandy a few minutes, and spoke rationally, then came into the drawing-room, and sat down on a chair for a few minutes to rest, and gave a short cry and never spoke again. We gave him ether for a little, but he could not swallow it. There was no warning, and apparently no pain. He expected it long, and described it so himself.

Do you think Uncle Robert could come and help a little ? Tell Dr. Bell and other people. As it is, it is better than if it had been when I was away. He would not let me stay. I was to go to Cambridge on Friday.—
Your aff. nephew,   J. C. Maxwell.

To Mrs. Blackburn of Killearn.

Glenlair, Thursday.

Dear Mrs. Blackburn—My father died suddenly to-day at 12 o'clock. He had been giving directions about the garden, and he said he would sit down and rest a little as usual. After a few minutes I asked him to lie down on the sofa, and he did not seem inclined to do so, and then I got him some ether, which had helped him before.

Before he could take any he had a slight struggle, and all was over. He hardly breathed afterwards.

He used often to talk to me about this, which has come at last, and he seemed fully to have made up his mind to it and to be prepared for it. His nights have sometimes been troubled, and last night I was with him the whole time trying to get him into a comfortable sleep, which did not come till light.

Otherwise we thought him better than when in Edin.^r He was very glad to get back here again.

I write to you that you may tell Mrs. Wedderburn. She ought to know, and I trust you will let her know, that not only was there no pain or distress about my father's death, but he had often been speaking of how glad he was that he had got everything put in order, and that he was home again.

I have written to ask my Uncle Robert Cay to come and help me in various things, as I am rather alone here. Of course I have written to Sir George, and will do so to other relatives as soon as I can.—Your affte. cousin,

JAMES CLERK MAXWELL.

FROM R. DUNDAS CAY, Esq., to MISS CAY, on MR. JOHN CLERK MAXWELL'S Death.

Glenlair, 8th April 1856.

I think you will be glad to hear how we are getting on. It is very nice to see how natural James is. There is no affectation of more feeling than he really has, but he talks away upon his own subjects when not busy with the necessary preparations for to-morrow. Fortunately these occupy him a good deal, and as I think the business is of use to him, I only assist him and keep him talking. For instance, he made out all the list and directed the letters himself; I sat by and sealed them. Then my health requires a walk every day, so we go out and talk away very much as usual all the time, discussing the thinning of plantations, etc.

It is beautiful to see the feeling of all the people towards him, all thinking for him, and trying to assist him in every way, and he trying to carry on everything as before :—or when he wants to make a change, his anxiety, lest people should think he disapproves of the former customs. For instance, he wished to have the servants in for prayers every evening, instead of our reading by ourselves and reading to them separately ; he was quite afraid they should think his doing so would look as if he thought it was wrong, it not having been done before.

To LEWIS CAMPBELL, Esq.

Trin. Coll., 22d April 1856.

I have had many things to attend to lately, which have kept me from writing to you. I am glad you wrote to me. I got a very kind letter from your mother and Bob, for which you must thank them meanwhile. My uncles, Robert and Albert, stayed with me till the 15th. That day I got a letter from Cambridge about college matters, and so I had to set to work at home more vigorously. George Wedderburn came in the afternoon, and we had two hard days' work of various kinds.

On Friday he and I left Glenlair, and I got here on Saturday, and since yesterday have been lecturing.

All things are as if I had been up after a common vacation, and I see them all the same as they used to be. I have got back among chapels and halls and scholarships, and all the regular routine, with now and then some expression of condolence, which is all that strangers can or ought to afford. Neither they nor I enter on a subject which must be misunderstood; but it seems to me that while all the old *subjects* are as interesting to me as ever, I talk about them without understanding the men I talk to.

I have two or three stiff bits of work to get through this term here, and I hope to overtake them. When the term is over I must go home and pay diligent attention to everything there, so that I may learn what to do.

The first thing I must do is carry on my father's work of personally superintending everything at home, and for doing this I have his regular accounts of what used to be done, and the memories of all the people, who tell me everything they know. As for my own pursuits, it was my father's wish, and it is mine, that I should go on with them. We used to settle that what I ought to be engaged in was some occupation of teaching, admitting of long vacations for being at home; and when my father heard of the Aberdeen proposition he very much approved. I have not heard anything very lately, but I believe my name is not yet put out

of question in the L^d. Advocate's book. If I get back to
Glenlair I shall have the mark of my father's work on everything I see. Much of them is still his, and I must be in
some degree his steward to take care of them. I trust that
the knowledge of his plans may be a guide to me, and never
a constraint.

I am glad to hear of your [Oxford] W[orking Men's]
Coll[ege]. The preparatory school here has at once got from
seventy to ninety scholars, all in earnest, and they have had
to migrate to a larger schoolhouse.

We might consider of you and Bob and W. Cay coming
for a quiet week or two to Glenlair in summer, if all goes
well. Bob is with Willie and Charlie now touring.

I am getting a new top turned to show my class the
motion of bodies of various forms about a fixed point. I expect to get very neat results from it, and agreeing with theory
of course.

FROM PROFESSOR J. D. FORBES.

Bridge of Allan, 30th April 1856.

MY DEAR SIR—I have just seen in the newspaper that
you have been appointed to the Chair in Marischal College,
on which I beg sincerely to congratulate you.

I regret much that it should at the same time be my
lot to express my sympathy on the occasion of the recent
death of your father. Such a loss occurs but once in a lifetime.
In your case I am sure that it has the greatest alleviation
which it admits of—I mean the consciousness that you have
been an affectionate and dutiful son, and that your excellent
conduct relieved your father's mind from every shade of
anxiety regarding you.—Believe me always, yours very
sincerely,   JAMES D. FORBES.

To R. B. LITCHFIELD, Esq.

Trinity, 4th June 1856.

On Thurs. evening I take the North-western route to the
North. I am busy looking over immense rubbish of papers,
etc., for some things not to be burnt lie among much com-
bustible matter, and some is soft and good for packing.

It is not pleasant to go down to live solitary, but it would not be pleasant to stay up either, when all one had to do lay elsewhere. The transition state from a man into a Don must come at last, and it must be painful, like gradual outrooting of nerves. When it is done there is no more pain, but occasional reminders from some suckers, tap-roots, or other remnants of the old nerves, just to show what was there and what might have been.

After his father's death, Maxwell set himself anew to the tasks before him, with a mingled sense of loss and responsibility. One of his first duties was to apply himself to the management of his estate. He remained at Glenlair during most of the summer, only making a short excursion to Belfast on account of his cousin, William Cay, who, in accordance with his advice, was about to study Engineering under James Thomson, the brother of the Glasgow Professor. In the autumn, besides entertaining Charles Hope Cay, then a boy of fifteen, in his school holidays, he had various Cambridge friends to stay with him, as in former years.[1]

1856.
Æt. 25.

In November he began his work at Aberdeen. A Scotch Professor has one advantage over a College lecturer at Cambridge. If his students are less advanced, he has the entire direction of their work in his own department. It is left to him, apart from any

---

1 With one of these, who happened to be " Carlyle-mad," he drove one day on pilgrimage to Craigenputtock. The enthusiast, in his rapture, harangued an old peasant, who was hoeing " neeps," on the glorious doings of the former tenant of the farm-house. The man listened, stooping over his work till the rhapsody was over, then looked up for a moment saying, " It is aye gude that mends," and resumed his labour. Maxwell was fond of relating this.

prescribed system, to determine the order in which the parts of his subject shall be developed. His selection of topics is not dominated by the Final Examination. This peculiarity of his position was fully appreciated by Clerk Maxwell, whose experience of the course in Edinburgh under Forbes, gave him a " standpoint" from which to arrange his great fund of scientific acquirement in presenting it to his students.

Had Maxwell the qualities of a teacher ? That he was not on the whole successful in oral communication is an impression too widespread to be contradicted without positive proof. Yet his letters bear sufficient evidence that in many respects he had a true vocation as an educator. The combination of keen sympathy with native authority and dignity, the intense interest in his subject, his endless power of taking trouble, his philanthropic enthusiasm, his critical study of mankind, his wide range of language and ideas, must have enabled him to make his mark as a public teacher, either at Aberdeen or Cambridge, if he had remained long enough at either place to wear off some superficial impediments, to adapt his methods to his environment, and to effect a thorough understanding with his pupils. As it was, his lectureship at Trinity lasted only for a year, and in the Scotch university he had only taught for three short sessions when Marischal College was on the point of being suppressed, and his reputation as a teacher was, under these circumstances, brought into comparison with that of others whose strength lay in exposition. To those who know what is implied in academical contests and controversies, the mention of these facts will be a sufficient caution against taking the lowest estimate of Maxwell's teaching powers; and in after years "at Cambridge, where his class consisted of picked students," we have good authority for saying "his lectures were listened to with an attention and pleasure similar to that with which his books are now read." But at this earlier time there were certainly drawbacks, of which he himself was imperfectly conscious. Between his students' ignorance and his vast knowledge, it was difficult to find a common measure. The advice which he once gave a friend whose duty it was to preach to a country congregation, "Why don't you give it them thinner?" must often have been applicable to himself. Another hindrance lay in the very richness of his imagination, and the swiftness of his wit. The ideas with which his mind was teeming were perpetually intersecting, and their interferences, like those of the waves of light, made "dark bands" in the place of colour, to the unassisted eye. Illustrations of *ignotum per ignotius*, or of the abstruse by some unobserved property of the familiar, were multiplied with dazzling rapidity. Then the spirit of indirectness and paradox, though he was aware of its dangers, would often take possession of him against his will, and either from shyness, or momentary excitement, or the despair of making himself understood, would land him in "chaotic statements," breaking off with some quirk of ironical humour. Add to this his occasional hesitation, his shortsightedness, and the long years of solitary intercourse with his father, who understood his meaning from the slightest hint and rather encouraged the family trick of " calling things out of their names," and the list of hindrances is sufficiently formidable.　But he was striving to overcome those of which he knew, and even if he had never done so completely, the weight of his character as well as the profundity of his genius, and his unvarying kindliness, must have won their way.

As marking his educational enthusiasm, it should not be forgotten here that he continued at Aberdeen the practice which he had commenced at Cambridge, of lecturing to working men.　This was entirely voluntary, and for aught I know may have been regarded as a piece of eccentricity.

A trivial incident may be recorded as throwing light on his relations to professors and students severally.　The professors had unlimited access to the library, and were in the habit of sometimes taking out a volume for the use of a friend.　The students were only allowed two volumes at a time.　Maxwell took out books for his students, and when checked for this by his colleagues explained that the students were his friends.

Amongst the human phenomena surrounding him, one which genuinely interested him was the religious " revival " which took place about that time in Scotland.　His intercourse with evangelical friends in England had prepared him to sympathise with such " experiences," and his Calvinistic reading had familiarised him with the language used.　And he was less jealous of Antinomianism than of a cut and dried morality.[1] But he was in no wise distracted from his professional duties by this or anything else, and although he referred to it in conversation, it has left no trace in any of his remaining letters which I have seen.

LETTERS, 1856-1857.

To R. B. LITCHFIELD, Esq.

Glenlair, 4th July 1856.

I have got some prisms and opticals from Edinbro', and I am fitting up a compendious colour-machine capable of transportation. I have also my top for doing dynamics and several colour-diagrams, so that if I come to Cheltenham I shall not be empty handed. At the same time I should like to hear from you soon.

I have been giving a portion of time to Saturn's Rings, which I find a stiff subject but curious, especially the case of the motion of a fluid ring. The very forces which would tend to divide the ring into great drops or satellites are made by the motion to keep the fluid in a uniform ring.

I find I get fonder of metaphysics and less of calculation continually, and my metaphysics are fast settling into the rigid high style, that is about ten times as far *above* Whewell as Mill is *below* him, or Comte or Macaulay *below* Mill, using above and below conventionally like *up* and *down* in Bradshaw.

Experiment furnishes us with the values of our arbitrary constants, but only suggests the form of the functions. Afterwards, when the form is not only recognised but understood scientifically, we find that it rests on precisely the same foundation as Euclid does, that is, it is simply the contradiction of an absurdity, out of which may we all get our legs at last!

---

[1] See p. 111.

To the Same.

Glenlair, 18th July 1856.

I can promise you milk and honey and mutton, with wind and water to match, a reasonable stock of natives of great diversity, and very unlike any natives I know elsewhere. I also expect a cousin [1] here, who carries a clear and active mind in a body ditto ditto, and I hope to make them stick closer together by the material above stated.

To the Same.

Glenlair, 9th September 1856.

My only hope for Pomeroy is that he may keep his health; if that remains I think it quite presumptuous to interfere with him by hopes or otherwise, for I would rather be interfered with by him (from which I am safe) than bother a man who steers so well himself. You must remember that besides the clerical shell of respectability which is to be put on, there is sometimes a lay shell of ανηριθμον γελασμα, which has in some measure to be put off, or perhaps more truly drawn in, for no one that has once known can ever forget that instead of two views there are three, good, bad, and grotesque; and tho' all things are full of jokes, that does not hinder them from being quite full, or even more so, of more solemn matters. It also strikes me that if we were to compare notes, the thing we would most differ about would be the notion we have of the " standpoint " (see religious prints, *passim*) of the men whom we know in common.

My own notion is that you see him where he ought to be according to principle, and I see him where he acts, as if he was, that is, in the position which would naturally produce his actual life. But I find on comparing notes with other people that a man always shows himself up differently according to the man who is with him. In fact I do it myself, so I must now show up the fishing side for the benefit of Charlie (Cay).

---

[1] Charles Hope Cay.

To C. J. Monro, Esq.

Glenlair, 14th October 1856.

. . . During September I had Lushington, Maclennan, and two cousins " Cay" here. Now I am writing a solemn address or manifesto to the Natural Philosophers of the North,[1] which I am afraid I must reinforce with coffee and anchovies, and a roaring hot fire and spread coat-tails to make it all natural. By the way, I have proved that if there be nine coefficients of magnetic induction, perpetual motion will set in, and a small crystalline sphere will inevitably destroy the universe by increasing all velocities till the friction brings all nature into a state of incandescence, or as H—— would say, Terrestrial all in Chaos shall exhibit efflorescence.

To Miss Cay.

129 Union Street, Aberdeen,
27th February 1857.

You are right about my being two letters in debt to you. I proceed to " post you up " to the most recent epoch. The weather is mild and sunny, but the winter has been severe. The planets Jupiter and Venus have been neighbours ; Saturn, Mars, and Mercury also visible.

To descend to particulars. I find everything going on very smoothly. I never passed an equal time with less trouble. I have plenty of work but no vexation as yet. In fact, I am beginning to fear that I must get into some scrape just to put an end to my complacency.

I will begin with the College. We are having public meetings and caucuses (that is, the students are) for the election of Lord Rector. Lord Stanley won't come. Lord Elgin is doubtful. They seem to prefer Elgin to Layard. We are to have a commission consisting of Col. Mure, Cosmo Innes, and Stirling of Keir.

To-morrow I hold my second *general* examination on the subject we have done. I hope that my men of science

---

[1] His Inaugural Lecture at Aberdeen.

won't have their heads turned with politics.   I have all the
squibs regularly presented to me.   They are not very good.

I have had 13 special examinations, and the two last
have been the best answered of any.   I send you my
paper for to-morrow to give it to Bob Campbell with my
profoundest esteem.

We have been at the theory of Heat and the Steam
Engine this month, and on Monday we begin Optics.   I have
a volunteer class who have been thro' astronomy, and we
are now at high Optics.   Tuesday week I give a lecture to
operatives, etc., on the Eye.   I have just been getting cods'
and bullocks' eyes, to refresh my memory and practise
dissection.   The size of the cod and the ox eye is nearly
the same.   As this was our last day of fluids, I finished off
with a splendid fountain in the sunlight.   We were not
very wet.

Out of College I have made the most of my time in
seeing the natives.   I used to walk *every day* with Professor
Martin, but he was not well for some time, and we broke
that habit.   I get on better with people of more decision
and less refinement, because they keep me in better order.

I have been keeping up friendly relations with the
King's College men, and they seem to be very friendly too.
I have not received any rebukes yet from our men for
so doing, but I find that the families of some of our
professors have no dealings, and never had, with those
of the King's people.   Theoretically we profess charity.

I had a glorious solitary walk to-day in Kincardineshire
by the coast—black cliffs and white breakers.   I took my
second dip this season.   I have found a splendid place,
sheltered and safe, with gymnastics on a pole afterwards.

To the Rev. Lewis Campbell.

129 Union Street,
Aberdeen, 6th February 1857.

I got your letter[1] this morning at breakfast.   I was
somewhat seedy from being up late, but the perusal seemed

---

[1] Announcing our marriage engagement.

to clear up everything, and I got on better with explaining the properties of elastic fluids than I had any reason to hope. So when I have doubts about the best mode of explaining anything, I must consult your letter, only it will not do for ever. I must have a new one now and then.

But I have not been so glad for long. Knowing you of old, I can see how things are by the way you write, and it is not always that similar announcements have given me similar satisfaction.

So I am glad that you do not know what "it" was. Avoid the neuter pronoun. "It" is unworthy of beasts that perish. "He" and "She" are for ever and ever. What the form of the pronoun may be after this I cannot tell, but I think more is meant in the distinction than is fully expressed in this life.

The Sadducees on the one side, and the ascetics on the other, point out the errors. Solomon, Prov. viii., *et passim,* and Eph. vi., indicate matter of contemplation not unallied to action, which in good ground bears good fruit.

But as Urania remarked to Melpomene, I am but displaying the fact of my belonging to a lower stage in the scale of things, so I must for the present go down beside my native rill.

With respect to this "northern hermitage," my cell is pretty commodious. In quitting the cœnobitic cloister of Trinity for the howling wilderness of Union Street, I have not been made an anchoret. It is quite consistent with the eremitic life to modify one's fast in friends' houses 4 days per week or so.

One thing I am thankful for, though perhaps you will not believe it.—Up to the present time I have not even been tempted to mystify any one.

I have made out who were most likely to excite my passion that way, and I have avoided some, and broken the ice with others. I am glad B—— is not here ; he would have ruined me. I once met him. I was as much astonished as he was at the chaotic statements I began to make. But as far as I can learn I have not been mis-

understood in anything, and no one has heard a single
oracle from my lips.   Of course I do not mean that my
class do not mistake my meaning sometimes.   That is found
out and remedied day by day.   I speak of professors,
ministers, doctors, advocates, matrons, maidens, and phenomenal existences (Chimeræ bombylantes in vacuo).   We
are through mechanics.   I had an ex$^{n.}$ on bookwork on
24th Jan.   I got answers to all the questions and riders,
though no one floored them all right.   I have now to be
brewing experiments on Heat, as well as determining the
form of doctrine to be presented to the finite capacities of
my men.

FROM C. J. MONRO, ESQ.

15th February 1857.

Have you seen the Pomeroy packet ?   It has much more
in it than any travels I ever read.   Lots of phenomena,
human and otherwise, on the way out : especially the waves
in a. storm. . . .

. . . They who deal in instruments of strings say that
if you strike a certain note you hear certain others above.
Is that because of the further terms in a Fourier's integral,
or because a sympathetic vibration is excited in certain other
of the strings of the same instrument ?   I observe Weber
says that it does not occur in wind instruments.

FROM PROFESSOR J. D. FORBES.

Edinburgh, 31st March 1857.

MY DEAR MAXWELL—I have often wished to ask you to
tell me how your first session had turned out ; consequently
I was exceedingly glad to get your letter this Evening, and
to find that you have not been disappointed in the results
of the step to which you kindly say that my assistance was
of some use.   In what you say about the monotony of re-
iteration, I can confidently assure you that your conclusions
are quite correct ; certain precautions being taken which an
active mind like yours is sure to fall upon.

We shall be delighted to see you at the R. S. on the 20th

and to have your paper, which, if convenient, please to put into my hands, as a matter of form, when ready.

I have been at several meetings of the Society, but am feeling a little just now the effects of the season and the winter's work, so I shall not be there on the 6th. On the whole, however, I have got through the winter well.

I shall like much to see your Top, of which I read the account in the *Athenæum*.

Have you observed in that same flippant paper for last Saturday an attack upon Faraday (as it seems to me) of a most presumptuous and ignorant kind? Though by no means as yet a convert to the views which Faraday maintains, yet I have so far a general appreciation of them as to believe that this conceited mathematician (some fifteenth Cambridge wrangler, I guess) is ignorant altogether of what Faraday wishes to prove.—Always yours sincerely,

JAMES D. FORBES.

To C. J. MONRO, Esq.

Glenlair, Springholm,
Dumfries, 20th May 1857.

I went to Old Aberdeen for Fourier, . . . but I have forgotten what was to be discovered out of him.

The session went off smoothly enough. I had Sun all the beginning of optics, and worked off all the experimental part up to Fraunhofer's lines, which were glorious to see with a water prism I have set up in the form of a cubical box, 5 inch side. The only things not generally done that I attempted last session were the undulatory medium made of bullets for advanced class, and Plateau's experiments on a sphere of oil in a mixture of spirits and water of exactly its own density.

I succeeded very well with heat. The experiments on latent heat came out very accurate. That was my part, and the class could explain and work out the results better than I expected. Next year I intend to mix experimental physics with mechanics, devoting Tuesday and THURSDAY (what would Stokes say?) to the science of experimenting accurately.

I got a glorified top made at Aberdeen. I think you saw the wooden type at Cambridge. I have made it the occasion of a short screed on rotation coming out in the Roy. Soc., Edinburgh, presently.

Last week I brewed chlorophyll (as the chemists word it), a green liquor, which turns the invisible light red. My pot of all the winter spinach that remained was portentous, so I exhibited the optical effects, which were allowed to be worth the potful.

My last grind was the reduction of equations of colour which I made last year. The result was eminently satisfactory.

To R. B. LITCHFIELD, Esq.

Glenlair, 29th May 1857.

It is with a profound feeling of pity that I write to a denizen of Hare Court after participating in the blessings of this splendid day. We had just enough of cloud to prevent scorching, and the grass seemed to like to grow just as much as the beasts to eat it.

I have not had a mathematical idea for about a fortnight, when I wrote them all away to Prof. Thomson, and I have not got an answer yet with fresh ones. But I believe there is a department of mind conducted independent of consciousness, where things are fermented and decocted, so that when they are run off they come clear.

By the way, I found it useful at Aberdeen to tell the students what parts of the subject they were *not* to remember, but to get up and forget at once as being rudimentary notions necessary to development, but requiring to be sloughed off before maturity.

I have no one with me but the domestics and dog. The valley seems deserted of its gentry; but we have one gentleman from Dumfriesshire, who is living in a hired house, and building with great magnificence an Episcopal Chapel in Castle-Douglas at his own expense. His own house is 20 miles off, a capital place, and this is perhaps the least

Episcopal part of Scotland by reason of the memory of the dragoons. One old family of the Stewartry is of that persuasion, and most of the persecutors' families are now Presbyterian and Whig, so that the congregation is but feeble.

It is very different at Aberdeen, where the Presbyterians persecuted far more than the Prelatists, so there I actually found a true Jacobite (female, I could not undertake to produce a male specimen), and there are three distinct Episcopal religions in Aberdeen, all pretty lively.

Can you tell me what the illustrated Tennyson is like? I shan't see it till I go to Edinbro'. I don't mean are the prints the best possible, or impervious to green spectacles; but are they nice diagrams as such things go? I should like to know before long about it, and whether the characters are of the Adamic type, and in reasonable condition, or pre-Rafaelitic in all but colour, and symbolising everything except the "Archetypal Skeleton" and the "Nature of Limbs."

To C. J. MONRO, Esq.

Glenlair, 5th June 1857.

I have not seen article seven, but I agree with your dissent from it entirely. On the vested interest principle, I think the men who intended to keep their fellowships by celibacy and ordination, and got them on that footing, should not be allowed to desert the virgin choir or neglect the priestly office, but on those principles should be allowed to live out their days, provided the whole amount of souls cured annually does not amount to £20 in the King's Book. But my doctrine is that the various grades of College officers should be set on such a basis that, although chance lecturers might be sometimes chosen from among fresh fellows who are going away soon, the reliable assistant tutors, and those that have a plain calling that way, should after a few years be elected permanent officers of the College, and be tutors and deans in their time, and seniors also, with leave to marry, or rather, never prohibited or asked any questions on that head, and with leave to retire after

so many years' service as seniors. As for the men of the
world, we should have a limited term of existence, and that
independent of marriage or " parsonage."

I saw a paragraph about the Female' Artists Exhibition,
and that Mrs. Hugh Blackburn had her Phaethon there. . . .
She has done a very small picture of a haystack making,
somewhat pre-Raphaelite in pose, but graceful withal, and
such that the Moidart natives know every lass on the stack,
whether seen behind or before. It was at the Edinburgh
Academy of Painters.

I have done a screed of introduction to optics, and am
at a sort of general summary of mechanical principles—
doctrines relating to absolute and relative motion, analysis
of the doctrine of Force into the smallest number of independent truths, theory of angular momentum and couples of
work done, and *vis viva*, of actual and potential energy, with
continual jaw on the doctrine of measurement by units all
through.

To REV. LEWIS CAMPBELL.

Glenlair, 7th August 1857.

Æt. 26.

I got your letter yesterday. I have oftener corresponded
with people I expected to see than with those I had just
left, so you must excuse my being rather more glad of
it than if I had expected it. So you were better than I
took you for; put that in the Logic-mill and grind it by
" Conversion of Props." Since you left I have been stirring up old correspondents. Poor W—— is "himself again,"
with not many to care about him. He could not keep the
A—— youths in order, and tried to get his authority backed
by the big authorities. Then I suppose ensued a struggle
between bodily weakness and hesitation, and mental stern-
ness, stubbornness, and conscientiousness. The result probably was something severe in substance and mild in manner,
or otherwise open to scorn from the youths. — I don't
know, but he has resigned his place. The youths then
proceeded to express their penitence, and the authorities their
regret. But he is now taking private pupils for that seat of
. . . learning, with not more friends and friendliness than

of old. Not exactly. I am glad to hear of his knowing some mathematical men, actuaries, etc., and corresponding with them, and he is much more friendly by the post than by speech and face.

Yesterday we did our Castle-Douglas, and round by Greenlaw (Gordon, Esq.) Old Greenlaw impounded us at once, and embarked us in his boat down to Threave Castle, where some falsified antiquity, and some apart behind thick woven thorns bathed in the black water of Dee.

Then back to dinner with another party of chance visitors, songs both of the drawing-room and the quire and the cotton fields, and, to conclude, the unpremeditated hop.

The thing was not destitute of its humours. Old Greenlaw, heir of entail, with charters in his bedroom belonging to " Young Lochinvar " his forbear, and various Douglases, with rights of pit and gallows, and other curious privileges, sending all his people and visitors neck and heels in the very best direction for themselves. Son and daughter—mild, indefatigable, generally useful, doing (at home) exactly as they are bid. One gay litter(ar)y widow, charming never so wisely, with her hair about her ears and her elbows on her knees, on a low stool, talking Handel, or Ruskin, or Macaulay, or general pathos of unprotected female, passing off into criticism, witticism, pleasantry, unmitigated slang, sporting, and betting.

One little Episcopal chaplain, a Celt, whom I see often, but do not quite fathom—that is, I don't know how far he respects and how far he is amused with his most patronising friends. One, mathematical teacher somewhere,—friend to chaplain. Voice. Mild, good fellow, like a grown up chorister, quite modest about everything except his voice— " What will they say in England," " The Standard Bearer," " Oh Susannah " (Chaplain leads chorus), " Courtin' down in Tenessee" (Chaplain obligato), " Yet once more " (Handel), " But who may abide " (do.), and so on.

One good old widow lady, with manners. One son to d?., —sanguine temperament, open countenance, very much run to nose, brain inactive, probably fertile in military virtue.

Two daughters to d°.,—healthy, physical force girls, brains more developed owing to their not having escaped in the form of nose.

Now, conceive the Voice set down beside one of the physical forces, and trying to interest her in the capacities of different rooms for singing in, she being more benevolent and horsefleshy than technically musical,—the Chaplain entertaining the other with an account of his solitary life in his rooms,—old Greenlaw hospitably entreating the mannerly widow, and trying to get the Nose to talk.

The young widow fixed on Colin, and informed him that if Solomon were to reappear with all his wisdom, as well as his glory, he would yet have to learn the polka; and that the mode of feasting adopted by the Incas of Peru reminded her strongly of a custom prevalent among the Merovingian race of kings of France.

Living in the Pampas she regarded as an enviable lot, and she was at a loss to know the best mode of studying Euclid for the advantage of being able to teach a young brother of six (years old).

So we did not get home till near 11, and I had to be up at Glenlair at 5 this morning, the result of which is that at 12 to-night I am a little sleepy. Johnnie[1] can swim across the big pool at the Chapel, all by himself. His taste of water through the nose did him great good. . . . I have had some races after stones down the water in Loch Roan. I have kept the stone in sight a good way, but it has always beaten me. I'll try some broken crockery to begin with.

I have succeeded in establishing the existence of an error in my Saturnian mazes, but I have not detected it yet. I have finished the first part of the Réligion Naturelle.[2] I am not a follower of those who believe they know what perfection must imply, and then make a deity to that pattern; but it is very well put, and carries one through, though if the book belongs to this age at all, it is eminently unlike most books of this century in England. But I only know

---

[1] His cousin John Cay, younger brother of William and Charles.
[2] By M. Jules Simon.

one other book of French argument on the positive (not positiviste) side, and that also worked by " demonstration." My notion is, that reason, taste, and conscience are the judges of all knowledge, pleasure, and action, and that they are the exponents not of a code, but of the unwritten law, which they reveal as they judge by it *in presence of the facts*. The facts must be witnessed to by the senses, and cross-examined by the intellect, and not unless everything is properly put on record and proved as fact, will any question of law be resolved at headquarters.

We are only going through our Lehrjahr in the knowledge of Perfection, and we may have a Wanderjahr to complete even after getting the first diploma, which is a certificate of having eyes to see the work, a conscience to feel after Right, and faith to believe in the Word, and to reach a station thereby where both those eyes and that conscience may be satisfied, or at least appeased. I do not think it is doing Reason, etc. any injustice to say that rough dead facts are the necessary basis on which to work in order to elicit the living truth, not from the facts, but either from the utterer of facts or the giver of Reason, which two are one, or Reason would never decipher facts.

> For know, whatever was created needs
> To be sustained and fed.   Of elements
> The grosser feeds the purer, etc.
> ———— various degrees
> Of substance, and in things that live, of life
>   Meanwhile enjoy
> Your fill what happiness this happy state
> Can comprehend, incapable of more.

T

CHAPTER X.

ABERDEEN—MARRIAGE—1857 TO 1860, ÆT. 26-29.

THE Glenlair letters of 1857 (see last chapter) sufficiently indicate Maxwell's mental condition in the
interval between his first and second sessions at Aberdeen. His expansive sociable spirit is putting forth
fresh feelers, and he has made a new beginning in his
observation of man in society. But he has not yet
recovered from the loss of the preceding year, and
those who read between the lines cannot fail to trace
here and there a touch of sadness peering from
beneath the habitual buoyancy of his style.

In September of this year another loss renewed the feeling of desolation which had haunted
him since his father's death. His friend Pomeroy,
whom he had nursed in illness, and of whose career in
India he had augured so highly, was carried off by a
second attack of fever, caused by a hurried journey
during the first outbreak of the Mutiny. Maxwell's
letters to Mr. Litchfield show how keenly he felt this
blow, and what deep thoughts on human life and
destiny were once more stirred up in him.

His original work on electricity was now for a
while interrupted by another laborious task, which

absorbed his best energies for more than a year. The examiners for the Adam's Prize, given by St. John's College in honour of the discovery of Neptune,[1] had set as a subject, "The Structure of Saturn's Ring." To frame and test an hypothesis which should account for the observed phenomena was a problem of no ordinary complexity, and one to which the speculative imagination and mathematical ingenuity of Clerk Maxwell were particularly adapted. It appears to have completely fascinated him for the time. The essay by which he gained the prize, and which he published after elaborately revising it, is well known to students, and the allusions to the subject in his letters at this time will be read with interest.

Such was the strain of feeling, and such the chief intellectual interest, with which he returned to Aberdeen, where he seems to have been once more destined, though in his native country, to understand more than he was understood; and in his letters, together with the deepening earnestness and the unfailing humour, there is now and then mingled for the first time a grain of bitterness, or what may be taken for such. But it is rather the cry of a spirit hungering for completion. And the phase of disharmony quickly passes off, and is followed by a song of triumph.

Of his new acquaintances at Aberdeen he had become most intimate with the family of Principal Dewar of Marischal College, and he was a frequent

---

[1] See above, p. 85.

visitor at their house. His deep and varied knowledge, not only of his own and kindred subjects, but
of history, literature, and theology, his excellence of
heart, and the religious earnestness which underlay
his humorous " shell," were there appreciated and
admired. He had been asked to join them in their
annual visit to Ardhallow, the home of the Principal's
son-in-law, Mr. M'Cunn, in the neighbourhood of
Dunoon, and had accepted the invitation. The time
of his stay there in September 1857 is marked by
letters which, unlike some others of this period,
reflect his brightest mood.

In February 1858 he announced his betrothal to
Katherine Mary Dewar, and they were married early
in the following June. In May he had made a
journey to the south of England to visit me in my
parish of Milford, in Hampshire, and to act as " best
man" on the occasion of our marriage, which took
place at Brighton. My wife and I found our way to
Aberdeen in time to be present at the wedding there,
and were shortly afterwards entertained at Glenlair.

The correspondence of these months and the poems
then written contain the record of feelings which in
the years that followed were transfused in action and
embodied in a married life which can only be spoken
of as one of unexampled devotion.[1]

He remained for two more sessions at Aberdeen.

---

[1] See the Poems of 1858 amongst the Occasional Pieces in
Part III.

But in 1860 came the fusion of the colleges, and
the Professorship of Natural Philosophy at Marischal
College was one of those suppressed. In the same
winter his old friend Professor James D. Forbes,
after struggling for eight years against ill-health,
resigned his Chair, and Maxwell became a candidate
for the vacant post. It is enough to have alluded
once for all [1] to the contest, which ended in the
appointment of Professor Tait. It only remains to
say that Maxwell's relations with that eminent
man, who had been his companion both in Edinburgh
and at Cambridge, always continued to be of the
most friendly kind; and their correspondence, often of
the quaintest description, would of itself fill a volume
of very entertaining reading for those possessed of a
clue to the labyrinth of science, learning, wit, and
frolicsome allusion, which it contains. The two men
looked over proof-sheets of each other's writings, and
when they most differed, Maxwell's criticism condensed
in humorous verse was always understood and welcomed by the Edinburgh Professor. In the summer

---

[1] See above, p. 258. As this candidature was the last occasion on
which Maxwell was compelled to collect testimonials, it may be well
to add here to what has been said above about his teaching powers,
that his success at Aberdeen was very strongly attested by his
colleagues, and in particular by Thomas Clark, the Professor of
Chemistry. And four years earlier, in February 1854, Professor
G. G. Stokes had given this important testimony :—" . . . One thing
more is wanted in a teacher, namely, a power of conveying clearly his
knowledge to others. That Mr. Maxwell possesses that power I feel
satisfied, having once been present when he was giving an account of
some of his geometrical researches to the Cambridge Philosophical
Society, on which occasion I was struck with the singularly lucid
manner of his exposition." (See p. 212.)

of 1860 the ex-professor of Aberdeen was appointed
to the vacant Professorship of Natural Philosophy in
King's College, London.

To LEWIS CAMPBELL, Esq.

Glenlair, 28th August 1857.

. . . I have been battering away at Saturn, returning
to the charge every now and then. I have effected several
breaches in the solid ring, and now I am splash into the
fluid one, amid a clash of symbols truly astounding. When
I reappear it will be in the dusky ring, which is something
like the state of the air supposing the siege of Sebastopol
conducted from a forest of guns 100 miles one way, and
30,000 miles the other, and the shot never to stop, but go
spinning away round a circle, radius 170,000 miles. . . .

To THE SAME.

Ardhallow, Dunoon, 4th Sept. 1857.

The road along Loch Eck is the most glorious for shape
and colour of hills and rocks that I have seen anywhere,
specially on a fine calm day, with clouds as well as sun, and
with large patches of withered bracken mixed with green on
the less steep parts of the hills. Then the crushing and
doubling up of the strata, and the slicing and cracking of
the already doubled up strata, quite without respect to
previous torment, gives a notion of active force, as well as
passive, even to ungeological minds. We inspected Duncan
Marshall, the Hermit of these parts, and wound up the day
with a pull in the boat till dark. . . .

Mrs. Wed[derburn] professes herself ready to " follow
follow South " when asked, so when Johnny and I have done
our Moidart and Loch Aylort, we shall hoist sail or get up
steam or something, and then very likely he may reappear
to his parent and aunt, and I shall continue my road with
my aunt to wait upon the faithful Tobs, and realise Saturn's
Rings, and probably feed a few natives of the valley with

the produce of its soil.   I was writing great screeds of
letters to Professor Thomson about those Rings, and lo! he
was a-laying of the telegraph which was to go to America,
and bringing his obtrusive science to bear upon the engineers,
so that they broke the cable with not following (it appears)
his advice.   However, I know nothing.   List to the new
words to a common song, which I conceived on the railway
to Glasgow.   As I have only a bizzing, loose, interruption-
to-talking-&-deathblow-to-general-conversation-memory of
the orthodox version, I don't know if the metre is correct;
but it is some such rambling metre anyhow, and contains
some insignificant though apparently treasonable remarks in
a perfect thicket of vain repetitions.   To avoid these let

$$(\text{U}) = \text{`` Under the sea,''}$$

so that 2(U), by parity of reasoning, represents two repetitions of that sentiment.   This being granted, we shall have
as follows :—

THE SONG OF THE ATLANTIC TELEGRAPH COMPANY.

I.

2(U)

Mark how the telegraph motions to me,

2(U)

Signals are coming along,

With a wag, wag, wag ;

The telegraph needle is vibrating free,

And every vibration is telling to me

How they drag, drag, drag,

The telegraph cable along,

II.

2(U)

No little signals are coming to me,

2(U)

Something has surely gone wrong,

And it's broke, broke, broke ;

What is the cause of it does not transpire,

But something has broken the telegraph wire

With a stroke, stroke, stroke,

Or else they've been pulling too strong.

III.

2(℧)

Fishes are whispering.   What can it be,

2(℧)

So many hundred miles long?

For it's strange, strange, strange,

How they could spin out such durable stuff,

Lying all wiry, elastic, and tough,

Without change, change, change,

In the salt water so strong.

IV.

2(℧)

There let us leave it for fishes to see ;

2(℧)

They'll see lots of cables ere long,

For we'll twine, twine, twine,

And spin a new cable, and try it again,

And settle our bargains of cotton and grain,

With a line, line, line,—

A line that will never go wrong.

Receive, etc.

To R. B. Litchfield, Esq.

Glenlair, 23d Sept. 1857.

I have just returned from the remote Highlands, and
have met all the Indian news on my way, and found your letter
at home. I suppose it is best to say what I think to you,
rather than what I feel, for that is confusion. You may
well ask " Why ?" I myself see a horrible despair waiting
for us if we knew or even paid enough attention to things
happening continually. Is it merely a reaction from our
animal life that makes us comfortable again ? or excitement
of some other kind ? or defective sympathy ? No; I think
real sympathy is the very thing we want, and we suffer
more from want of union than from any other cause. I
cannot make the thing clearer either to you or to myself;
but as I was coming home and expecting bad news, I
thought of dead and absent friends, and how they endea-

voured when alive to make themselves known to us, and
how the impression they had left in us remained untouched
and sacred during their absence. Then I thought of those
who had left the clearest impression on me,—how some were
dead, and their character never known or proved to the
world, and their deeds never done as they would have been
if they had lived. But this secret knowledge is strengthening as well as sad, if our brother's life is an inheritance to us
when he falls, and we rise (like Triamond) to fight his battle
as well as our own.

Do not understand all this as a theory. I wish to say
that it is in personal union with my friends that I hope to
escape the despair which belongs to the contemplation of
the outward aspect of things with human eyes. Either be a
machine and see nothing but "phenomena," or else try to
be a man, feeling your life interwoven, as it is, with many
others, and strengthened by them whether in life or in death.
You will say that this is what a man writes after a course of
healthy exercise and boisterous health, when he suddenly
feels a pull on his soul, but his body goes on as before.
But though my knowledge of our friend does not reach so
far back as yours, there is a great part of all my thoughts
which bears the mark of his honest handling, for to me he
was most liberal in communication, so that all manner of
thought became our common property. When he was ill at
Cambridge, my father, who was then rather better, was very
much concerned about him, so that afterwards, when he was
worse himself, he would speak of him from recollections of
what I had told him before. So I used to think of them
together,—the one guiding me along by wise plans in the ways
of freedom, and the other supplying the energy of speculative honesty and the freshness of a younger mind. And is
all gone? Certainly not, look at it as you will. I am not
trying to persuade myself into hope. I find I cannot do
otherwise. I have been miserable about these very things
when there seemed no particular reason at the time, and
then, when the time was worst, felt well; but that must be
one's own personal affair. If we can hear the General's voice

we will rise in our tears and go to our places, having
conquered ourselves.   Long ago I felt like a peasant in a
country overrun with soldiers, and saw nothing but carnage
and danger.   Since then I have learned at least that some
soldiers in the field die nobly, and that all are summoned
there for a cause.

I am very sorry for India, and for you and poor Mrs.
Pomeroy.   She had a stake in him that none of us could
have.

To Miss Cay.

Glenlair, 28th September 1857.

My dear friend Pomeroy died at Ghazeepore about
1st August, from overwork and forced marches.   He was a
civilian, and appointed assistant magistrate at Azinghur.
He and the rest of the civilians proceeded thither with about
400 men.   When they got there they fought about
2000 insurgents in two feet of water, and beat them off
three times.   Pomeroy volunteered to take in a wounded
man (Lieut. Lewis, 65 N. I.) to Ghazeepore, when no one
else would venture to go.   He stayed there a few days, but
was unwell.   Then he went back to his post at Azinghur
for about a week.   On the news of the Dinapore mutiny
they were all ordered in, and made a forced march on
Ghazeepore, forty miles.

He died soon after getting in.

Of all the men I have known he was the most likely
to have done something for India.   I never knew a man
more able to see his way through difficulties, more respected
by men of all classes, or more determined that duty should
be done whatever might happen.   We have one comfort for
ourselves, that few men have made themselves·more open to
their friends, so that many men may receive something of
his spirit, though he is cut off before strangers could take
example by his deeds.

To R. B. Litchfield, Esq.

Glenlair, 15th October 1857.

I was glad S. sent me the letter.   Remember that

besides all the danger and distance from friends, there was the fever, of which he had already long experience, and which in such times he knew to be as inconvenient to his friends as himself. But it is no use saying this and that. Some men redeem their characters by their deeds, and we praise them. Those that merely show their character by their deeds should be remembered, not praised; and a complete true man will live longest in the memory, and I cannot but think will be less changed in reality, than one who has doubtfully struggled with duplicity in his constitution, and has walked with hesitation, though along a good path. I know that both do deny and renounce themselves in favour of duty and truth, as they come to see them; and as they come to see how goodness, having the knowledge of evil, has passed through sorrow to the highest state of all, they accept it as a token that they have found their true head and leader, and so, with their eyes on him, they complete the process called the knowledge of Good and Evil, which they commenced so early and so ignorantly.

Now, what the " completion of the process " is I cannot conceive, but I can feel the difference of Good and Evil in some degree, and I can conceive the perception of that difference to grow by contemplating the Good till the confusion of the two becomes an impossibility. Then comes the mystery. I have memory and a history, or I am nothing at all. That memory and history contain evil, which I renounce, and must still maintain that I was evil. But it contains the image of absolute good, and the fight for it, and the consciousness that all this is right.

So there the matter lies, a problem certain of solution. . . .

I am grinding hard at Saturn, and have picked many holes in him, and am fitting him up new and true. I am sure of most of him now, and have got over some stumbling-blocks which kept me niggling at calculations two years.

I am to have some artisans as weekly students this winter.

To the Same.

Lauriston Lodge,
Edinburgh, 25th October 1857.

. . . As to collecting memorials, that is a thing to do faithfully if at all. No man can write himself down wholly at once, so no one thing will give a complete autograph of the man. As for anecdotes, they are to be tolerated as the roughest way of giving to the public sketches of public men; but they will satisfy no private friend, not even him from whose memory they are drawn.

But if any essays (I know some that were not read aloud to any particular people, but were written by himself for himself) or anything else of his are to be had, they might be given to Mrs. Pomeroy with what explanation we could give, and so (not hastily but) when times occur, she might continue to learn more of him and his honestly fought and thoroughly conquered and well secured path through a land of shadows, where friends and foes seemed to exchange appearances, so that honesty and sincerity had parted with order and reverence, and indecision had passed for conscientiousness, and enquiry after truth had been taken for infidelity. A sad world to seek for truth in under any man's guiding. The exact point at which this progress has been interrupted may be disputed by men, but they have no jurisdiction, and are ignorant even of their own true faith till something bring it into action.

I could not conceive of any one undertaking to *write* a *memoir*. I am fully satisfied of the impossibility of that, with respect to any *young* man. The journal of voyage, which was meant to be read, is far the best memorial, and it is a very good one indeed for his mother and friends, for it is so wonderfully open-hearted and unaffected. I appreciate these qualities on account of the force which it requires for me to say or do anything according to nature, especially in ordinary circumstances.

This spring I read a great deal of MS. at home, which gave me much light about my father and his dealings with various people. There were journals of travels and adventures from twelve years of age to thirty, and many other things. Well, I have found that reading profitable in making me remember him better, and honour him more, and understand better how he ordered his life, and who were his friends. That is the use and intent of keeping anything belonging to our friends. I do not intend to advertise these things for public sale, for I doubt whether the public would be better, and I would be worse.

The following letter from Pomeroy to his mother, of which a copy was found in Maxwell's handwriting, throws further light on the character of the man whose loss was felt so deeply :—

Azingurh, July 28, 1857.

MY DEAR MOTHER—I sit down to write to you in rather a solemn mood, partly owing to my having received yesterday a note from Mrs. W. H. T., which I enclose, and partly because that, in this part of India, and particularly in these outlying stations, with no European soldiers, the lives of Englishmen more clearly lie in God's hand than on most occasions in our sojourn on earth.

I have told you in former letters that I volunteered to go with any civilians who reoccupied a station in the Benares district. I did this completely of my own responsibility, and whether I was right or wrong God alone knows. I was then in excellent health, and I could not bear to think that I should be absolutely doing nothing in a safe station, when an additional European, who could ride or use firearms, even if he could do nothing else, was not utterly useless to the brave men who were going out in order to make, persuade, or enable the natives to cultivate their fields, during the three important months of the year, and so avoid a famine.

Mr. Tucker responded immediately to my volunteering, and appointed me " Assistant Magistrate " of Azingurh, with nearly —— a month. You know the delay in going, and that I was ill during that time. We left (Benares) on the 16th, and arrived here on the 18th. The march, which I performed partly on horseback and partly in my buggy, did not do me any harm, but the fatigue, excitement, and varied feelings caused by an affray with a body of natives, led by Oude Zemindars, in which we were finally victorious, after five hours of half fighting (not that I fought much, but I was on horseback most of the time), rather weakened me. Then there was a wounded man to be taken into Ghazipoor, and of those whose duty it was to accompany him nobody who could be spared could be induced to go, so I volunteered, whereby I pleased him, poor fellow, and his fellow-officers very much, and I believe I was of use to him. I hoped to have got three days rest at Ghazipoor, but the excitement of seeing new faces, and giving all my news over and over again, and trying to soothe angry officers whose corps had been abused, and explaining mistakes, etc. etc., made the *rest* very ambiguous ; and the journey back knocked me up. I was very feverish last night, but Home (the magistrate) took tremendous care of me, put my bed next his, and got me limes and honey, and everything he could get for cough or fever. To-day I am a great deal better, but still on rice-milk and chicken-broth, and as I have no duties but a share in the mess management, I get lots of lying down, but no absolute rest, as the nine of us, civil and military, live in one six-roomed bungalow. Don't imagine that nine (only) is the number walking about inside : everybody's bearer except mine and another's, I believe, walk about like tame dogs, bringing their masters' slightest wants, for few think of getting anything for themselves, even from across the room.

We have a very grave piece of intelligence from the neighbourhood of Legowlie, the headquarters of the 12th Irregular Cavalry. A native brought a report to a gentleman living 8 miles from Legowlie, that on the 24th of July the men of the 12th who were there left the station, and that the commanding officer (Major Holmes) and his wife, and the Doctor (Gardner), had been killed. This news has at length reached us, whom it concerns nearly, as one hundred of the same regiment are here. On the other side, it is to be said that all the principal native officers were absent from Legowlie, two being here now, and one else-where, so that the mutiny may not spread, and we have no treasure here to tempt them, and the Hindu Sepoys are not likely to join them, though their hatred of Musselmans has two or three times given way to their desire for plunder or love of Christian blood. Our position is precarious, but I can look it straight in the face, and in four days or less the crisis will, I hope, have passed.

Herne is a Poole man, with an uncle there of the name of Biddel. His wife is in the hills, and he hears from her now pretty often, but there is a month's interval between the departure and arrival of her letters. He is a decidedly religious man of a very genuine stamp. He read us last Sunday that grand sermon the Bishop of Calcutta preached towards the end of June. You should get it if it is to be had in England.

We are daily expecting to hear of the safety of Lucknow. Poor Mrs. Cooper ! She and her children must have been living in cellars for the last month, to be out of the way of cannon balls : and to live in a cellar in India must be dreadfully trying. If any children survive it will be a mercy,—a mercy indeed if they escape the awful, horrible fate of the ladies of Cawnpore. Mrs. Cave, whom I mentioned in my Journal, was, I fear, at the latter place. As a Madras fusilier said to me (in a broad Irish accent), " I don't care what they do to the likes of me—but the women—and Ladies too ! "

Poor T—— writes me letters intended to be cheery, but really awfully dismal. She bears up bravely, and thinks no one sees her sorrow.

We have a picture in the house that belonged to some one who fled from this on the 3d June. It is called "Woman's Mission." You must have seen it in the print-shop windows,—two Scutari volunteer nurses attending a wounded man. A pathetic picture suits our feelings. All but two, I believe, have either wife or lover to think of; but sorry should I or any be that any woman should be with us. Poor Batty married a very short time ago, and had to part from his Bride almost immediately, and send her to Calcutta. He, too, has lost a Brother.

I hope the English papers will have better and truer accounts than the Indian generally have ; but you cannot follow events without some good map, like Allen's £2 : 2s. one. So don't try ; it will only puzzle you, and the *Times* will give you the pith of the matter. The fire is now really being extinguished, though it will be long before stray points cease to flare up and burn those within their range.

There are two hundred of the King of Nepaul's Goorkas on the march to form part of our garrison, and on those we think we can really depend. They are very brave, and more than a match for Sepoys, and are more different in race and manners from the Hindus than we are, and very different in religion.

Urgent requisitions have been sent to Ghazipoor and Benares for Europeans, which may now at length be answered.

With regard to Mrs. T——, whom I was nearly forgetting in my egotism, I hope—if she has by the time this reaches you visited Cheltenham—that you have seen her. What a comfort it must be to her, poor thing, to think that she made him face those thoughts he had shunned and shirked so long. She left by last mail, so she will have been a fortnight in England or Ireland when this reaches you. I hope to write *you* a letter by Bombay and Marseilles of a later date, and to Editha by Calcutta and Marseilles ; so that this letter, which I confess is an alarmist one, will not arrive till you know the Event. I have written what I have that you may see I know my danger, and have, I hope, right feelings under it.—I remain, my very dear mother, your most affectionate son,   ROBERT HENRY POMEROY.

To Prof. Maxwell from Prof. G. G. Stokes.

School of Mines,
Jermyn Street, 7th November /57.

I have just received your papers on a dynamical top, etc., and the account of experiments on the perception of colour. The latter, which I missed seeing at the time when it was published, I have just read with great interest. The results afford most remarkable and important evidence in favour of the theory of three primary colour-perceptions, a theory which you, and you alone, so far as I know, have established on an exact numerical basis.

To Prof. J. C. Maxwell from Prof. Faraday.

Albemarle Street, 7th November 1857.

I have just read and thank you heartily for your papers. I intended to send you copies of two of mine. I think I have sent them, but do not find them ticked off. So I now send copies, not because they are assumed as deserving your attention, but as a mark of my respect, and desire to thank you in the best way that I can.

To Prof. J. C. Maxwell from Prof. Tyndall.

Royal Institution, 7th November 1857.

I am very much obliged to you for your kind thoughtfulness in sending me your papers on the Dynamical Top and on the Perception of Colour, as also for your memoir on Lines of Force, received some time ago. I never doubted the possibility of giving Faraday's notions a mathematical form, and you would probably be one of the last to deny the possibility of a totally different imagery by which the phenomena might be represented.[1]

To Prof. Maxwell from Prof. Faraday.[2]

Albemarle Street,
London, 13th November 1857.

If on a former occasion I seemed to ask you what you

---

[1] For confirmation of this, see Maxwell's (fragmentary) preface to the smaller treatise on electricity, published posthumously in 1881 ; especially these words : " In the larger treatise I sometimes made use of methods which I do not think the best in themselves, but without which the student cannot follow the investigations of the founders of the Mathematical Theory of Electricity. I have since become aware of the superiority of methods akin to those of Faraday, and have therefore adopted them from the first."

[2] This letter has already been published in the *Life of Faraday*.

thought of my paper, it was very wrong; for I do not think any one should be called upon for the expression of their thoughts before they are prepared, and wish to give them. I have often enough to decline giving an opinion because my mind is not ready to come to a conclusion, or does not wish to be committed to a view that may by further consideration be changed. But having received your last letter, I am exceedingly grateful to you for it, and rejoice that my forgetfulness of having sent the former paper on conservation has brought about such a result. Your letter is to me the first intercommunication on the subject with one of your mode and habit of thinking. It will do me much good, and I shall read and meditate it again and again.

I daresay I have myself greatly to blame for the vague use of expressive words. I perceive that I do not use the word "force" as you define it, "the tendency of a body to pass from one place to another." What I mean by the word is the *source* or *sources* of all possible actions of the particles or materials of the universe; these being often called the *powers* of nature when spoken of in respect of the different manners in which their effects are shown. In a paper which I have received at the moment from the *Phil. Mag.*, by Dr. Woods, they were called the "forces, such as electricity, heat, etc." In this way I have used the word "force" in the description of gravity which I have given as that expressing the received idea of its nature and source; and such of my remarks as express an opinion or are critical apply only to that sense of it. You may remember I speak to labourers like myself, experimentalists on force generally, who receive that description of gravity as a physical truth, and believe that it expresses all, and no more than all, that concerns the nature and locality of the power. To these it limits the formation of their ideas, and the direction of their exertions, and to these I have endeavoured to speak; showing how such a thought, if accepted, pledged them to a very limited and probably erroneous view of the cause of the force, and to ask them to consider whether they should not look (for a time at least) to a source in part external to

U

the particles.   I send you two or three old printed lines *marked*, relating to this point.   To those who *disown* the definition or description as imperfect, I have nothing to urge, as there is then probably no real difference between us.

I hang on to your words because they are to me weighty, and where you say, " I for my part can not realise your dis- satisfaction with the law of gravitation, provided you conceive it according to your own principles," they give me great comfort.   I have nothing to say against the law of action of gravity.   It is against the law which measures its total strength as an inherent force that I venture to oppose my opinion ; and I must have expressed myself badly (though I do not find the weak point) or I should not have conveyed any other impression.   All I wanted to do was to move men (not No. I. but No. II.) from the unreserved acceptance of a principle of physical action which might be opposed to natural truth.   The idea that we may possibly have to con- nect *repulsion* with the lines of gravitation force (which is going far beyond anything my mind would venture on at present, except in private cogitation), shows how far we may have to depart from the view I oppose.

There is one thing I would be glad to ask you.   When a mathematician engaged in investigating physical actions and results has arrived at his conclusions, may they not be expressed in common language as fully, clearly, and definitely as in mathematical formulæ ?   If so, would it not be a great boon to such as I to express them so ?—translating them out of their hieroglyphics, that we also might work upon them by experiment.   I think it must be so, because I have always found that you could convey to me a perfectly clear idea of your conclusions, which, though they may give me no full understanding of the steps of your process, give me the results neither above nor below the truth, and so clear in character that I can think and work from them.   If this be possible, would it not be a good thing if mathematicians, work- ing on these subjects, were to give us the results in this popular, useful, working state, as well as in that which is their own and proper to them ?

To H. R. Droop, Esq.

129 Union Street,
Aberdeen, 14th November 1857.

I am very busy with Saturn on the top of my regular work. He is all remodelled and recast, but I have more to do to him yet, for I wish to redeem the character of mathematicians, and make it intelligible. I have a large advanced class for Newton, physical astronomy, the electric sciences, and high optics. What is your department by the way?

I have also a mechanics' class in the evening, once a week, on mechanical principles, such as doctrine of lever work done by machines, etc. So I have 15 hours a week, which is a deal of talking straight forward.

I am getting several tops (like the one I had at Cambridge) made here for various parties who teach rigid dynamics.

To C. J. Monro, Esq.

127 Union Street,
Aberdeen, 26th November 1857.

The enclosed letters came from Mrs. Pomeroy to me. I think they are, one of them at least, yours. I doubt of the smaller one, but you will know. They seem dropping in still on poor Mrs. Pomeroy. Even ordinary returned letters are strange things to read again, as if you had been talking on when everybody had gone away.

I got your letter of the 6th. I have been grinding so hard ever since I came here that I left many letters unanswered. When I have time I shall write to you, and meanwhile only thank you for your letter.

I am at full college work again. A small class with a bad name for stupidity, so there was the more field for exciting them to more activity. So I have got into regular ways, and have every man *viva voce'd* once a week, and the whole class examined in writing on Tuesdays, and roundly and sharply abused on Wednesday morning; and lots of exercises which I find it advantageous to brew myself overnight.

Public Opinion here says that what our colleges want is inferior professors, and more of them for the money. Such men, says P. O., would devote their attention more to what would pay, and would pay more deference to the authority of the local press than superior or better paid men. Therefore, although every individual but one who came before the Commission was privately convinced that the best thing in itself would be to fuse our two institutions into one, with one staff of teachers, yet they all agreed that the public opinion of the whole was the opposite of the private opinion of each, and that more harm than good would result from adopting the course which seemed good to the members, but not to the body, of the public. So the battle rages hot between Union (of Universities only) and Fusion (of classes and professors). Almost all the professors in Arts are fusionists, and all the country south of the Dee, together with England and the rest of Europe and the world, but Aberdeen (on the platform), is Unionist, and nothing else is listened to at public meetings, though perhaps a majority of those present might be fusionists. Such is Public Opinion ; but we are all quiet again now, and I am working at various high matters, for I have a very good class for Physical Astronomy, Electricity, and Undulations, etc., and I want to do them justice.

I have had a lot of correspondence about Saturn's Rings, Electric Telegraph, Tops, and Colours. I am making a Collision of Bodies' machine, and a model of Airy's Transit Circle (with lenses), and I am having students' teas when I can. Also a class of operatives on Monday evening, who do better exercises than the University men about false balances, Quantity of Work, etc.

To Miss Cay.

129 Union Street, Aberdeen,
28th Nov. 1857.

I had a letter from Willy to-day about jet pumps to be made for real drains, but not saying anything about the Professorship of Engineering.

I have been pretty steady at work since I came. The

class is small and not bright, but I am going to give them plenty to do from the first, and I find it a good plan. I have a large attendance of my old pupils, who go on with the higher subjects. This is not part of the College course, so they come merely from choice, and I have begun with the least amusing part of what I intend to give them. Many had been reading in summer, for they did very good papers for me on the old subjects at the beginning of the month. Most of my spare time I have been doing Saturn's Rings, which is getting on now, but lately I have had a great many long letters to write,—some to Glenlair, some to private friends, and some all about science. . . . I have had letters from Thomson and Challis about Saturn—from Hayward of Durham University about the brass top, of which he wants one. He says that the Earth has been really found to change its axis regularly in the way I supposed. Faraday has also been writing about his own subjects. I have had also to write Forbes a long report on colours, so that for every note I have got I have had to write a couple of sheets in reply, and reporting progress takes a deal of writing and spelling. . . .

I have had two students' teas, at which I am becoming expert. I have also indulged in long walks, and have seen more of the country. The evenings are beautiful at this season. There have been some very fine waves on the cliffs south of the Dee.

To the Rev. L. Campbell.

(On taking Priest's Orders).

129 Union Street, Aberdeen,
22d Dec. 1857.

I take for granted that sometime on Sunday last you entered the second of the ecclesiastical transformations. May your life and doctrine set forth God's glory, and be the means of setting forward the salvation of all men! Some of my friends think that the separation to a " holy function " puts a man into an artificial position with respect to the conduct of his thoughts, words, and acts, and that he is immersed in a professional atmosphere,—a *world*, in fact, differing from the world of business or of fashion only in the general colouring of its scenery.   It has always seemed to me that men who have fallen into this " religious world" have completely failed in getting into the Church, seeing that the Church professes to be an escape from the world, and the only escape.   And what holds of the Church ought to hold of the clergy preëminently.   So far my theory of the Church not being a clerical world.   Now I believe it not only as a theory, but as a fact, that a man will find the thing so if he will try it himself.

The restraints and professional stiffness of sentiment are not made for lawful members, but for those whom the truth has not yet made entirely free.   I have to tell my men that all they see, and their own bodies, are subject to laws which they cannot alter, and that if they wish to do anything they must work according to those laws, or fail, and therefore we study the laws,   You have to say that what men are and the nature of their actions depends on the state of their wills, and that by God's grace, through union with Christ, the contradictions and false action of those wills may be settled and solved, so that one way lies perfect freedom, and the other way bondage under the devil, the world, and the flesh, and therefore you entreat them to give heed to the things which they have heard.

Now, no man accuses me of being stiff, because I try to make what I say precise.   Everybody knows that I believe it, whether I am stating a well-established law or only a half-verified conjecture.   And I think that you have fully more right to be respected, inasmuch as the nature of your message implies the duty of *preaching* it, and the convictions that may be arrived at are as cogent, being more *clear*, if less *distinct*, than scientific truths.

I have been reading Butler's *Analogy* again, specially with reference to obscurities in style and language, and also to distinguish the merits of the man, and what habits of thought they depended on.   Also Herschel's Essays, of which read that on Kosmos, and Froude's History.   One night I read

160 pages of Buckle's *History of Civilisation*—a bumptious book, strong positivism, emancipation from exploded notions, and that style of thing, but a great deal of actually original matter, the true result of fertile study, and not mere brain-spinning. The style is not refined, but it is clear, and avoids fine writing. Froude is very good that way, though you can see the sort of pleasure that a University man takes in actually realising what he has talked over at Hall about showing what England was in the middle ages, and transfusing himself, style and all, thereinto, that his friends may see. A solitary student never does that sort of thing, nor can he appreciate the graces of imitation. I wish Froude would state whether he translates, and from what language, in each document.

I am still at Saturn's Rings. At present two rings of satellites are disturbing one another. I have devised a machine to exhibit the motions of the satellites in a disturbed ring; and Ramage is making it, for the edification of sensible image worshippers. He has made four new dynamical tops, for various seats of learning.

I have set up a model of Airy's Transit Circle, and described it to my advanced class to-day. That institution is working well, with a steady attendance of fourteen, who have come of their own accord to do subjects not required by the College, *and the dryest first.*

To the present time we have been on Newton's *Principia* (that is, Sects. i. ii. iii., as they are, and a general view of the Lunar Theory, and of the improvements and discoveries founded on such inquiries). Now we go on to Magnetism, which I have not before attempted to explain.

The other class is at two subjects at once. Theoretical and mathematical mechanics is the regular subject, but two days a week we have been doing principles of mechanism, and I think the thing will work well. We now go on to Friction, Elasticity, and Breakage, considered as subjects for experiment, and as we go on we shall take up other experimental subjects germane to the regular course. I am happy in the knowledge of a good tinsmith, in addition to a smith, an optician, and a carpenter.   The tinsmith made the Transit Circle.

College Fusion is holding up its head again under the fostering care of Dr. David Brown (father to Alexander of Queen's).[1]   Know all men I am a Fusionist, and thereby an enemy of all the respectable citizens who are Unionists (that is, unite the three learned faculties, and leave double chairs in Arts).   But there is no use writing out their theory to you.   They want inferior men for professors—men who will find it their interest to teach what will pay to small classes, and who will be more under the influence of parents and the local press than more learned or better paid men would be in a larger college.

I send you a description of the Murtle Lecture delivered in our Public School :—

To those who admire the genius of the bard who sang of The Dee, the Don, Balgownie Brigg's black wall, the following lines will be welcome from their resemblance to the opening of one of his poems :—

> Know ye the Hall where the birch and the myrtle
>   Are emblems of things half profane, half divine,
> Where the hiss of the serpent, the coo of the turtle,
>   Are counted cheap fun at a sixpenny fine ?
> Know ye the Hall of the pulpit and form,
> With its air ever mouldy, its stove never warm ;
> Where the chill blasts of Eurus, oppressed with the stench,
> Wax faint at the window, and strong at the bench ;
> Where Tertian and Semi are hot in dispute,
> And the voice of the Magistrand never is mute ;
> Where the scrape of the foot and the audible sigh
> In nature though varied, in discord may vie,
> Till the accents of Wisdom are stifled and die ;
> Where the Bajuns are dense as the cookies they chew,
> And all save the Regents have something to do :—

---

[1] See a little book entitled " Crushed Hopes Crowned in Death," London, 1862.

'Tis our Hall of Assembly, our high moral School,
Must its walls never rest from the bray of the fool ?
Oh vain as the prospect of summer in May
Are the lessons they learn and the fines that they pay.

All the public discipline, fines, etc., are arranged and levied at
the Public School. The Bajuns, Semis, Tertians, and Magis-
trands are the four years of men. The Regents are the four
Professors—Greek, Nat. Hist., Nat. Phil., and Mor. Phil.

Gaiety is just beginning here again. Society is pretty
steady in this latitude,—plenty of diversity, but little of
great merit or demerit,—honest on the whole, and not
vulgar. . . . No jokes of any kind are understood here. I
have not made one for two months, and if I feel one coming I
shall bite my tongue. I shall write as soon as I hear again
from you.

TO THE SAME.

129 Union Street, 20th Jan. 1858.

I should have written to thank you for the little book,[1]
but I think you will prefer my thanks now that I have
read it.

What is the English book that says *dasz Wahrheit
Offenbarung mache, nicht Offenbarung Wahrheit* ?

The marrow of the book lies in the man's being a
Fremdling, conscious of the shell that surrounds him and
divides him from others, and able neither to live in it nor
to break it. But the shell is only the outer surface of a
minute drop of fluid, impenetrable because minute, and
arising from molecular forces in himself. But he meets with
another drop, confined to one spot it is true, but a great deal
larger than himself. He coalesces, and both are now larger
and more fluid than before. Note that though he was
movable and active, she fixed and ill, she knew the
Hofrath; he knew no one, unless perhaps his mother.
People differ in their need of knowing others, and in their
power of conquering that knowledge. I have been reading

---

[1] *Deutche Liebe.* (By Prof. Max Müller.)

Lavater and his life. He needed to know people; he was a man of a refined and tender spirit, but it was vigorous, although not very massive or powerful, and he came to know people, made friends, a few enemies, stuck to his work, and lived happy. Other men have lived well and done good without even wishing to burst the shell of separate existence, feeling it like the natural garment of a personal being.

Now I find that the transfusive tendency is not identical with personal attraction (using the last two words in anything but their newspaper sense). There are some people whom I feel disposed to love, honour, and obey, though in many things I may dislike them, and may have no wish to have a complete fusion of thought and feeling with them. There are others who are easily sympathised with, and open out willingly, but do not thereby acquire the power and authority which the first have without seeking it. Both is best, but of the two the first is more permanent than the second.

To return to the book, the different sex of the parties is treated as an accident, but by the effect on the man it certainly is not, neither is the effect of it insensible on the lady. She treats his statements with more reverence than is their due, as coming from a man, and, I think, fails entirely in framing a scheme for coalescing, without entering on that state of which marriage is the symbol, even though by accident there may be checks which may enable or compel the parties to stop short. It is not society that does it; it is a law in us. Now I must stop, or I shall be teaching my grandmother to suck eggs.

Let me try my hand on that worthy relative in her professional as well as private capacity, with respect to the Lilleshall sermon. I am sure you will be able, with pains, to put anything you have sure hold of before your hearers; but there are certain subjects which, after being handled by some of our writers, get coated over with language so tenacious that it is difficult to recognise them in plain clothes, so that you become like the "lovely song of one that hath a pleasant voice."

Now it is good to learn wisdom wherever it is to be found, but in teaching it, it must be made light, wholesome, and digestible, by being stripped of all vagueness and wordiness, and refitted with illustrations and conceptions carefully adapted to the hearers. Not but what the other method is pleasant to listen to, and not without profit (if taken with salt); but there are good books, out of which you may preach very bad sermons, with which your people may be as delighted as Mary Anne was with Faraday's lecture, of which she gave an account to *Punch.*

I find my principal work here is teaching my men to avoid vague expressions, as " a certain force," meaning uncertain ; *may* instead of *must; will be* instead of *is; proportional* instead of *equal.* . . .

As to yourself, I do not know whether college or parish work is best for you to be set to. I am not sorry about the rich people. They require you as much as the poor. But you must find out your own spirit, and what you were made for, and not steer by men either towards one man or from another.

Now I sent you a libellous description of our public school here last letter. Last week I was walking with the tune of the " Lorelei " running in my head, and it set itself into a " kind of allegory." The words are very crude, but that is the way they came together. I profess myself responsible for most things I say, but less for this than for most :—

> Alone on a hillside of heather,
>   I lay with dark thoughts in my mind,
> In the midst of the beautiful weather
>   I was deaf, I was dumb, I was blind.
> I knew not the glories around me,
>   I thought of the world as it seems,
> Till a spirit of melody found me,
>   And taught me in visions and dreams.
>
> For the sound of a chorus of voices
>   Came gathering up from below,
> And I heard how all Nature rejoices,
>   And moves with a musical flow.

O strange ! we are lost in delusion,
　　Our ways and doings are wrong,
We are drowning in wilful confusion
　　The notes of that wonderful song.

But listen, what harmony holy
　　Is mingling its notes with our own !
The discord is vanishing slowly,
　　And melts in that dominant tone.
And they that have heard it can never
　　Return to confusion again ;
Their voices are music for ever,
　　And join in the mystical strain.

No mortal can utter the beauty
　　That dwells in the song that they sing ;
They move in the pathway of duty,
　　They follow the steps of their King.
I would barter the world and its glory,
　　The vision of joy to prolong,
Or to hear and remember the story
　　That lies in the heart of their song.

To the Same.

129 Union Street, 31st January 1858.

Thank you for your letter, so kind and so speedy. Now there are two of us, and I have that knowledge which is better than all advice. Not that I undervalue the advice at all, only the sense of unity between us is the main thing, whether we keep up correspondence or not. And I know that my friends are *ipso facto* your's, and your's mine, so that we are a large and influential body. . . .

But don't suppose that I intend to make you my confessor. It would not be just to you, for I do not like being confessed to myself; and I think every one should bear his own burden, though willing to lighten that of others. Besides, confession brings into set words and distinct outlines, doubtful suspicions and half-formed thoughts, which would fade at once if they were not stirred up.

To do the thing adequately, without extenuation, voluntary exaggeration, or colouring of any kind, is far beyond human power. The consciousness of the presence of God is the only guarantee for true self-knowledge. Everything else is mere fiction, fancy portraiture,—done to please one's friends or self, or to exhibit one's moral discrimination at the expense of character.

.   .   .   .   .   .

And now be assured that I feel like Spenser's " Diamond," who had " Priamond's " spirit in him as well as his own. There is another human aid that is with me. The memory of my father is a great help to being practical and active. The more I think of him the better I get on, and I am the less tempted to absurdity and eccentricity in thought.

As for outward act, no one here seems to think me odd or daft. Some did at Cambridge, but here I have escaped. My rule is to avoid the company of young men whom I do not respect, unless I have the control 'of them.

King's College has its Senior Wrangler this year. I announced it to my class yesterday morning. Lightfoot has commissioned some more from me to be sent to Trinity.

To R. B. LITCHFIELD, Esq.

129 Union Street,
Aberdeen, 7th February 1858.

When I last wrote I was on my way here. Since then I have been at work, Statics and Dynamics ; two days a week being devoted to Principles of Mechanism, and afterwards to Friction, Elasticity and Strength of Materials, and also Clocks and Watches, when we come to the pendulum. We have just begun hydrostatics. I have found a better text-book for hydrostatics than I had thought for,—the run of them are so bad, both Cambridge and other ones,—Galbraith and Haughton's *Manual of Hydrostatics* (Longmans, 2s.) There are also manuals of Mechanics and Optics of the same set. There is no humbug in them, and many practical matters are introduced instead of mere intricacies. The only defect is a somewhat ostentatious resignation of the demonstrations of certain truths, and a leaning upon feigned experiments instead of them. But this is exactly the place where the students trust most to the professor, so that I care less about it. I shall adopt the Optics, which have no such defect, and possibly the Mechanics, next year.

My students of last year, to the number of about fourteen, form a voluntary class, and continue their studies. We went through Newton i. ii. iii., and took a rough view of the Lunar Theory, and of the present state of Astronomy. Then we have taken up Magnetism and Electricity, static and current, and now we are at Electro-magnetism and Ampere's Laws. I intend to make Faraday's book the backbone of all the rest, as he himself is the nucleus of everything electric since 1830.

So much for class work. Saturn's Rings are going on still, but this month I am clearing out some spare time to work them in. I have got up a model to show the motions of a ring of satellites, a very neat piece of work, by Ramage, the maker of the "top."

For other things—I have not much time in winter for improving my mind. I have read Froude's History, *Aurora Leigh*, and Hopkins's *Essay on Geology*, also Herschel's collected Essays, which I like much, also Lavater's life and *Physiognomy*, which has introduced me to him pleasantly though verbosely. I like the man very much, quite apart from his conclusions and dogmas. They are only results, and far inferior to methods. But many of them are true if properly understood and applied, and I suppose the rest are worth respect as the statements of a truth-telling man.

Well, work is good, and reading is good, but friends are better. I have but a finite number of friends, and they are dropping off, one here, one there. A few live and flourish. Let it be long, and let us work while it is day, for the night is coming, and work by day leads to rest by night.

To Rev. Lewis Campbell.

129 Union Street,
Aberdeen, 17th February 1858.

. . . I have not been reading much of late. I have

been hard at mathematics. In fact I set myself a great arithmetical job of calculating the tangential action of two rings of satellites, and I am near through with it now. I have got a very neat model of my theoretical ring, a credit to Aberdeen workmen. Here is a diagram, but the thing is complex and difficult to draw :—

Two wheels turning on parallel parts of a cranked axle; thirty-six little cranks of same length between corresponding points of the circumferences; each carries a little ivory satellite.[1]

To Miss Cay.

129 Union Street,
18th February 1858.

Dear Aunt—This comes to tell you that I am going to have a wife.

I am not going to write out a catalogue of qualities, as I am not fit; but I can tell you that we are quite necessary to one another, and understand each other better than most couples I have seen.

Don't be afraid; she is not mathematical; but there are other things besides that, and she certainly won't stop the mathematics. The only one that can speak as an eye-witness is Johnnie, and he only saw her when we were both trying to act the indifferent. We have been trying it since, but it would not do, and it was not good for either.

So now you know who it is, even Katherine Mary Dewar (hitherto). I have heard Uncle Robert speak (second-hand) of her father, the Principal. Her mother is a first-rate lady, very quiet and discreet, but has stuff in her to go through anything in the way of endurance. . . . So there is the state of the case. I settled the matter with her, and the rest of them are all conformable.

I hope some day to make you better acquainted. I can hardly admit that Johnnie saw her at all,—not as he will when she appears in a true light. . . . For the present you

---

[1] The sketch which follows corresponds to the model which is preserved in the Cavendish Laboratory. See Part II.

must just take what I say on trust.   You know that I am
not given to big words.   So have faith and you shall
know.

. . . I don't write separately to Uncle Robert, seeing
he is with you, and I am very busy, and just now I should
just write the same thing over again, and I have not a
copying press.   So good-bye.—Your affectionate nephew.

To R. B. LITCHFIFLD, Esq.

129 Union Street,
Aberdeen, 5th March 1858.

My " lines " are so pleasant to me that I think that everybody ought to come to me to catch the infection of happiness.
This college work is what I and my father looked forward
to for long, and I find we were both quite right—that it was
the thing for me to do.   And with respect to the particular
college, I think we have more discipline and more liberty,
and therefore more power of useful work, than anywhere else.
It is a great thing to be the acknowledged " regent " of one's
class for a year, so as to have them to one's self except in
mathematics and what additional classes they take.   Then
the next year I get those that choose to come, which makes
a select class for the higher branches.   They have all great
power of work.

In Aberdeen I have met with great kindness from all sects
of people, and you now know of my greatest achievement in
the way of discovery, namely, the method of converting
friendship and esteem into something far better.   We are
following up that discovery, and making more of it every day ;
getting deeper and deeper into the mysteries of personality,
so as to know that we ourselves are united, and not merely
attracted by qualities or virtues, either bodily or mental.—
(Don't suppose we talk metaphysics.) . . .

You will easily see that my " confession of faith " must
be liable to the objection that Satan made against Job's
piety.   One thing I would have you know, that I feel as free
from compulsion to any form of compromised faith as I did
before I had any one to take care of, for I think we both

believe too much to be easily brought into bondage to any set of opinions.

With respect to the " material sciences," they appear to me to be the appointed road to all *scientific* truth, whether metaphysical, mental, or social. The knowledge which exists on these subjects derives a great part of its value from ideas suggested by analogies from the material sciences, and the remaining part, though valuable and important to mankind, is not *scientific* but aphoristic. The chief *philosophical* value of physics is that it gives the mind something distinct to lay hold of, which, if you don't, Nature at once tells you you are wrong. Now, every stage of this conquest of truth leaves a more or less presentable trace on the memory, so that materials are furnished here more than anywhere else for the investigation of *the* great question, " How does Knowledge come ? "

I have observed that the practical cultivators of science (*e.g.*, Sir J. Herschel, Faraday, Ampère, Oersted, Newton, Young), although differing excessively in turn of mind, have all a distinctness and a freedom from the tyranny of words in dealing with questions of Order, Law, etc., which pure speculators and literary men never attain.

Now, I am going to put down something on my own authority, which you must not take for more than it is worth. There are certain men who write books, who assume that whatever things are orderly, certain, and capable of being accurately predicted by men of experience, belong to one category ; and whatever things are the result of conscious action, whatever are capricious, contingent, and cannot be foreseen, belong to another category.

All the time I have lived and thought, I have seen more and more reason to disagree with this opinion, and to hold that all want of order, caprice, and unaccountableness results from interference with liberty, which would, if unimpeded, result in order, certainty, and trustworthiness (certainty of success of predicting). Remember I do not say that caprice and disorder are not the result of free will (so called), only I say that there is a liberty which is not disorder, and that this is by no means less free than the other, but more.

In the next place, there are various states of mind, and schools of philosophy corresponding to various stages in the evolution of the idea of liberty.

In one phase, human actions are the resultant (by par^{m} of forces) of the various attractions of surrounding things, modified in some degree by internal states, regarding which all that is to be said is that they are subjectively capricious, objectively the " RESULT OF LAW,"—that is, the wilfulness of our wills feels to us like liberty, being in reality necessity.

In another phase, the wilfulness is seen to be anything but free will, since it is merely a submission to the strongest attraction, after the fashion of material things. So some say that a man's will is the root of all evil in him, and that he should mortify it out till nothing of himself remains, and the man and his selfishness disappear together. So said Gotama Buddha (see Max Muller), and many Christians have said and thought nearly the same thing.

Nevertheless there is another phase still, in which there appears a possibility of the exact contrary to the first state, namely, an abandonment of wilfulness without extinction of will, but rather by means of a great development of will, whereby, instead of being consciously free and really in subjection to unknown laws, it becomes consciously acting by law, and really free from the interference of unrecognised laws.

There is a screed of metaphysics. I don't suppose that is what you wanted. I have no nostrum that is exactly what you want. Every man must brew his own, or at least fill his own glass for himself, but I greatly desire to hear some more from you, just to get into *rapport*.

As to the Roman Catholic question, it is another piece of the doctrine of Liberty. People get tired of being able to do as they like, and having to choose their own steps, and so they put themselves under holy men, who, no doubt, are really wiser than themselves. But it is not only wrong, but impossible, to transfer either will or responsibility to another;

and after the formulæ have been gone through, the patient
has just as much responsibility as before, and feels it too.
But it is a sad thing for any one to lose sight of their work,
and to have to seek some conventional, arbitrary treadmill-
occupation prescribed by sanitary jailors. . . . .

With respect to the class, I send you the paper they did
last week. Five floored it approximately, two first-rate. I
got half-a-dozen correct answers to questions on the effects of
mixtures of ice and steam in various proportions, and on the
effect of heating and cooling on the thrust of iron beams
(numerical). From the higher class I have essay on Vision
(construction of eye, spectacles, stereoscopes, etc.) So the
work done is equivalent to the work spent.

To Rev. Lewis Campbell.

Aberdeen, 15th March 1858.

When we had done with the eclipse to-day, the next
calculation was about the conjunction. The rough approximations bring it out early in June. . . .

The first part of May I will be busy at home. The
second part I may go to Cambridge, to London, to Brighton,
as may be devised. After which we concentrate our two
selves at Aberdeen by the principle of concerted tactics.
This done, we steal a march, and throw our forces into the
happy valley, which we shall occupy without fear, and we
only wait your signals to be ready to welcome reinforce-
ments from Brighton. . . . Good night.—Your affectionate
friend.   J. C. Maxwell.

N.B.—We are going to do optical experiments together
in summer. I am getting two prisms, and our eyes are so
good as to see the *spot* on the sun to-day without a telescope.

To Prof. J. C. Maxwell from Prof. Forbes.

Edinburgh, 16th March 1858.

I was much obliged to you for your letter, and the
announcement of your marriage, which I have not the smallest

doubt will add to your happiness, while at the same time I
do not fear its abstracting you from science.

Your notice of Saturn will be very acceptable. But it
should not run to too great a length, as the 19th will probably
be our last meeting, and is always a crowded billet. Give me
an idea of the least time requisite to give an idea of your
subject; and more particularly try to send me a piece of the
MS., so that I may legally hold it as a MS. delivered, and
take precedence of some trivial communications of which I
stand rather in dread.

I duly received the *Top*. I suppose it is all right, but
my energies have been absorbed in Electrical experiments
merely for lecture, and which have been very heavy upon
me. I ought to have paid for the Top ere now, but I will
soon.

We saw the Eclipse very badly, and it seems that in
England it was no better. I had arranged to give a tele-
graphic account to R. Soc. last night, from no less than 3
points on the central line. . . .

To REV. LEWIS CAMPBELL.

Glenlair, Springholm,
Dumfries, 28th April 1858.

. . . I wish you great joy, now and always! I hope to
certify myself ere long what sort of "friend's wife" I am to
have. I have faith already, but sight is better, and you
will have some pleasure in getting my verdict, though you
don't need anything of the kind.

I have been very happy in observing the very admirable
frame of mind in which all my friends seem disposed to
regard my affairs, and yet I would rather that their opinions
and sentiments had a more distinct basis of observation.
But I suppose they observe me, and see I am "all right and
no mistake." . . .

I tell you this . . . because you are our friend for
better for worse.

I shall bring you a small pen-wiper that Katherine made
for you the last day I was in Aberdeen. If you are careful

in using it as it ought to be used, you will get rid of all the
"odium theologicum" and other bitter principles sometimes
occurring in parsons' ink, and your heart will indite good
matter with the pen of a ready writer. . . .

To C. J. MONRO, Esq.

Glenlair, 29th April 1858.

. . . I displayed my model of Saturn's Ring at the Edinburgh Royal Society on the 19th. The anatomists seemed to
take most interest in the construction of it. We are going
to do some experiments on colour this summer, if my prisms
turn out well. I have got a beautiful set of slits made by
Ramage, to let in the different pencils of light at the proper
places, and of the proper breadths.

To MISS K. M. DEWAR.

2d May 1858.

Now you must remember that all I say about texts and
matters of that sort is only a sort of help to being together
when we read, for I am not skilful to know what is the
right meaning of anything so as to tell other people, only I
have a right to try to make it out myself, and what I say
to myself I may say to you.

To THE SAME.

6th May 1858.

Isaiah li. and Gal. v. I suppose the leaven in v. 9 is
the little bit of Judaism that they were going to adopt on
the plea that it is "safer" to do and believe too much than
too little, and yet these little things altered the character of
the whole of their religion by making it a thing of labour
and wages, instead of an inward growth of faith working by
love, which purifies the heart now, and encourages us to
wait for the hope of righteousness. But still the desire of
the spirit is contrary to the desire of the flesh, the one tending towards God, and the other towards the elements of the
world, so that we are kept stretched as it were, and this is
our training in this life. Our flesh is God's making, who

made us part of His world; but then He has given us the power of coming nearer to Himself, and so we ought to use the world and our bodies as means towards the knowledge of Him, and stretch always as far as our state will permit towards Him. If we do not, but wilfully seek back again to the elements as the Israelites to Egypt, then we are not like infants or even brutes, but far worse, as recoiling from God and His blessedness. Here are manifest the works of the flesh, which are not only not those of righteousness, but opposed to them; but the fruit of the spirit comes when, like good trees, we stretch our best affections upwards till we see the sun, and breathe the air and drink the rain, and receive all free gifts, instead of sending our branches after our roots, down among things that once had life but now are decaying, and seeking there for nourishment that can only be had from above.

See the order of ripening of the fruit. Now, love brings joy to ourselves, and this, peace with others, and this, long-suffering of their attacks, for why should we be angry? Gentleness is a higher degree of this, being active. Goodness is used in a less general sense than we use it. It seems something like " good nature," only better and more manly, and refers to the good disposition of a man among men. Faith is put in here as the *result* of good living; which is true, for it is nourished thereby. Then see what comes of adding faith to a good disposition,—meekness, which we cannot afford to have without faith; and lastly, temperance or moderation, which is also founded on faith, and is a virtue that can never be perfect till all the rest are so.

TO THE SAME.

Milford, Lymington (Hants),
9th May 1858.

To-day we were called at seven; were down soon thereafter, and had everything leisurely and comfortably till ten, when we went to school. I had a class of youths just beginning to read, and some of them knew not what swine were, still less what a herd was. At eleven to church—

Lewis read prayers and lessons very well and distinctly, and Chester preached on James i. 15. Sin when it is finished bringeth forth death. He showed up sin as the universal poison, and showed how it might be seen working death in several instances, and also in all, good and bad, in this life, and then turned to the next, and finally indicated the remedy, though not so clearly as Paul in Rom. viii. 2, which he should have read.

In the afternoon Chester read prayers and Lewis preached on " Ye must be born again," showing how respectable a man it was addressed to, and how much he, and all the Jews, and all the world, and ourselves, needed to be born *from above* (for that is the most correct version of the word translated *again*). Then he described the changes on a man new-born, and his state and privileges. I think he has got a good hold of the people, and will do them good and great good.

To the Same.

10th May 1858.

Eph. iii. 19.—Paul can express no more, but read the last two verses and you will see this is not the crown, but only what can be asked or thought. What a field for ambition there is,—for climbing up, or rather, being drawn up, into Christ's love, and receiving into our little selves all the fulness of God. Let us bless God even now for what He has made us capable of, and try not to shut out His spirit from working freely.

To the Same.

13th May 1858.

I have been reading again with you Eph. vi. Here is more about family relations. There are things which have meanings so deep that if we follow on to know them we shall be led into great mysteries of divinity. If we despise these relations of marriage, of parents and children, of master and servant, everything will go wrong, and there will be confusion as bad as in Lear's case. But if we reverence them, we shall even see beyond their first aspect a spiritual meaning, for God speaks to us more plainly in these bonds of our life than in anything that we can understand. So we find a great deal of Divine Truth is spoken of in the Bible with reference to these three relations and others.

TO THE SAME.

16th May 1858.

Phil. iii.—There is great wisdom in v. 13. Never look back with complacency on anything done, or attained, or possessed. See the description of those who mind earthly things, and let us depart from their ways. Conversation in v. 20 means going backwards and forwards, and refers to the walking of the preceding verses. What a description of the power of Christ in the last verse, over "all things," and our vile bodies among the rest, and what a day it will be when He has done all His work and is satisfied.

I think the more we enter together into Christ's work He will have the more room to work His work in us. For He always desires us to be one that He may be one with us. Our worship is social, and Christ will be wherever two or three are gathered together in His name.

I have been vexed that I could not speak better to ——— I had a long walk with him, talking of what people have believed, and what was necessary to be believed. I hope we may come to understand each other, but more that he may come to the clear light. I wish I could speak to him wise words. He is so anxious to hear, and I to speak, and then the words are all wind after all.

TO PROF. J. C. MAXWELL FROM VERNON LUSHINGTON, ESQ.

Ockham, 31st May 1858.

Next Wednesday is <i>your</i> second of June, after which we shall no longer be able to think of you as one of ourselves —the youthful wanderers and seekers of the earth. So how can I better employ the end of this Sunday evening than by bidding you Farewell and God speed ? . . .

When Wednesday comes I hope I shall think of you. I like thinking of you,—what you were, what you are, and what you may be. All happiness be with you and yours!

To C. J. MONRO, Esq.

Glenlair, 24th July 1858.

. . . We are no great students at present, preferring various passive enjoyments, resulting from the elemental influences of sun, wind, and streams. This week I have begun to make a small hole into Saturn, who has slept on his voluminous ring for months.

To MRS. MAXWELL.

16th September 1859.

Mrs. Sabine learnt mathematics of her husband after she was married, so she was not married for it. Murchison knew no geology when he was married, but his wife did a little ; and there was a fall of a cliff in the morning early, and her maid told her of it, so she was for up ; [1] so Murchison got up too, and there were the great bones of an Icthyosaurus in the broken cliff, and he was interested and took to geology. Before that he was an idle young officer.

---

[1] " was for up," *i.e.* wished to get up.

CHAPTER XI.

FROM this point onward the interest of Maxwell's life (save things "wherewith the stranger inter- meddles not") is chiefly concentrated in his scientific career.   As some account of his labours in science will be given in the second portion of this book, what re- mains of the present narrative is comparatively brief.

1860-1865.   The work at King's College was more exacting
Æt. 29-34. than that in Aberdeen.   There were nine months of lecturing in the year, and evening lectures to artisans, etc., were recognised as a part of the Professor's regular duties.   Maxwell retained the post until the spring of 1865, when he was succeeded by Professor W. G. Adams, but continued lecturing to the working men during the following winter.

In June 1860 Maxwell attended the British Association's meeting at Oxford, where he exhibited his box for mixing the colours of the spectrum.   He also presented to Section A a most important paper on Bernoulli's Theory of Gases ; a theory which sup- poses that a gas consists of a number of independent particles moving about among one another without mutual interference, except when they come into collision. Maxwell showed that the apparent viscosity of gases, their low conductivity for heat, and Graham's laws of diffusion, could be satisfactorily explained by this theory, and gave reasons for believing that in air at ordinary temperature each particle experiences on an average more than 8,000,000,000 collisions per second. It is probable that the contemplation of the "flight of brick-bats" (his own vivid phrase for the constitution of Saturn's rings) led him on to his far-reaching investigations in this field of molecular physics.

On the 17th of May 1861 he delivered his first lecture before the Royal Institution. The subject was " On the Theory of the three Primary Colours."

All this while Maxwell was quietly and securely laying the foundations, deep and wide, of his great work on Electricity and Magnetism, but he had not the leisure that was requisite for bringing it to completion.

The period of his King's College Professorship was far, however, from being scientifically unfruitful. The colour-box was perfected, and many series of observations were made with it. Mrs. Maxwell's observations were found to have a special value. Through a striking discrepancy between her readings and C. H. Cay's, Maxwell discovered that the blindness of the *Foramen Centrale* to blue light, which was strongly marked in his own dark eyes, was either altogether absent from hers, or present in a very low degree. The comparison of J. C. M.'s (J.'s) eyes, and Mrs. M.'s (K.'s) is referred to in Part II.

The experimental measurements by which the present standard of electrical resistance (the Ohm) was first determined, were made at King's College by a sub-committee of the B.A., consisting of Maxwell, Balfour Stewart, and Fleeming Jenkin, in 1862-63, in accordance with a method proposed by Sir Wm. Thomson. A further experimental measurement was made next year by Maxwell, Fleeming Jenkin, and Charles Hockin (Fellow of St. John's). The importance of the work may be estimated by the fact that the system of units then determined by the B.A. Committee was, in the main, adopted by the Electrical Congress which met last year (1881) in Paris, and an International Commission has been appointed by the European Governments to make a redetermination of the standard of resistance first measured by the B.A. Committee. Maxwell's papers on this subject, with those of his fellow-workers, were republished in 1873 in a volume edited by Fleeming Jenkin.[1] Many are the references to successful or fruitless "spins" in the home letters of this period. The following quotation will suffice :—

28th January 1864.

We are going to have a spin with Balfour Stewart tomorrow. I hope we shall have no accidents, for it puts off time so when anything works wrong, and we cannot at first find out the reason, or when a string breaks, and the whole spin has to begin again. . . . However, we hope to bring out our standards by September, and Becker [2] makes them up excellently.

---

[1] *Reports of the Committee on Electrical Standards, appointed by the British Association for the advancement of Science.* Spon, London and New York, 1873.   [2] Of Messrs. Elliott Brothers.

A mass of correspondence, containing numerous suggestions made by Maxwell from day to day in 1863-4, has been preserved by Professor Jenkin. Two of the least technical passages will be found amongst the letters in this chapter (pp. 337, 340).

Another very important experimental investigation conducted by Maxwell about this period was the determination of the ratio of the electromagnetic and electrostatic units of electricity, for the purpose of comparing this quantity with the velocity of light. As this investigation will be again referred to in Part II., it is only necessary to say here that the experiment amounts to a comparison between the attractions of two electric currents flowing in coils of wire, and the attraction or repulsion between two metal plates which have each received a charge of electricity. Maxwell had pointed out that, in accordance with his theory, the ratio of the units should be equal to the velocity of light, and the value obtained by him was intermediate between the extreme values obtained for that velocity by previous observers. The experiment was the outcome of his theory of the constitution of the space in the neighbourhood of magnetic and electric currents, by which he accounted for all the then known phenomena of magnetism and electricity, and which he published in a semi-popular form in the *Philosophical Magazine* in 1861 and 1862.

During most of the King's College time Maxwell resided at 8 Palace Gardens Terrace, Kensington, where he carried on many of his experiments in a large garret which ran the whole length of the house. When experimenting at the window with the colour-box (which was painted black, and nearly eight feet long), he excited the wonder of his neighbours, who thought him mad to spend so many hours in staring into a coffin. This was also the scene of his well-known experiments on the viscosity of gases at different pressures and temperatures. For some days a large fire was kept up in the room, though it was in the midst of very hot weather. Kettles were kept on the fire, and large quantities of steam allowed to flow into the room. Mrs. Maxwell acted as stoker, which was very exhausting work when maintained for several consecutive hours. After this the room was kept cool, for subsequent experiments, by the employment of a considerable amount of ice.

During Maxwell's residence in London his brother-in-law, the Rev. Donald Dewar, came and stayed in his house in order to undergo a painful operation at the hands of Sir James Ferguson. Maxwell gave up the ground floor of his house to Mr. Dewar and his nurse. He himself, meanwhile, used to take his meals in a very small back room, where frequently he breakfasted (on porridge) on his knees, because there was no room for another chair at the table. Maxwell acted frequently in the capacity of nurse to Mr. Dewar, who would always look out anxiously for his return from college, and whose face would light up with a smile of pleasure and relief when he saw him coming, because he said he knew he should be comfortable when Maxwell returned.

One pleasant incident of his stay in London was the improvement of his acquaintance with Faraday, with whom he seems to have dined on the occasion of his lecture before the Royal Institution in 1861.

On one occasion he was wedged in a crowd attempting to escape from the lecture theatre of the Royal Institution, when he was perceived by Faraday, who, alluding to Maxwell's work among the molecules, accosted him in this wise — " Ho, Maxwell, cannot you get out ? If any man can find his way through a crowd it should be you."

He also renewed his personal intercourse with Litchfield, Droop, and other Cambridge friends.

His habit at this time was to do his scientific work chiefly in the mornings, unless when entertaining friends, when he would give up his days to them and take hours for work out of the night. In the afternoons he would ride with Mrs. Maxwell. She had been recommended horse exercise in 1860, when the pony " Charlie," called after Charles Hope Cay, was bought at the Rood fair. He was a high-bred, spirited, light bay Galloway, with arched neck and flowing tail. Maxwell himself broke him in, riding side-saddle, with a piece of carpet to take the place of a habit. This pony was a great favourite until the end in 1879.

About this time (between 1860 and 1865) the endowment of Corsock Church was completed, and the Manse built. Maxwell gave largely to both objects, which were promoted mainly by his zeal and energy.

At the beginning and at the close of the King's College period Maxwell suffered from two severe

illnesses, both of a dangerously infectious nature, and
in both of them he was nursed by Mrs. Maxwell. In
September 1860 he had an attack of smallpox at
Glenlair, which he was supposed to have caught at the
fair, where " Charlie " was bought. During this illness
his wife was left quite alone with him—the servants
only coming to the door of the sick-room. He has
been heard to say that by her assiduous nursing on
this occasion she saved his life.

The second illness was in September 1865, also at
Glenlair. Maxwell had been riding a strange horse,
and got a scratch on the head from a bough of a tree ;
this was followed by an attack of erysipelas, which
brought him very low. Mrs. Maxwell was again his
nurse, and to listen, as he insisted on doing, to her
quiet reading of their usual portion of Scripture every
evening, was the utmost mental effort which he could
bear.

1866-1870.   The years which followed the resignation of his
Æt. 35-39.
post at King's College were spent, for the most part,
at Glenlair, the house being at this time enlarged
in general accordance with his father's plan. And
Maxwell took advantage of this retirement to embody some of the results of his investigations in
substantive books. The great work on Electricity
and Magnetism, although not published till 1873, was
now taking definite shape, and the treatise on Heat,
which appeared in 1870, had been undertaken as a
by-work during the same period.

His scientific and other correspondence also took

up a good deal of energy. Some measure of it is afforded by the fact that a "pillar"-box was let into the rough stone wall on the roadside, across the Urr, for the sole use of Glenlair House. Maxwell would himself carry the letters to and from this rustic post-office in all weathers, at the same time giving the dogs a run.

Both now and afterwards, his favourite exercise —as that in which his wife could most readily share —was riding, in which he showed great skill. Mr. Fergusson remembers him in 1874, on his new black horse, "Dizzy," which had been the despair of previous owners, "riding the ring," for the amusement of the children at Kilquhanity, throwing up his whip and catching it, leaping over bars, etc.

A considerable portion of the evening would often be devoted to Chaucer, Spenser, Milton, or a play of Shakespeare, which he would read aloud to Mrs. Maxwell.

On Sundays, after returning from the kirk, he would bury himself in the works of the old divines. For in theology, as in literature, while reckoning frankly with all phases, his sympathies went largely with the past. Not that he would have checked the real progress of thought on the subject of religion, but he did not share the sanguine hopes of some who have sought to hasten these "slow-paced" changes ; nor did he believe in progress by ignoring differences, or by merging the sharp outlines of traditional sys-tems in the haze of a "common Christianity." He was one of those in whom physical studies seem to

Y

have the effect of leading the mind to dwell on the permanent aspects of thought as well as of things, thus reinforcing the instincts of conservatism. No mind ever delighted more in speculation, and yet none was ever more jealous of the practical application or the popular dissemination of what appeared to him as crude and half-baked theories about the highest subjects. He preferred resting on the great thoughts of other ages, though no man knew better wherein they (and scientific theories likewise) fell short of certainty; and while he was anything rather than a formalist or a dogmatist, and still clung to the belief that love remains while knowledge vanishes away, he was the enemy of indefiniteness and indifferentism, as well as of a style of preaching which, as he used to say, "dings ye wi' mere morality." His theological attitude, which it would be rash to develop further here, is indicated to some extent in his letter to Bishop Ellicott, and in his reply to the Secretary of the Victoria Institute, both of which will be found in Chapter XII (pp. 393, 404).

But he was far, indeed, from judging men by their opinions. "I have no nose for heresy," he used to say. His sympathy pierced beneath the outer shell of circumstance and association, and he hardly ever failed to discover what was best and strongest in those with whom he had to do.

His kindly relations with his neighbours and with their children may be passed without further notice after what has been said above. But it may be mentioned that he used occasionally to visit any sick person in the village, and read and pray with them in cases where such ministrations were welcomed.

One who visited at Glenlair between 1865 and 1869 was particularly struck with the manner in which the daily prayers were conducted by the master of the household. The prayer, which seemed *extempore*, was most impressive and full of meaning.[1]

It is right also to record briefly his continued intercourse with his cousins of the Cay family. Mr. William Dyce Cay, who had now entered on his profession as a civil engineer, was employed by him to build the bridge over the Urr, and has a vivid recollection of their intercourse, both then (1861-2) and in

---

[1] The following fragments have been found amongst his papers :—

" Almighty God, who hast created man in Thine own image, and made him a living soul that he might seek after Thee and have dominion over Thy creatures, teach us to study the works of Thy hands that we may subdue the earth to our use, and strengthen our reason for Thy service ; and so to receive Thy blessed Word, that we may believe on Him whom Thou hast sent to give us the knowledge of salvation and the remission of our sins. All which we ask in the name of the same Jesus Christ our Lord."

" O Lord, our Lord, how excellent is Thy name in all the earth, who hast set Thy glory above the heavens, and out of the mouths of babes and sucklings hast perfected praise. When we consider Thy heavens, the work of Thy fingers, the moon and the stars which Thou hast ordained, teach us to know that Thou art mindful of us, and visitest us, making us rulers over the works of Thy hands, showing us the wisdom of Thy laws, and crowning us with honour and glory in our earthly life ; and looking higher than the heavens, may we see Jesus, made a little lower than the angels for the suffering of death, crowned with glory and honour, that He, by the grace of God, should taste death for every man. O Lord, fulfil Thy promise, and put all things in subjection under His feet. Let sin be rooted out of the earth, and let the wicked be no more. Bless Thou the Lord, O my soul, praise the Lord."

former years. In particular he remembers how, on
one occasion, Maxwell spent the whole time during a
walk of several miles over the hill from Glenlair to
Parton, "giving one example after another to explain
by illustration the principle of virtual velocities." . . .
"The feeling I had," says Mr. Cay, "was that before
I got to the bottom of one example he had rushed off
to another."

And the reader will find in the correspondence
two of Maxwell's letters to my friend Charles Hope
Cay, and in another letter a few words of bright description of him. He died in 1869, at the early age of
twenty-eight, the most devoted of teachers, one of the
purest-hearted and most amiable of men. If he *could*
have listened to his cousin's gentle warnings against
excessive zeal, perhaps his services to Clifton College,
if less vividly remembered, might have been continued
longer. But who knows? "They whom the gods
love die young."

Maxwell's retirement was not by any means unbroken. There was a visit to London in the spring of
every year. And in the spring and early summer of
Æt. 35. 1867 he made a tour in Italy with Mrs. Maxwell.
They had the misfortune to be stopped for quarantine
at Marseilles, and his remarkable power of physical
endurance and of ministration were felt by all who
shared in the mishap. True to the associations of his
early days (see above, pp. 28, 121), he became the
general water-carrier, and in other ways contributed
greatly to the alleviation of discomforts that were by
no means light.

We met accidentally at Florence, and I remember his mentioning two things as having particularly struck him amongst the innumerable objects of interest at Rome. He had looked at the dome of St. Peter's with an eye of sympathetic genius,[1] and his ear for melody had been satisfied by "the Pope's band." He acquired Italian with great rapidity, and amused himself with noticing the different phonetic values of the letters in Italian and English.[2] One of his chief objects in learning the language was to be able to converse with Professor Matteucci, whose bust now stands in the Campo Santo at Pisa. During the same tour he took special pains to improve his acquaintance with French and German. The only language he had any difficulty in mastering was Dutch.

In the years 1866, 1867, 1869, and 1870, he was either Moderator or Examiner in the Mathematical Tripos at Cambridge, where his influence was more and more felt. His work on these occasions was, indeed, a principal factor in the movement, to be hereafter described, which led ultimately to important changes in the Examination system; to the creation of the Cavendish Laboratory; and to the foundation of the Chair of Experimental Physics.

His paper on the Viscosity of Gases, printed in the *Phil. Trans.* for 1866, had been delivered by him as the Bakerian Lecture for that year.

---

[1] The tone in which he spoke of this brought home to me, more than anything I have seen in books, the joy of Michael Angelo in etherealising the work of Brunelleschi.

[2] On learning from our teacher, Sign. Briganti, the pronunciation of *suolo*, he said, "That is the English for *rondinella*."

He also attended several meetings of the British Association, and, in 1870, at the Liverpool meeting, was President of Section A (Mathematics and Physics). His Presidential Address was on the relation of Mathematics and Physics to each other—a theme suggested by Professor Sylvester, who had been president of the same section in the previous year. The opening passage, in which he alludes to other recent scientific addresses, is characteristic, and may be quoted here :—

Æt. 39.

I have endeavoured to follow Mr. Spottiswoode, as with far-reaching vision he distinguishes the systems of science into which phenomena, our knowledge of which is still in the nebulous stage, are growing. I have been carried, by the penetrating insight and forcible expression of Dr. Tyndall, into that sanctuary of minuteness and of power, where molecules obey the laws of their existence, clash together in fierce collision, or grapple in yet more fierce embrace, building up in secret the forms of visible things. I have been guided by Professor Sylvester towards those serene heights

> " Where never creeps a cloud or moves a wind,
> Nor ever falls the least white star of snow,
> Nor ever lowest roll of thunder moans,
> Nor sound of human sorrow mounts, to mar
> Their sacred everlasting calm."

But who will lead me into that still more hidden and dimmer region where Thought weds Fact,—where the mental operation of the mathematician and the physical action of the molecules are seen in their true relation ? Does not the way to it pass through the very den of the metaphysician, strewed with the remains of former explorers and abhorred by every man of science ? . . .

Two important papers read by Maxwell at the

same meeting were that "On Hills and Dales," to which reference will be found in the correspondence (pp. 382, 383), and that "On Colour Vision at different Points of the Retina."

The Cambridge examinations were the only cause which separated him for more than a day or two from Mrs. Maxwell. When most pressed with the load of papers to be read, he would write to her daily— sometimes twice a day—in letters full of "*enfantill-ages*," as in his boyish endeavours to amuse his father, telling her of everything, however minute, which, if she had seen it, would have detained her eye, small social phenomena, grotesque or graceful (including the dress of lady friends), together with the lighter aspects of the examinations ; College customs, such as the "grace-cup ;" his dealings with his co-examiners, and marks of honour to himself which he knew would please her, though they were indifferent to him. And sometimes he falls into the deeper vein, which was never long absent from his communion with her, commenting on the portion of Scripture which he knew that she was reading, and passing on to general meditations on life and duty.

In November 1868 his old teacher, James D. Forbes, had resigned the principalship of the United College in the University of St. Andrew's, and an effort was made by several of the professors[1] to induce Maxwell to stand for the vacant post, which was in the gift of the Crown, and had been held by Brewster and

---

[1] It is right that I should add that the suggestion did not proceed from me.—*L. C.*

Forbes successively. He was touched by the kindness, and travelled a whole day from Galloway to confer with us, but, on mature consideration, relinquished the idea.

LETTERS, 1860 TO 1870.

To REV. LEWIS CAMPBELL.

Marischal College,
Aberdeen, 5th January 1860.

. . . I have been publishing my views about Elastic Spheres in the *Phil. Mag.* for Jany., and am going to go on with it as I get the prop$^{ns.}$ written out. I have also sent my experiments on Colours to the Royal Society of London, so I have two sets of irons in the fire, besides class work. I hope you get on with Plato, and that your pupils are all Theætetuses, and that wisdom soaks like oil into their inwards. There is a man here who is striving after a general theory of things, but he has great difficulty in so churning his thoughts as to coagulate and solidify the vague and nebulous notions which wander in his head. He has been applying to me very steadily whenever he can pounce on me, and I have prescribed for him as I best could, and I hope his abstract of his general theory of things will be palatable to the readers of the *British Ass. Reports* for 1859.

To HIS WIFE.

Edinburgh, 13th April 1860.

Now let us read (2 Cor.) chapter xii., about the organisation of the Church, and the different gifts of different Christians, and the reason of these differences that Christ's body may be more complete in all its parts. If we felt more distinctly our union to Christ, we would know our position as members of His body, and work more willingly and intelligently along with all the rest in promoting the health and growth of the body, by the use of every power which the spirit has distributed to us.

14th April 1860.

Let us read about charity,—that love which is so perfect that it remains when that which is in part shall be done away. May God purify our love, and make it fit for eternity, by grafting upon it the love of Himself, that so both the human root and the engrafted branches and the divine fruit may be holy to Him!

FROM C. J. MONRO, Esq.

Hadley, Barnet, N.
23d October 1861.

Thank you much for the papers. That about vortices I had skimmed already in the magazine. I shall now be able to do more than skim it. The coincidence between the observed velocity of light and your calculated velocity of a tranverse vibration in your medium seems a brilliant result. But I must say I think a few such results are wanted before you can get people to think that, every time an electric current is produced, a little file of particles is squeezed along between rows of wheels. But the instances of bodily transfer of matter in the phenomena of galvanism look like it already, and I admit that the possibility of convincing the public is not the question.

To H. R. DROOP, Esq. (of the Equity Bar).

Glenlair, Dalbeattie, N.B., 28th December 1861.

I enclose a short statement of the scheme of endowing the chapel which was built near us in 1838 for this district, which is very far from any parish church. If we can raise £1000, there is a fund already raised which will contribute £2000, so as to give a salary of £120 to the minister permanently, and as the people are too poor to support the minister themselves, we hope to make the chapel independent of chance contributions in this way. Great part of the funds for building the church were subscribed in London by all kinds of people who were friends of an English gentleman who then had property here; but we have no longer any such means of drawing on the metropolis.

If you can put us in the way of diminishing the deficit we shall be grateful, and I will see that the money goes to the fund, and that the names are duly entered, however small the contributions.

. . . I have nothing to do in King's College till Jany. 20, so we came here to rusticate. We have clear hard frost without snow, and all the people are having curling matches on the ice, so that all day you hear the curling-stones on the lochs in every direction for miles, for the large expanse of ice vibrating in a regular manner makes a noise which, though not particularly loud on the spot, is very little diminished by distance. I am trying to form an exact mathematical expression for all that is known about electro-magnetism without the aid of hypothesis, and also what variations of Ampère's formula are possible, without contradicting his expressions. All that we know is about the action of *closed* currents—that is, currents through closed curves. Now, if you make a hypothesis (1) about the mutual action of the elements of two currents, and find it agree with experiment on closed circuits, it is not proved, for—

If you make another hypothesis (2) which would give *no action* between an element and a *closed* circuit, you may make a combination of (1) and (2) which will give the same result as (1). So I am investigating the most general hypothesis about the mutual action of elements, which fulfils the condition that the action between an element and a closed circuit is null. This is the case if the action between two elements can be reduced to forces between the extremities of those elements depending only on the distance and + or − according as they act between similar or opposite ends of the elements. If the force is an attraction

$$= \phi\,(r)\; ss'\, (\cos \omega + 2 \cos \theta \cos \theta')$$

where ω is the angle between s and s', r the distance of s and s' and θ and θ' the angles s and s', the elements, make with r, then the condition of no action will be fulfilled.

To the Same.

8 Palace Gardens Terrace, W.,
24th January 1862.

. . . When I wrote to you about closed currents, it was partly to arrange my own thoughts by imagining myself speaking to you. Ampère's formula containing n and k is the most general expression for an attractive or repulsive force in the line joining the elements; and I now find that if you take the most general expression consistent with symmetry for an action transverse to that line, the resulting expression for the action of a closed current on an element gives a force not perpendicular to that element. Now, experiment 3d (Ampère) shows that the force on a movable element is perp. to the directions of the current, so that I see Ampère is right.

But the best way of stating the effects is with reference to "lines of magnetic force." Calculate the magnetic force in any plane, arising from every element of the circuit, and from every other magnetising agent, then the force on an element is in the line perp. to the plane of the element and of the lines of force.

But I shall look up Cellerier and Plann, and the long article in Karsten's *Cyclopædia*. I want to see if there is any evidence from the mathematical expressions as to whether element acts on element, or whether a current first produces a certain effect in the surrounding field, which afterwards acts on any other current.

Perhaps there may be no mathematical reasons in favour of one hypothesis rather than the another.

As a fact, the effect on a current at a given place depends solely on the direction and magnitude of the magnetic force at that point, whether the magnetic force arises from currents or from magnets. So that the theory of the effect taking place through the intervention of a medium is consistent with fact, and (to me) appears the simplest in expression; but I must prove either that the direct action theory is completely identical in its results, or that in some conceivable case they may be different. My theory of the rotation

of the plane of polarised light by magnetism is coming out
in the *Phil. Mag.* I shall send you a copy.

To the Same.

8 Palace Gardens Terrace,
Kensington, London, W., 28th January 1862.

Some time ago, when investigating Bernoulli's theory
of gases, I was surprised to find that the internal friction
of a gas (if it depends on the collision of particles) should
be independent of the density.

Stokes has been examining Graham's experiments on
the rate of flow of gases through fine tubes, and he finds
that the friction, if independent of density, accounts for
Graham's results, but, if taken proportional to density, differs
from those results very much. This seems rather a curious
result, and an additional phenomenon, explained by the
"collision of particles" theory of gases. Still one
phenomenon goes against that theory—the relation between
specific heat at constant pressure and at constant volume,
which is in air = 1·408, while it ought to be 1·333.

My brother-in-law, who is still with us, is getting better,
and had his first walk on crutches to-day across the room.

To C. J. Monro, Esq.

8 Palace Gardens Terrace,
London W., 18th February 1862.
(Recd. 3d March.)

I got your letter in Scotland, whither we had gone
for the Christmas holidays. I have been brewing Platonic
suds, but failed, owing I suppose to a too low temperature.
I had not read Plateau's recipe then. Some of the
bubbles on the surface lasted a fortnight in the air, but
they were scummy and scaly and inelastic. I shall take
more care next time. Elliot of the Strand (30) is going to
produce colour-tops, with papers from De La Rue, and directions for use by me; and so I shall be put in competition
with the brass Blondin and the Top on the top of the Top.

. . . With regard to Britomart's nurse—I have not Spenser here, but I think Spenser was not a magician himself, and got all his black art out of romances and not out of the professional treatises,—the notions to be brought out were:—1st, The unweaving any web in which B. had been caught; 2d, Doing so in witch-like fashion; 3d, Not like a wicked witch, but like a well-intentioned nurse, unused to the art, and therefore blunderingly. She believes in the number three and in contrariety, and therefore says everything thrice and does everything thrice, saying inversions of sentences, and doing reversions of her revolutions, which are described in similar language. The revolutions begin by $+ 3 (2\pi)$ against the visible motion of the sun, then by a revolution $- 6 \pi$ she returns all contrary and unweaves the first. Then she goes round $+ 6\pi$, to make the final result contrary to the natural revolution, and to make a complete triad. Withershins is, I believe, equivalent to wider die Sonne in High Dutch, which I am not aware is a modern or ancient idiom in that language, but it may be one in a cognate language. If the "phamplets" have not turned up in Madeira yet, let me know, that I may "replace" them.

I suppose in your equations, when the numbers do not amount to unity, Black has been present.

$$\cdot 841 \text{ Brunsw.·G} + \cdot 159 \text{ W} = \cdot 200 \text{ V} + \cdot 423 \text{ U} + \cdot 377 \text{ Black.}$$

That is, green a little palish and dark mauve, your last equation by the young eyes. It is something like a colour-blind eq^n, but all those I know say 100 Brunswick G $= 100$ Vermillion, so that this person sees the green darker than the Vermillion, or in other words sees much more of the second side of the equation than a colour-blind person would. But in twilight U comes out strong, while G does not; so that I think the apparent equality arises from suppression of all colours but blue (in U and W) in the twilight, so that you may write—

$$\cdot 841 \text{ Black} + \cdot 159 \text{ W} = \cdot 577 \text{ Black} + \cdot 423 \text{ U.}$$

There is no use going to the 3rd place of decimals, unless you spend a good while on each observation, and have first-

rate eyes.   But if you can get observations to be consistent to the 3$^{rd.}$ place of decimals, glory therein, and let me
know what the human eye can do.

Donkin gave me tea in Oxford, July 1, 1860.

I find that my belief in the reality of State affairs is no
greater in London than in Aberdeen, though I can see the clock
at Westminster on a clear day.   If I went and saw the parks
of artillery at Woolwich, and the Consols going up and down
in the city, and the Tuscarora and Mr. Mason, I would
know what like they were, but otherwise a printed statement
is more easily appropriated than experience is acquired by
being near where things are being transacted.

I am getting a large box made for mixture of colours.
A beam of sunlight is to be divided into colours by a prism,
certain colours selected by a screen with slits.   These
gathered by a lens, and restored to the form of a beam by
another prism, and then viewed by the eye directly.   I
expect great difficulties in getting everything right adjusted,
but when that is done I shall be able to vary the intensity
of the colours to a great extent, and to have them far purer
than by any arrangement in which white light is allowed to
fall on the final prism.

I am also planning an instrument for measuring
electrical effects through different media, and comparing
those media with air.   A and B are two equal metal discs,
capable of motion towards each other by fine screws ; D is
a metal disc suspended between them by a spring, C; E is
a piece of glass, sulphur, vulcanite, gutta-percha, etc.   A
and B are then connected with a source of + electricity,
and D with − electricity.   If everything was symmetrical, D
would be attracted both ways, and would be in unstable
equilibrium, but this is rendered stable by the elasticity of
the spring C.   To find the effect of the plate E, you work
A further or nearer till there is no motion of D consequent
on electrification.   Then the plate of air between A and D
is electrically equivalent to the two plates of air and one of
glass (say) between D and B, whence we deduce the coeff$^{rt.}$
for E.

To REV. LEWIS CAMPBELL.

8 Palace Gardens Terrace,
Kensington, W., 21st April 1862.

It is now a long time since I wrote half a letter to you, but I have never since had time to write or to find the scrap. I suppose, as it was more than a good intention, but less than a perfect act, it may be regarded as destined to paper purgatory. This is the season of work to you, when folks visit shrines in April and May, but I get holiday this week. I have been putting together a large optical box, 10 feet long, containing two prisms of bisulphuret of carbon, the largest yet made in London, five lenses and two mirrors, and a set of movable slits. Everything requires to be adjusted over and over again if one thing is not quite right placed, so I have plenty of trial work to do before it is perfect, but the colours are most splendid.

I think you asked me once about Helmholtz and his philosophy. He is not a philosopher in the exclusive sense, as Kant, Hegel, Mansel are philosophers, but one who prosecutes physics and physiology, and acquires therein not only skill in discovering any desideratum, but wisdom to know what are the desiderata, e.g., he was one of the first, and is one of the most active, preachers of the doctrine that since all kinds of energy are convertible, the first aim of science at this time should be to ascertain in what way particular forms of energy can be converted into each other, and what are the equivalent quantities of the two forms of energy.

The notion is as old as Descartes (if not Solomon), and one statement of it was familiar to Leibnitz. It was wholly unknown to Comte, but all sorts of people have worked at it of late,—Joule and Thomson for heat and electricals, Andrews for chemical combinations, Dr. E. Smith for human food and labour. We can now assert that the power of our bodies is generated in the muscles, and is not conveyed to them by the nerves, but produced during the transformation of substances in the muscle, which are supplied fresh by the blood.

We can also form a rough estimate of the efficiency of

a man as a mere machine, and find that neither a perfect heat engine nor an electric engine could produce so much work and waste so little in heat. We therefore save our pains in investigating any theories of animal power based on heat and electricity. We see also that the soul is not the direct moving force of the body. If it were, it would only last till it had done a certain amount of work, like the spring of a watch, which works till it is run down. The soul is not the mere mover. Food is the mover, and perishes in the using, which the soul does not. There is action and reaction between body and soul, but it is not of a kind in which energy passes from the one to the other,— as when a man pulls a trigger it is the gunpowder that projects the bullet, or when a pointsman shunts a train it is the rails that bear the thrust. But the constitution of our nature is not explained by finding out what it is not. It is well that it will go, and that we remain in possession, though we do not understand it.

Hr. Clausius of Zurich, one of the heat philosophers, has been working at the theory of gases being little bodies flying about, and has found some cases in which he and I don't tally, so I am working it out again. Several experimental results have turned up lately, rather confirmatory than otherwise of that theory.

I hope you enjoy the absence of pupils. I find that the division of them into smaller classes is a great help to me and to them; but the total oblivion of them for definite intervals is a necessary condition of doing them justice at the proper time.

To FLEEMING JENKIN, Esq.[1]

27th Aug. 1863.

. . . To compare electromagnetic with electrostatic units :—

1*st*, Weber's method.—Find the capacity of a condenser in electrostatic measure (meters).

---

[1] Now Professor of Engineering in Edinburgh.

Determine its potential when charged, and measure the charge of discharge through a galvanometer.

2*d*, Thomson's.— Find the electromotive force of a battery by electromagnetic methods, and then weigh the attraction of two surfaces connected with the two poles.

3*d*, (Not tried, but talked of by Jenkin).—Find the resistance of a very bad conductor in both systems—

(1) By comparison with (4th June),

(2) By the log. decrement of charge per second.

All the methods require a properly graduated series of steps. The 1st and 2d determine V, a velocity $=$ 310,740,000 meters per second.

The 3d method determines V^2.

The first method requires a condenser of large capacity, and the measurement of this capacity and that of the discharge by a galvanometer.

I think this method looks the best; but I would use a much larger condenser than Weber, and determine its capacity by more steps.

The chief difficulty of Thomson's method is the measurement of a very small force and a very small distance. I think these difficulties may be overcome by making the force act on a comparatively stiff spring and magnifying optically the deflection.

On the third method we require a very large condenser indeed, also a series of resistances in steps between 4th June and that of the insulating substance of the condenser, and a galvanometer (or electrometer) to measure discharge (or tension). . . .

To C. H. CAY, Esq.

8 Palace Gardens Terrace,
18th November 1863.

We hope to hear how you are. A little literature helps to chase away mathematics from the mind. I have read *Paracelsus* in parts, but concluded that there was a great deal of poetry in it; but Mr. Browning has written much better poems with half the quantity of poetry at his disposal.

Have you seen *Pessimus, a Prose Poem in Paradox,* from Oxford, and *Sketch from Cambridge* by a Don who imagines that mathematical men are safer not to talk shop than classical. I know several men who see all nature in symbols, and express themselves conformably whether in Quintics or Quantics, Invariants or Congruents. I send you the electric scheme.

To HIS WIFE.

22d June 1864.

May the Lord preserve you from all evil, and cause all the evil that assaults you to work out His own purposes, that the life of Jesus may be made manifest in you, and may you see the eternal weight of glory behind the momentary lightness of affliction, and so get your eyes off things seen and temporal, and be refreshed with the things eternal! Now love is an eternal thing, and love between father and son or husband and wife is not temporal if it be the right sort, for if the love of Christ and the Church be a reason for loving one another, and if the one be taken as an image of the other, then, if the mind of Christ be in us, it will produce this love as part of its complete nature, and it cannot be that the love which is first made holy, as being a reflection of part of the glory of Christ, can be any way lessened or taken away by a more complete transformation into the image of the Lord.

I have been back at 1 Cor. xiii. I think the description of charity or divine love is another loadstone for our life—to show us that this is one thing which is not in parts, but perfect in its own nature, and so it shall never be done away. It is nothing negative, but a well-defined, living, almost acting picture of goodness; that kind of it which is human, but also divine. Read along with it 1 John iv., from verse 7 to end; or, if you like, the whole epistle of John and Mark xii. 28.

To THE SAME.

23d June 1864.

Think what God has determined to do to all those who

submit themselves to His righteousness and are willing to
receive His gift. They are to be conformed to the image of
His Son, and when that is fulfilled, and God sees that they
are conformed to the image of Christ, there can be no more
condemnation, for this is the praise which God Himself gives,
whose judgment is just. So we ought always to hope in
Christ, for as sure as we receive Him now, so sure will we
be made conformable to His image. Let us begin by taking
no thought about worldly cares, and setting our minds on the
righteousness of God and His kingdom, and then we shall
have far clearer views about the worldly cares themselves,
and we shall be continually enabled to fight them under
Him who has overcome the world.

To the Same.

26th June 1864.

Note in (2 Cor.) ver. 10 that the judgment is according
to what we have done, so that if we are to be counted
righteous, we must really get righteousness and do it. Note
also that we are to receive the things done in the body, not
rewards or punishments merely, but the things themselves
are to be brought back to us, and we must meet them in the
spirit of Christ, who bore our sins and abolished them, or
else we must be overwhelmed altogether.

. . . I have come from Mr. Baptist Noel. The church
was full to standing, and the whole service was as plain as
large print. The exposition was the Parable of Talents, and
the sermon was on John iii. 16. The sermon was the text writ
large, nothing ingenious or amusing, and hardly any attempt
at instruction, but plain and very serious exhortation from a
man who evidently believes neither more nor less than what
he says.

To the Same.

28th June 1864.

I can always have you with me in my mind—why
should we not have our Lord always before us in our minds,
for we have His life and character and mind far more clearly
described than we can know any one here ? If we had seen

Him in the flesh we should not have known Him any better, perhaps not so well. Pray to Him for a constant sight of Him, for He is man that we may be able to look to Him, and God, so that He can create us anew in His own image.

To C. Hockin, Esq.

Glenlair, Dalbeattie, September 7th 1864.

. . . I have been doing several electrical problems. I have got a theory of " electric absorption," *i.e.* residual charge, etc., and I very much want determinations of the specific induction, electric resistance, and absorption of good dielectrics, such as glass, shell-lac, gutta-percha, ebonite, sulphur, etc.

I have also cleared the electromagnetic theory of light from all unwarrantable assumption, so that we may safely determine the velocity of light by measuring the attraction between bodies kept at a given difference of potential, the value of which is known in electromagnetic measure.

I hope there will be resistance coils at the British Association.

To Professor Lewis Campbell.[1]

8 Palace Gardens Terrace,
London, W., 22d November 1864.

It was very kind of you to think of me at this time, and write to me. I shall always remember your mother's kindness to me, beginning more than twenty-three years ago, and how she made me the same as you two when I came to see you. To you her memory is what you can share with none, so I can say no more except that you will continue to find that to have had a mother so devoted to her duty gives you a consciousness of your own obligations which will be strengthened whenever you think of her.

---

[1] Mrs, Morrieson died on the 17th of November 1864.

To C. H. CAY, Esq.

Glenlair, 5th January 1865.

We are sorry to hear you cannot come and see us, but you seem better by your letter, and I hope you will be able for your travels, and be better able for your work afterwards, and not take it too severely, and avoid merimnosity and taking over too much thought, which greatly diminishes the efficiency of young teachers.   We have been here since 22d ult., and are in the process of dining the valley in appropriate batches. We have had very rough weather this week, which, combined with the dining, has prevented our usual airings.   The ordinary outing is to the Brig of Urr, Katherine on Charlie and I on Darling.   Charlie has got a fine band on his forehead, with his name in blue and white beads.

The Manse of Corsock is now finished; it is near the river, not far from the deep pool where we used to bathe.

I set Prof. W. Thomson a prop. which I had been working with for a long time.   He sent me 18 pages of letter of suggestions about it, none of which would work; but on Jan 3, in the railway from Largs, he got the way to it, which is all right; so we are jolly, having stormed the citadel, when we only hoped to sap it by approximations.

The prop. was to draw a set of lines like this

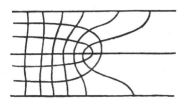

so that the ultimate reticulations shall all be squares.

The solution is exact, but rather stiff.   Now I have a

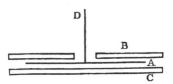

disc A hung by a wire D, between two discs B, C, the interval being occupied by air, hydrogen, carbonic acid, etc., the friction of which gradually brings A to rest. In order to calculate the thickdom or viscosity of the gas, I require to solve the problem above mentioned, which is now done, and I have the apparatus now ready to begin. We are also intent on electrical measurements, and are getting up apparatus, and have made sets of wires of alloy of platinum and silver, which are to be sent all abroad as standards of resistance. I have also a paper afloat, with an electro-magnetic theory of light, which, till I am convinced to the contrary, I hold to be great guns.

Spice[1] is becoming first-rate: she is the principal patient under the ophthalmoscope, and turns her eyes at command, so as to show the tapetum, the optic nerve, or any required part. Dr. Bowman, the great oculist, came to see the sight, and when we were out of town he came again and brought Donders of Utrecht with him to visit Spice.

To H. R. DROOP, Esq.

Glenlair, Dalbeattie, 19th July 1865.

There are so many different forms in which Societies may be cast, that I should like very much to hear something of what those who have been thinking about it propose as the plan of it.

There is the association for publishing each other's productions; for delivering lectures for the good of the public and the support of the Society; for keeping a reading room or club, frequented by men of a particular turn; for dining together once a month, etc.

I suppose W———'s object is to increase the happiness of men in London who cultivate physical sciences, by their meeting together to read papers and discuss them, the publication of these papers being only one, and not the chief end of the Society, which fulfils its main purpose in the act of meeting and enjoying itself.

---

[1] The Scotch terrier of the period.

The Royal Society of Edinburgh used to be a very sociable body, but it had several advantages. Most of the fellows lived within a mile of the Society's rooms. They did not need to disturb their dinner arrangements in order to attend.

Many of them were good speakers as well as sensible men, whose mode of considering a subject was worth hearing, even if not correct.

The subjects were not limited to mathematics and physics, but included geology, physiology, and occasionally antiquities and even literary subjects. Biography of deceased fellows is still a subject of papers. Now those who cultivate the mathematical and physical sciences are sometimes unable to discuss a paper, because they would require to keep it some days by them to form an opinion on it, and physical men can get up a much better discussion about armour plates or the theory of glaciers than about the conduction of heat or capillary attraction.

The only man I know who can make everything the subject of discussion is Dr. Tyndall. Secure his attendance and that of somebody to differ from him, and you are all right for a meeting.

If we can take the field with a plan in our head, I dare say we could find a good many men who would co-operate.

We ride every day, sometimes both morning and evening, and so we consume the roads. I have made 68 problems, all stiff ones, not counting riders.

I am now getting the general equations for the motion of a gas considered as an assemblage of molecules flying about with great velocity. I find they must repel as inverse fifth power of distance.

To C. H. CAY, Esq.

Glenlair, 14th October 1865.

... I hope you keep your conscience in good order, and do not bestow more labour on erroneous papers than is useful to the youth who wrote it. Always set him to look for the mistake, if he prefers that to starting fresh, for to find your

own mistake may sometimes be profitable, but to seek for
another man's mistake is weariness to the flesh.

There are three ways of learning props.—the heart, the
head, and the fingers ; of these the fingers is the thing for
examinations, but it requires constant practice.   Nevertheless the fingers have a fully better retention of methods than
the heart has.   The head method requires about a mustard
seed of thought, which, of course, is expensive, but then it
takes away all anxiety.   The heart method is full of anxiety,
but dispenses with the thought, and the finger method
requires great labour and constant practice, but dispenses
with thought and anxiety together.

We have had very fine weather since you went away,
but I was laid up for more than three weeks with erysipelas
all over my head, and got very shaky on my pins.   But I
have been out for a fortnight, and riding regularly as of old,
which is good for Katherine after the nursing, and I eat
about double what any man in Galloway does, and know
nothing of it in half an hour ; but my legs are absorbing the
beef as fast as it is administered.

To the Rev. C. B. Tayler.

8 Palace Gardens Terrace, W.,
2d February 1866.

I was very glad to get your kind letter, and to be assured
that you still remembered me.   I thought of you when I
was in Cambridge, and made up my mind to write to you
and hear of you and Mrs. Tayler, and your nephew George.
A nephew of yours was for a short time in my class in King's
Coll., and I asked him about you, but he had not seen you
lately.   Is George still in Hull ?

You ask for my history since I wrote to you before my
marriage.   We remained in Aberdeen till 1860, when the
union or fusion of the Colleges took place, and I went to
King's Coll., London, where I taught till last Easter, when I
was succeeded by W. G. Adams, brother of the astronomer.
I have now my time fully occupied with experiments and
speculations of a physical kind, which I could not undertake

as long as I had public duties. These are the chronological data.   It is 13 years nearly since I was with you, and you carried me about when I could not move myself, but I remember everything about you and Otley much better than most things before and after that time.   I got advantage from your nursing when my father was ill, and many other things have since brought you and Mrs. Tayler to mind.   If you and Mrs. Tayler are to be in London during the spring, we shall be exceedingly glad to see you here, or if you ever go to Scotland in summer or autumn, we hope you will try and stay with us some time.   My wife knows you quite well,—that is, as well as I do,—all but what can only be got by seeing and hearing directly, and it would do us both great good to see you, and open up our minds a little.

Many people's minds seem to be shut up with solemn charms, so that though they seem Christians, and know what they mean to speak about, they can say nothing.   At Cambridge I heard several sermons from excellent texts, but all either on other subjects or else right against the text.   There is a Mr. Offord in this street, a Baptist [1] who knows his Bible, and preaches as near it as he can, and does what he can to let the statements in the Bible be understood by his hearers. We generally go to him when in London, though we believe ourselves baptized already.

Pray let me hear from you occasionally.   We shall be here till the end of March, and after that address Glenlair, Dalbeattie, N.B., which is my permanent address, and is sure at all times to find me.

Mrs. Maxwell joins me in kind regards to you and Mrs. Tayler, and I remain your afft. friend,

J. CLERK MAXWELL.

TO PROFESSOR LEWIS CAMPBELL.

Glenlair, Dalbeattie, 3d November 1868.

I have given considerable thought to the subject of the

---

[1] Whilst in London, Mr. and Mrs. Maxwell occasionally attended Nonconformist services, partly led, perhaps, by recollections of the simple Presbyterian worship, to which Mrs. Maxwell had been accustomed.

candidature, and have come to the decision not to stand. The warm interest which you and other professors have taken in the matter has gratified me very much, and the idea of following Principal Forbes had also a great effect on my feelings, as well as the prospect of residing among friends; but I still feel that my proper path does not lie in that direction.—Your afft. friend,

J. CLERK MAXWELL.

To C. J. MONRO, Esq.

Glenlair, Dalbeattie, 6th July 1870.

My question to the Mathematical Society bore fruit in various forms. . . . It would give my mind too great a wrench just now to go into elliptic integrals, but I will do so when I come to revise about circular conductors. . . . I can cut the subject short with an easy conscience, for I have no scruple about steering clear of tables of double entry, especially when, in all really useful cases, convergent series may be used with less trouble, and without any knowledge of elliptic integrals.   On this subject see a short paper on Fluid Displacement in next part of the *Math. Soc. Trans.*, where I give a picture of the stream lines, and the distortion of a transverse line as water flows past a cylinder.

Mr. W. Benson, architect, 147 Albany Street, Regent Park, N.W., told me that you had been writing to *Nature*, and that yours was the only rational statement in a multitudinous correspondence on colours.   Mr. Benson considers that Aristotle and I have correct views about primary colours.   He has written a book, with coloured pictures, on the science of colour, and he shows how to mix colours by means of a prism.   He wants to publish an elementary book with easy experiments, but gets small encouragement, being supposed an heretic.   No other architect in the Architect's Society believes him.   This is interesting to me, as showing the chromatic condition of architects.   I made a great colour-box in 1862, and worked it in London in '62 and '64.   I have about 200 equations each year, which are reduced but not published. I have set it up here this year, and have just got it in working order. I expect to get some more material, and work up the whole together. In particular, I want to find any change or evidence of constancy in the eyes of myself and wife during eight years. I can exhibit the yellow spot to all who have it,— and all have it except Col. Strange, F.R.S., my late father-in-law, and my wife,—whether they be Negroes, Jews, Parsees, Armenians, Russians, Italians, Germans, Frenchmen, Poles, etc. Professor Pole, for instance, has it as strong as me, though he is colour-blind; Mathison, also colour-blind, being fair, had it less strongly marked.

One J. J. Müller in *Pogg. Ann.* for March and April 1870, examines compound colours, and finds the violet without any tendency to red, or the red to blue. He also selects a typical green out of the spectrum.

CHAPTER XII.

CAMBRIDGE—1871 TO 1879.

THE Chair of Experimental Physics in the University of Cambridge was founded by a Grace of the Senate on the 9th of February 1871.

In October 1870 the Duke of Devonshire, who was Chancellor of the University, had signified his desire to build and furnish a Physical Laboratory for Cambridge. In acting as a member of the Royal Commission on Scientific Education, he had perceived how useful such an institution might be made. It was in connection with the acceptance of this munificent offer that the new professorship was established by the Senate.

The question, who should be the first professor? was for some time attended with anxiety. It was understood that Sir William Thomson had declined to stand, and it was thought uncertain whether Clerk Maxwell could be persuaded to leave the retirement of his country-seat. After some hesitation, arising chiefly from genuine diffidence, he was induced to become a candidate, on the understanding that he might retire at the end of a year, if he wished to do so. His candidature was announced on the 24th of

February.[1]   There was no opposition, and he was appointed on the 8th of March.

The following letters indicate the part taken by various persons in bringing about this result :—

FROM THE HON. J. W. STRUTT (Lord Rayleigh).

Cambridge, 14th February 1871.

When I came here last Friday I found every one talking about the new professorship, and hoping that you would come.   Thomson, it seems, has definitely declined. . . . There is no one here in the least fit for the post.   What is wanted by most who know anything about it is not so much a lecturer as a mathematician who has actual experience in experimenting, and who might direct the energies of the younger Fellows and bachelors into a proper channel. There must be many who would be willing to work under a competent man, and who, while learning themselves, would materially assist him. . . .   I hope you may be induced to come ; if not, I don't know who it is to be.   Do not trouble to answer me about this, as I believe others have written to you about it.

FROM THE REV. E. W. BLORE, M.A. (now Vice-Master of Trinity).

14th February 1871.

Many residents of influence are desirous that you should occupy the post, hoping that in your hands this University would hold a leading place in this department.   It has, I believe, been ascertained that Sir W. Thomson would not accept the professorship.   I mention this lest you should wish to avoid the possibility of coming into the field against him.

---

[1] On 23d February, Professor Stokes (who had been urgent in pressing Maxwell to stand) wrote to him :—" I am glad you have decided to come forward."

Maxwell's usual modesty is apparent in the draft of his reply to this letter:—

Glenlair, Dalbeattie, 15th February 1871.

My DEAR BLORE—Though I feel much interest in the proposed Chair of Experimental Physics, I had no intention of applying for it when I got your letter, and I have none now, unless I come to see that I can do some good by it.

. . . I am sorry Sir W. Thomson has declined to stand. He has had practical experience in teaching experimental work, and his experimental corps have turned out very good work. I have no experience of this kind, and I have seen very little of the somewhat similar arrangements of a class of real practical chemistry. The class of Physical Investigations, which might be undertaken with the help of men of Cambridge education, and which would be creditable to the University, demand, in general, a considerable amount of dull labour which may or may not be attractive to the pupils.

In the Grace of Senate of 9th February, it had been enacted that it should be "the principal duty of the professor to teach and illustrate the laws of Heat, Electricity, and Magnetism; to apply himself to the advancement of the knowledge of such subjects; and to promote their study in the University."

For some time after his appointment, Maxwell's principal work was that of designing and superintending the erection of the Cavendish Laboratory.

He inspected the Physical Laboratories of Sir William Thomson at Glasgow and of Professor Clifton at Oxford, in order to embody in the new structure the best features of both of these institutions. But many of the most important arrangements were of his own invention.   An account of the Laboratory itself will be found in *Nature* (vol. x. p. 139) ; it is sufficient here to say that it would be difficult to imagine a building better adapted to its purpose, or one in the construction of which more provision should be made for possible requirements.   In no case was convenience sacrificed to architectural effect, but in both respects the building is a decided success.   The architect was Mr. W. M. Fawcett of Cambridge, who appears to have fully appreciated and thoroughly carried out all Professor Maxwell's suggestions.   The contract was given to Mr. Loveday of Kibworth, his tender being recommended by the report of the Syndicate appointed to superintend the building, dated 1st March 1872.

The work of arranging and furnishing the Cavendish Laboratory occupied a considerable time.   It was not completed until the spring of 1874, when the practical work of experimenting commenced, and on the 16th of June in that year, the Chancellor formally presented his gift to the University.   Sir Charles Lyell and the French astronomer Leverrier were among those who visited the Laboratory and received the honorary degree of LL.D. from the University on that occasion.

The following draft of a letter from Maxwell to the Vice-Chancellor in the previous year affords an interesting illustration of the thorough and business-like manner in which he had addressed himself to these preliminary labours.

To the Vice-Chancellor, Cambridge.

(Draft of a Letter.)

Glenlair, 5th July 1873.

I enclose a provisional list of fixtures and apparatus required for the Laboratory.

At present I am not able to estimate the prices of many of the articles.

Some of them are in the market, and have simply to be ordered ; others require to be constructed specially for the Laboratory.

I have begun with a list arranged according to the places and rooms in the Laboratory, but, of course, all small things must be kept in cases, either in the apparatus room, or in the special rooms.

The special duty of the professor of experimental physics is to teach the sciences of heat and electricity, and also to encourage physical research. The Laboratory must therefore contain apparatus for the illustration of heat and electricity, and also for whatever physical research seems most important or most promising.

The special researches connected with heat which I think most deserving of our efforts at the present time are those relating to the elasticity of bodies, and in general those which throw light on their molecular constitution ; and the most important electrical research is the determination of the magnitude of certain electric quantities, and their relations to each other.

These are the principles on which I have been planning the arrangement of the Laboratory. But if in the course of years the course of scientific research should be deflected, the plans of work must vary too, and the rooms must be allotted differently.

I agree with you that the income of the Museums must be largely increased in order to meet the demands of this and other new buildings, and I am glad that the University is able to increase it.

It is impossible to procure many of the instruments, as

they are not kept in stock, and have to be made to order. Some of the most important will require a considerable amount of supervision during their construction, for their whole value depends on their fulfilling conditions which can as yet be determined only by trial, so that it may be some time before everything is in working order.

Even in 1874, however, there were still manifold desiderata, and the Duke expressed his wish to furnish the Laboratory completely with the necessary apparatus. To carry out this wish was again a work of time, for the Professor would never order an important instrument until he was satisfied that its design and construction were the best that could be obtained. In his annual report to the University in 1877 Professor Maxwell announced that the Chancellor had now "completed his gift to the University, by furnishing the Cavendish Laboratory with apparatus suited to the present state of science;" but at the same time he wrote to the Vice-Chancellor stating that he should reserve to himself the privilege of presenting to the Laboratory such apparatus as the advancement of science might render it desirable for the University to possess. And during the short remainder of his tenure of the professorship he expended many hundreds of pounds in this manner. And already, in the spring of 1874, he had presented to the Laboratory all the apparatus in his own possession. The apparatus provided by the British Association for their Committee on Electrical Standards (see p. 316), was also deposited in the Laboratory, in accordance with a resolution passed at the Edinburgh Meeting of 1871

2 A

—the apparatus remaining the property of the Association, and subject to the control of the Committee.

While the Laboratory was thus gradually made available, the other work of the professorship went on uninterruptedly from the first. Maxwell gave annual courses of lectures on the subjects prescribed in his commission,[1] commencing with October 1871, when he delivered his inaugural lecture. This and the lecture "On Colour Vision," given at the Royal Institution shortly after his appointment in the preceding spring, are perhaps the happiest of his literary efforts. Philosophic grasp, scientific clearness, and poetic imagination could hardly be more successfully combined.

The Cambridge lecture (October 1871) sets forth in luminous outline the meaning and tendency of the moment in the evolution of the University of Cambridge, which was marked by the institution of the

---

[1] Throughout the tenure of his Cambridge Chair Maxwell annually delivered a course of lectures on Heat and the Constitution of Bodies during the October Term ; on Electricity in the Lent Term ; and on Electro-Magnetism in the Easter Term. The character of these lectures very much resembled that of the early chapters in the *Elementary Treatise on Electricity*, which he wrote before taking the Cavendish papers in hand, and which was published in a fragmentary form by the Delegates of the Clarendon Press in October 1881. During the first four or five years that Maxwell lectured in Cambridge, candidates for the Ordinary or Poll Degree were compelled to attend professors' lectures, and not unfrequently they would appear at the Cavendish Laboratory. Maxwell's lectures were the delight of those who could follow him in his brilliant expositions and rapid changes of thought.

course of Experimental Physics, and the erection of the Devonshire Laboratory.

The following passage is especially characteristic :—

Science appears to us with a very different aspect after we have found out that it is not in lecture-rooms only, and by means of the electric light projected on a screen, that we may witness physical phenomena, but that we may find illustrations of the highest doctrines of science in games and gymnastics, in travelling by land and by water, in storms of the air and of the sea, and wherever there is matter in motion.

This habit of recognising principles amid the endless variety of their action can never degrade our sense of the sublimity of nature, or mar our enjoyment of its beauty. On the contrary, it tends to rescue our scientific ideas from that vague condition in which we too often leave them, buried among the other products of a lazy credulity, and to raise them into their proper position among the doctrines in which our faith is so assured that we are ready at all times to act on them. Experiments of illustration may be of very different kinds. Some may be adaptations of the commonest operations of ordinary life; others may be carefully arranged exhibitions of some phenomenon which occurs only under peculiar conditions. They all, however, agree in this, that their aim is to present some phenomenon to the senses of the student in such a way that he may associate with it some appropriate scientific idea. When he has grasped this idea, the experiment which illustrates it has served its purpose.

In an experiment of research, on the other hand, this is not the principal aim. . . . Experiments of this class—those in which measurement of some kind is involved—are the proper work of a physical laboratory. In every experiment we have first to make our senses familiar with the phenomenon; but we must not stop here,—we must find out which of its features are capable of measurement, and what measurements are required in order to make a complete specification of the phenomenon. We must, then, make these measurements, and deduce from them the result which we require to find.

This characteristic of modern experiments—that they consist principally of measurements—is so prominent that the opinion seems to have got abroad that, in a few years, all the great physical constants will have been approximately estimated, and that the only occupation which will then be left to men of science will be to carry these measurements to another place of decimals.

If this is really the state of things to which we are approaching, our Laboratory may perhaps become celebrated as a place of conscientious labour and consummate skill; but it will be out of place in the University, and ought rather to be classed with the other great workshops of our country, where equal ability is directed to more useful ends.

But we have no right to think thus of the unsearchable riches of creation, or of the untried fertility of those fresh minds into which these riches will continually be poured. . . . The history of science shows that even during that phase of her progress in which she devotes herself to improving the accuracy of the numerical measurement of quantities with which she has long been familiar, she is preparing the materials for the subjugation of new regions, which would have remained unknown if she had been contented with the rough methods of her early pioneers.

The movement which was now to receive so great an impulse may be roughly dated from Sir William Thomson's first appearance as a Public Examiner in Cambridge; and Maxwell's own influence, as Examiner and Moderator, had been mainly instrumental in promoting it. The nature of the change has been described as follows by one whose University experience reaches back into the previous time :—

The style of mathematics which was popular in Cambridge for some time before was, to say the least, one-sided, and one-sided in a somewhat unproductive direction. There were many complaints that Cambridge was behind the rest of the scientific world, and that, whereas the students of so many other Universities were introduced to the splendid discoveries of such subjects as Electricity and Heat, the Wranglers of Cambridge spent their time upon mathematical trifles and problems, so-called, barren alike of practical results and scientific interest. Maxwell's questions (as Moderator in 1866) infused fresh life into the Cambridge Tripos, and, therefore, into the University studies, by the number of original ideas and new lines of thought opened up by them, thus preparing for the change of system in 1873, when so many interesting subjects were added to the Examination.

Sir William Thomson gives the following important testimony to the same effect :—

The University, Glasgow,
21st January 1882.

The influence of Maxwell at Cambridge had undoubtedly a great effect in directing mathematical studies into more fruitful channels than those in which they had been running for many years. His published scientific papers and books, his action as an examiner at Cambridge, and his professorial lectures, all contributed to this effect; but above all, his work in planning and carrying out the arrangements of the Cavendish Laboratory. There is, indeed, nothing short of a revival of Physical Science at Cambridge within the last fifteen years, and this is largely due to Maxwell's influence.

Evidence might easily be multiplied, but it is enough to quote the weighty words of Lord Rayleigh at a recent public meeting at Cambridge in support of the proposed Devonshire Memorial :—

It was no little thing to have had Professor Maxwell so

closely connected with Cambridge, for by his genius effects
were produced which could hardly have been produced in
any other way. Before coming there to occupy the position
he then held, he (Lord Rayleigh) had not given any particular attention to electricity, but he found Cambridge to be so
saturated with the subject that he quickly came to the conclusion that it would be best to make it his particular study.
All this was owing to the influence of Maxwell.[1]

While speaking of his work in lecturing, it may be
well briefly to advert to the famous " Discourse on
Molecules," delivered before the British Association at
Bradford in September 1873, which has been more
often quoted than, perhaps, any other of his writings.
This address was extremely rich in scientific matter,
but its chief interest lay in the concluding paragraphs,
which may be said to indicate more clearly than any
other of Maxwell's writings the position of his mind
towards certain doctrines maintained by scientific
men :—

In the heavens we discover by their light, and by
their light alone, stars so distant from each other that no
material thing can ever have passed from one to another ;

---

[1] Professor Westcott's utterance on the same occasion, though less
immediately relevant, ought not to be omitted :—" It was impossible
to think of him whom they had so lately lost, to whom first the charge
of the Cavendish Laboratory had been committed, Prof. Clerk Maxwell, and to recollect his genius and spirit, his subtle and profound
thought, his tender and humble reverence, without being sure that that
close connection between Physics and Theology which was consecrated
by the past was still a living reality among them.   That was an omen
for the future.   He felt, as probably all present felt, that he owed a
deep debt of gratitude to him, both for his researches, and for the
pregnant words in which he gathered up their lessons."

and yet this light, which is to us the sole evidence of the existence of these distant worlds, tells us also that each of them is built up of molecules of the same kinds as those which we find on earth. A molecule of hydrogen, for example, whether in Sirius or in Arcturus, executes its vibrations in precisely the same time.

Each molecule therefore throughout the universe bears impressed upon it the stamp of a metric system as distinctly as does the metre of the Archives at Paris, or the double royal cubit of the temple of Karnac.

No theory of evolution can be formed to account for the similarity of molecules, for evolution necessarily implies continuous change, and the molecule is incapable of growth or decay, of generation or destruction.

None of the processes of Nature, since the time when Nature began, have produced the slightest difference in the properties of any molecule. We are therefore unable to ascribe either the existence of the molecules or the identity of their properties to any of the causes which we call natural.

On the other hand, the exact equality of each molecule to all others of the same kind gives it, as Sir John Herschel has well said, the essential character of a manufactured article, and precludes the idea of its being eternal and self-existent.

Thus we have been led, along a strictly scientific path, very near to the point at which Science must stop,—not that Science is debarred from studying the internal mechanism of a molecule which she cannot take to pieces, any more than from investigating an organism which she cannot put together. But in tracing back the history of matter, Science is arrested when she assures herself, on the one hand, that the molecule has been made, and, on the other, that it has not been made by any of the processes we call natural.

Science is incompetent to reason upon the creation of matter itself out of nothing. We have reached the utmost limits of our thinking faculties when we have admitted that because matter cannot be eternal and self-existent it must have been created.

It is only when we contemplate, not matter in itself, but the form in which it actually exists, that our mind finds something on which it can lay hold.

That matter, as such, should have certain fundamental properties,—that it should exist in space and be capable of motion,—that its motion should be persistent, and so on,—are truths which may, for anything we know, be of the kind which metaphysicians call necessary. We may use our knowledge of such truths for purposes of deduction, but we have no data for speculating as to their origin.

But that there should be exactly so much matter and no more in every molecule of hydrogen is a fact of a very different order. We have here a particular distribution of matter—a *collocation*—to use the expression of Dr. Chalmers, of things which we have no difficulty in imagining to have been arranged otherwise.

The form and dimensions of the orbits of the planets, for instance, are not determined by any law of nature, but depend upon a particular collocation of matter. The same is the case with respect to the size of the earth, from which the standard of what is called the metrical system has been derived. But these astronomical and terrestrial magnitudes are far inferior in scientific importance to that most fundamental of all standards which forms the base of the molecular system. Natural causes, as we know, are at work, which tend to modify, if they do not at length destroy, all the arrangements and dimensions of the earth and the whole solar system. But though in the course of ages catastrophes have occurred and may yet occur in the heavens, though ancient systems may be dissolved and new systems evolved out of their ruins, the molecules out of which these systems are built—the foundation-stones of the material universe—remain unbroken and unworn. They continue this day as they were created—perfect in number and measure and weight; and from the ineffaceable characters impressed on them we may learn that those aspirations after accuracy in measurement, and justice in action, which we reckon among our noblest attributes as men, are ours because they are essential constituents of the image of Him who in the beginning created, not only the heaven and the earth, but the materials of which heaven and earth consist.

In 1875 he read before the Chemical Society a paper " On the Dynamical Evidence of the Molecular Constitution of Bodies."

The lecture on Thermodynamics at the Loan Exhibition of Scientific Apparatus in London in 1876 (to which he had contributed his real-image Stereoscope, etc.), was illustrated by his own model of the Thermodynamic Surface.[1]

The last of his public lectures was the Rede Lecture " On the Telephone," delivered at Cambridge in 1878, and illustrated with the aid of Mr. Gower's Telephonic Harp.

After pointing out the extreme simplicity as well as the absolute novelty of the invention, he made it the text of a discourse which is remarkable both for suggestiveness and discursiveness.

I shall . . . consider the telephone as a material symbol of the widely separated departments of human knowledge, the cultivation of which has led, by as many converging paths, to the invention of this instrument by Professor Graham Bell.

. . . In a University we are especially bound to recognise not only the unity of Science itself, but the communion of

---

[1] See Part II. In the official handbook to the collection, the articles entitled " General considerations respecting Scientific Apparatus " and " Molecular Physics," were written by Professor Maxwell. When her Majesty the Queen visited the collection Professor Maxwell, at the invitation of the Lords of the Committee of Council on Education, attended as the representative of Molecular Physics.

the workers of Science.   We are too apt to suppose that we
are congregated here merely to be within reach of certain
appliances of study, such as museums and laboratories,
libraries and lectures, so that each of us may study what he
prefers.   I suppose that when the bees crowd round the
flowers it is for the sake of the honey that they do so,
never thinking that it is the dust which they are carrying
from flower to flower which is to render possible a more
splendid array of flowers and a busier crowd of bees in the
years to come.

We cannot therefore do better than improve the shining hour in helping forward the cross-fertilisation of the
Sciences.

One great beauty of Professor Bell's invention is that
the instruments at the two ends of the line are precisely
alike. . . . The perfect symmetry of the whole apparatus—
the wire in the middle, the two telephones at the ends of
the wire, and the two gossips at the ends of the telephones,
may be very fascinating to a mere mathematician, but it
would not satisfy the evolutionist of the Spenserian
type, who would consider anything with both ends alike,
such as the Amphisbæna, or Mr. Bright's terrier, or Mr.
Bell's telephone, to be an organism of a very low type, which
must have its functions differentiated before any satisfactory
integration can take place.

Accordingly many attempts have been made, by differ-
entiating the function of the transmitter from that of the
receiver, to overcome the principal limitation of the power
of the telephone.   As long as the human voice is the sole
motive power of the apparatus, it is manifest that what is
heard at one end must be fainter than what is spoken at the
other.   But if the vibration set up at one end is used no
longer as the source of energy, but merely as a means of
modulating the strength of a current supplied by a voltaic
battery, then there will be no necessary limitation of the
intensity of the resulting sound, so that what is whispered
to the transmitter may be proclaimed *ore rotundo* by the
receiver.

He then briefly referred to Edison's loud-speaking telephone, and went on to exhibit and explain the microphone of Professor Hughes.

I have said the telephone is an instance of the benefit to be derived from the cross-fertilisation of the sciences. . . . Professor Graham Bell . . . is the son of a very remarkable man, Alexander Melville Bell, author of a book called *Visible Speech,* and of other works relating to pronunciation. In fact his whole life has been employed in teaching people to speak. He brought the art to such perfection that, though a Scotchman, he taught himself in six months to speak English, and I regret extremely that when I had the opportunity in Edinburgh I did not take lessons from him.[1] Mr. Melville Bell has made a complete analysis and classification of all the sounds capable of being uttered by the human voice, from the Zulu clicks to coughing and sneezing; and he has embodied his results in a system of symbols, the elements of which are not taken from any existing alphabet, but are founded on the different configurations of the organs of speech.

. . . Helmholtz, by a series of daring strides, has effected a passage for himself over that untrodden wild between acoustics and music—that Serbonian bog where whole armies of scientific musicians and musical men of science have sunk without filling it up.

We may not be able even yet to plant our feet in his tracks and follow him right across—that would require the seven league boots of the German Colossus; but to help us in Cambridge we have the Board of Musical Studies vindicating for music its ancient place in a liberal education. On the physical side we have Lord Rayleigh laying our foundation deep and strong in his *Theory of Sound.* On the æsthetic

---

[1] Maxwell had profited not a little by his own studies in this direction. But the Gallowegian tones are hard to modify, and even in his verse such rhymes as "hasn't" = "pleasant" recall to those who knew him his peculiar mode of speech.

side we have the University Musical Society doing the
practical work, and, in the space between, those conferences
of Mr. Sedley Taylor, where the wail of the Siren draws
musician and mathematician together down into the depths
of their sensational being, and where the gorgeous hues of the
Phoneidoscope are seen to seethe and twine and coil like
the

> Dragon boughts and elvish emblemings

on the gates of that city, where

> An ye heard a music, like enow
> They are building still, seeing the city is built
> To music, therefore never built at all
> And therefore built for ever.

The special educational value of this combined study of
music and acoustics is that more than almost any other
study it involves a continual appeal to what we must observe
for ourselves.

The facts are things which must be felt; they cannot be
learned from any description of them.

All this has been said more than 200 years ago by one
of our own prophets, William Harvey of Gonville and
Caius College :—" For whosoever they be that read authors,
and do not, by the aid of their own senses, abstract true
representations of the things themselves (comprehended
in the author's expressions) they do not represent true ideas,
but deceitful idols and phantasmas; by which means they
frame to themselves certaine shadows and chimæras, and all
their theory and contemplation (which they call science)
represents nothing but waking men's dreams and sick men's
phrensies."

After the opening of the Cavendish Laboratory in
1874, the most continuous, as well as the most important, work of the Chair was the superintendence of
various courses of experiments, undertaken by young
aspirants for scientific distinction. With character-

istic loyalty and humility, Maxwell seems often to
have taken more pride in their researches than in his
own.   To enumerate the men who were thus favoured
would be to name many who are now amongst the
most efficient teachers of science in the United Kingdom. But there can be nothing invidious in making
particular mention of those who are named by Maxwell
himself in his correspondence, although the omission
of other names may be accidental.   Besides Mr. W.
Garnett, who was his demonstrator in the Laboratory
from first to last, he refers with especial satisfaction to
the work of Mr. George Chrystal, now Professor of
Mathematics in Edinburgh, and to that of Mr. W. D.
Niven.

Mr. Chrystal was encouraged by him to undertake
a series of experiments for verifying Ohm's Law respecting the relation between the current and the
electro-motive force in a wire, on which some doubt
had been thrown by Weber's theories, and, in an opposite direction, by a series of experiments reported to
the British Association by Dr. Schuster in 1874.

In consequence of these doubts a committee was
appointed by the British Association consisting of Professor Maxwell, Professor Everitt, and Dr. Schuster,
and the report of this committee was presented to the
Association at their annual meeting in Glasgow in
1876. The report consists mainly of an account of
two experimental investigations planned by Professor
Maxwell and carried out in the Cavendish Laboratory
by Mr. Chrystal. To this report Mr. Chrystal added
a brief account of his experiments on the unilateral

and bilateral deflection of a galvanometer, affording a possible explanation of Dr. Schuster's result. The investigation proved that when a unit current passes through a conductor of a square centimetre section, its resistance does not differ from its value for indefinitely small currents by 0·000,000,001 per cent.

1873-9.   The scene of these congenial labours was surrounded with manifold associations, which his love for Cambridge intensified. He had pleasant intercourse with many persons there, and after a while resumed the habit of occasional essay writing. Under the name of Erănus (or pic-nic) a club of older men was formed, differing little apparently from the "Apostles," except in the greater seriousness of the discussions. Dr. Lightfoot (now Bishop of Durham) and Professors Hort and Westcott were members of this little circle of congenial spirits. Maxwell's contributions, containing his matured thoughts on various speculative questions, will be found in Chapter XIII. It may be remarked generally that the most marked feature of his later life was an ever-increasing soberness of spirit, and a deepening inward repose, which took nothing from the brightness of his companionship, but rather kept fresh the inexhaustible springs of cheerfulness and humorous mirth in him. The beginnings of such "life in earnest" may be traced far back, but are most obviously perceptible in his third year at Cambridge (1853),[1] in the summer

---

[1] How readily his thoughts took a serious turn, even in the earlier

of 1856, after his father's death, and in the crisis of
his life at Aberdeen (1857-8).

This graver tone by no means checked the playful impulses that burst forth from time to time in
sparkling *jeux d'esprits*.   It rather fledged his arrows,
while it loaded them, giving them a steadier aim,
so that his lightest effusions carried an unsuspected
weight of meaning.   His wit was never more brilliant,
more incisive, or (it may be added) more perfectly
good-humoured, than in the verses on Professor Cayley's portrait, and the "Notes of the President's
Address."   He found time also to indulge his old
taste for reading and writing in cypher, and thus, on
one occasion, considerably disconcerted a contributor
to the second column of the *Times*.

His outward appearance in these later years has
been well described by one who saw him first in
1866 :—

A man of middle height, with frame strongly knit, and

---

undergraduate days, may be seen in a letter (not given above) of
26th March 1852 :—   *Æt.* 20.
    "A. was sent for by telegraph to his sister : he found her past re-
covery, and she is since dead.   The family is large, and till now was
entire, so that the grief is great and new.
    "The attributes of man, as one of a family, seem to be more highly
developed in large families.   The pronoun 'we' acquires a peculiar
significance.   The family man has an idea of a *living* home, to which
he can in imagination retreat, and which gives him a steadiness and
force not his own.   He is one member of a naturally constituted
society ; he has protected his juniors and been protected by his seniors ;
and now he has the consciousness that he is but one of the arrows in
the quiver of the Mighty, and that it is the interest of others as well
as his own that he should succeed."

a certain spring and elasticity in his gait; dressed for comfortable ease rather than elegance; a face expressive at once
of sagacity and good humour, but overlaid with a deep shade
of thoughtfulness; features boldly but pleasingly marked;
eyes dark and glowing; hair and beard perfectly black, and
forming a strong contrast to the pallor of his complexion.
. . . He might have been taken, by a careless observer, for
a country gentleman, or rather, to be more accurate, for a
north country laird.   A keener eye would have seen, however, that the man must be a student of some sort, and one
of more than ordinary intelligence.

In later years his hair had turned to iron gray, but
until a few years before his death he retained his elasticity
of step.

The picture of Maxwell, as he appeared in 1866, became afterwards perfectly familiar to residents in Cambridge.
They will remember his thoughtful face as he walked in the
street, revolving some of the many problems that engaged
him, Toby lagging behind, till his master would suddenly
turn, as if starting from a reverie, and begin calling the
dog.

The same authority continues—

. . . He had a strong sense of humour, and a keen
relish for witty or jocose repartee, but rarely betrayed enjoyment by outright laughter.   The outward sign and conspicuous manifestation of his enjoyment was a peculiar
twinkle and brightness of the eyes.   There was, indeed,
nothing explosive in his mental composition, and as his
mirth was never boisterous, so neither was he fretful or
irascible.   Of a serenely placid temper, genial and temperate
in his enjoyments, and infinitely patient when others would
have been vexed or annoyed, he at all times opposed a solid
calm of nature to the vicissitudes of life.

In performing his private experiments at the laboratory,
Maxwell was very neat-handed and expeditious.   When
working thus, or when thinking out a problem, he had a habit
of whistling, not loudly, but in a half-subdued manner, no

particular tune discernible, but a sort of running accompaniment to his inward thoughts. . . . He could carry the full
strength of his mental faculties rapidly from one subject to
another, and could pursue his studies under distractions
which most students would find intolerable, such as a loud
conversation in the room where he was at work. On these
occasions he used, in a manner, to take his dog into his confidence, and would say softly, "Tobi, Tobi," at intervals, and
after thinking and working for a time, would at last say (for
example), " It must be so : Plato (*i.e.* Plateau), thou reasonest
well." He would then join in the conversation.

. . . His acquaintance with the literature of his own
country, and especially with English poetry, was remarkable
alike for its extent, its exactness, and the wide range of his
sympathies. His critical taste, founded as it was on his
native sagacity, and a keen appreciation of literary beauty,
was so true and discriminating that his judgment was, in
such matters, quite as valuable as on mathematical writings.
. . . As he read with great rapidity, and had a retentive
memory, his mind was stored with many a choice fragment
which had caught his fancy. He was fond of reading aloud
at home from his favourite authors, particularly from Shak-
speare, and of repeating such passages as gave him the
greatest pleasure.[1]

Maxwell was rarely seen walking without a dog
accompanying him, and, when visiting the Laboratory
for a short time, Toby or Coonie, or both, would always
attend him. Toby (II. or III.) came to Cambridge with
Professor Maxwell in 1871, and was thoroughly conversant with the details of the Laboratory and some of

---

[1] There was found amongst his papers a scrap on which he had
written, in pencil, *the whole* of Shelley's " Ode to the West Wind," in
all probability from memory, and as a distraction from anxiety or from
severer study. His note-books, one of which he always carried with
him, are full of the most miscellaneous jottings, plans of works, solutions of problems, extracts in prose and verse, etc.

its apparatus. He always betrayed signs of uneasiness
when he heard electric sparks, but when summoned
to his post he would sit down between his master's
feet and allow the electrophorus to be excited upon
his back, growling all the time in a peculiar manner,
as though to relieve his mind, but not evidencing any
signs of real discomfort. On one occasion Toby sat
quietly on an insulating support, and allowed himself
to be rubbed with a cat's skin, when it was found that
the dog became positively electrified, contrary to the
general belief that a cat's skin is positive to everything; whereupon Professor Maxwell remarked that
" a live dog is better than a dead lion." It remains
for a future physicist to determine the electric relations of a live cat and dog.

One great charm of Maxwell's society was his
readiness to converse on almost any topic with those
whom he was accustomed to meet, although he always
showed a certain degree of shyness when introduced
to strangers. He would never tire of talking with
boyish glee about the d——l on two sticks and similar
topics, and no one ever conversed with him for five
minutes without having some perfectly new ideas set
before him; sometimes so startling as to utterly
confound the listener, but always such as to well
repay a thoughtful examination. Men have often
asked, after listening to a conversation on some
scientific question, whether Maxwell were in earnest
or joking.[1] The charm of his conversation rendered it

---

[1] For an instance of humorous mystification, see the letter to Mr.
Garnett of 4th January 1877.

very difficult to carry on any independent work when he was present, but his suggestions for future work far more than compensated for the time thus spent.

On one occasion, after removing a large amount of calcareous deposit which had accumulated in a curiously oolitic form in a boiler, Maxwell sent it to the Professor of Geology with a request that he would identify the formation. This he did at once, vindicating his science from the aspersion which his brother professor would playfully have cast on it.

Maxwell still found occasional recreation in riding at Cambridge as well as more frequently at Glenlair, where he resided as much as he could consistently with his professional duties.[1] He always arranged to leave Cambridge at the end of the Easter term in time to officiate at the midsummer communion in the kirk at Parton, where he was an elder. His liberality in his own neighbourhood was very great. Besides the endowment of the church, and building of the manse at Corsock, he had planned a large contribution to the cause of primary education. When the School Board was instituted in the district, Maxwell was very anxious to keep up the school established in the reign of George III. at Merkland, in the immediate neighbourhood of the village of Kirkpatrick-Durham, in addition to the Board school at Corsock, five miles away. When this offer was refused, he set apart a site and had plans made for a school to be

---

[1] He kept up the old habit of regulating the clocks at Glenlair by the sun, which, when on the meridian, threw the shadow of a stick upon a notch cut in the stone outside the door.

erected and supported at his own expense upon his estate, but failing health prevented the accomplishment of his purpose.

The last few years of Maxwell's life were saddened by the serious and protracted illness of Mrs. Maxwell. Notwithstanding the inexhaustible freshness of his spirit, his work could not but be somewhat modified by a cause so grave. He was an excellent sick-nurse, and we have already seen how he attended upon Pomeroy when attacked with fever in college, how he devoted himself to his father during his illness, and how he cared for his brother-in-law when in London. On one occasion during Mrs. Maxwell's illness he did not sleep in a bed for three weeks, but conducted his lectures and other work at the Laboratory as usual. While attending on his wife he would continue working at his manuscripts, or would arrange a series of experiments to be carried out by one of the workers at the Cavendish Laboratory; but the time which he could personally devote to his own experiments was very limited. The same cause prevented his attendance at meetings in London and at the British Association, for which, however, he retained his affection. His wonderful devotion to his wife, and the almost mystical manner in which he regarded the marriage tie, are sufficiently apparent from his letters.

The meeting of the British Association, held at Belfast in 1874 when Professor Tyndall was President, was the last which Maxwell attended. Before Section A he read a note " On the Application of Kirchhoff's Rules for Electric Circuits to the Solution of a Geometrical Problem;" but his attendance at this meeting will be remembered chiefly on account of his paraphrase of the President's address, which was published in *Blackwood's Magazine*, and, together with the late Mr. Shilleto's Greek translation of it, will be found reprinted in Part III. His verses on the Red Lions, a social club consisting of members of the Association, were also written at this meeting.

In university politics Maxwell was regarded as a Conservative, and, as such, in November 1876, he was elected a member of the Council of the Senate of the University. His views respecting various questions of university reform are sufficiently indicated by his letters, especially those addressed to Mr. Monro (see p. 269). He was also a member of the Mathematical Studies and Examinations Syndicate, which was appointed on 17th May 1877, and which sat every week during term for a whole year for the purpose of reorganising the Mathematical Tripos.

In 1873 and 1874 Professor Maxwell was one of the examiners for the Natural Sciences Tripos, and in 1873 he was the first "Additional Examiner" in the Mathematical Tripos under the new regulations which then came into force. This was the fifth time that he had examined in the Mathematical Tripos in the course of seven years. He was president of the Cambridge Philosophical Society during the session 1876-7.[1]

---

[1] An account of the last years of Maxwell's life would not be complete without a reference to his acquaintance with Professor H. A. Rowland, formerly of Troy, and now of the Johns Hopkins University,

Besides many contributions to *Nature* and other similar publications during his residence in Cambridge, Maxwell wrote several articles for the Ninth Edition of the *Encyclopædia Britannica.* The last scientific paper he ever wrote was the very brief article on Harmonic Analysis, the proof of which was sent for correction when its author was too weak to read it.

Although the publication of the *Treatise on Heat* and of the *Electricity and Magnetism* falls within this period, they were mainly written during the time of his retirement at Glenlair. The "small book on a great subject," entitled *Matter and Motion,* was merely the concise expression of his most habitual thoughts. But his chief literary work during the last seven years of his life was the editing of the *Electrical Researches of the Hon. Henry Cavendish, F.R.S.*

Henry Cavendish was son of Lord Charles Cavendish

Baltimore. Professor Rowland visited Maxwell more than once, and on these occasions much time was spent in comparing notes on electrical questions. Some instruments which Professor Rowland designed were not only identical with Maxwell's in the relative dimensions of the several parts, but their absolute dimensions also were very nearly the same. After Maxwell's death Professor Rowland pointed out some sources of error in the experimental determination of the Ohm as carried out at King's College, and in the recent redetermination made by Lord Rayleigh in the Cavendish Laboratory these sources of error have been removed. Maxwell's opinion of Professor Rowland was very high, and he frequently alludes to him in his correspondence, and more than once "Rowland of Troy, that doughty knight," appears in his verses, where, as the American investigator in a certain branch of magnetic science studied here by Professor Oliver Lodge and Mr. Oliver Heaviside, he is in one place referred to as "One Rowland for two Olivers." I well remember the interest with which Maxwell looked forward to Mr. Rowland's first visit, and the meeting of "Greek and Trojan" on that occasion at Glenlair.

and great uncle to the present Duke of Devonshire.
He published only two papers relating to electricity—
"An Attempt to Explain some of the Phenomena of
Electricity by means of an Elastic Fluid" (*Phil. Trans.*
1771) and "An Account of some Attempts to Imitate
the Effects of the Torpedo by Electricity" (*Phil. Trans.*
1776).   He had prepared, however, some twenty
packets of manuscript on Mathematical and Experimental Electricity.   These, after his death, were placed
by the then Earl of Burlington, now Duke of Devonshire, in the hands of the late Sir William Snow Harris,
who appears to have made an abstract of them, with
a commentary of great value on their contents.   Of
this abstract and commentary Professor Maxwell was
unable to gain possession, but the Cavendish Manuscripts were placed in his hands by the Duke of
Devonshire in 1874.   The manner in which the contents of these manuscripts were investigated by
Professor Maxwell, and the series of experiments he
conducted in order to test Cavendish's results, will be
referred to in Part II.   The final proof-sheets were
returned to press during the summer of 1879, and
the book was published in October of the same year.
The letters on this subject, which will be found below,
are types of very many that were written by Maxwell
respecting the Cavendish papers.

The title of the book as published in October 1879
(one thick volume, 8vo) is *An Account of the Electrical Researches of the Honourable Henry Cavendish,
F.R.S., between* 1771 *and* 1781.   Few or none could
have performed that task as he has performed it.

And yet some may wish that these precious years
had been given rather to the unimpeded prosecution
of his own original researches.[1]

At my last meeting with him,—it was in his house
at Cambridge, in the year 1877,—in the midst of
some discursive talk, he took the MS. of this book out
of a cabinet, and began showing it to me and dis-
coursing about it in the old eager, playful, affectionate
way, just as with the magic discs in boyhood, or the
register of the colour-box observations at a later time,
in the little study at Glenlair.   " And what," I said,
" of your own investigations in various ways ?"   " I
have to give up so many things," he answered, with a
sad look, which till then I had never seen in his eyes.
Even before this, as it now appears, he had felt the
first symptoms of the inexorable malady, which in
the spring of 1879 assumed a dangerous aspect, and
killed him in the autumn of that year.

LETTERS, 1871 TO 1879—ÆT. 39-48.

FROM C. J. MONRO, Esq.

Hadley, Barnet, 3d March 1871.

The Hon. J. W. Strutt, son of Lord Rayleigh, and
senior wrangler in 1865, has been meddling with your
colours, and has given occasion also to me to do so again.
I send a selection of *Natures* containing him and me, and
my old contributions of last year, which, or one of which,
you say met Mr. Benson's approval.   Strutt's last letter
ends with a sentence which obliged me to write to him

---

[1] An unfinished fragment of a new work on Electricity, in which
he treads more closely than ever in the steps of Faraday, has been
edited since his death by Mr. Garnett and published in 1881.

personally; and I could not help saying, with regard to the sentence which begins it, p. 264, that I thought you would object to inferences founded on comparison by *contrast*, and that the proper way was to compare by matching recognised browns with a compound.

Listing's paper, mentioned in p. 102, was to me rather a paradox,—I had got to regard the subdivisions of the colour-scale which are assumed in language, as something so arbitrary. If you cared to see it, I have that number of Poggendorff; I think he would hardly agree with your J. J. Müller. I wish you or Benson could eradicate the insane trick of reasoning about colours *as identified by their names*. People seem to think that blue is blue, and one blue as good as another. Benson's book I have seen (since I heard of it from you), but not read. His way of mixing by means of a prism is very happy. . . . I wish, with your new set up box, you would just put the prism observations into relation with the disk ones. It would be very easy. White we have got; and it would only be *strictly* necessary to determine *two* other standard colours, such as vermilion or emerald, by reference to the spectrum primaries. You don't say whether your dwellers in Mesopotamia and elsewhere agree on the whole better or worse than "J" and "K," who, I suppose, agree for better and for worse. To judge by their case, the discrepancy would be a little diminished by taking as units of colour co-ordinates for any given pair of eyes, not the intensities of the primaries as they appear in the spectrum, but their intensities as they appear in the combination which to that pair of eyes makes (say) *white*. This amounts to transforming from trilinear to Hamilton's anharmonic co-ordinates with white for the fourth point,— in the language of "scientific metaphor." On the other hand, ought not all your co-ordinates to be cooked by multiplying by $\dfrac{d \text{ . scale, page } 68}{d \text{ . wave-length}}$?

You know where I learnt *scientific metaphor*. I have read the address in Section A more than once with much pleasure, and, I hope, profit in proportion. The pleasure, I

confess, was with me, as I found it was with Litchfield,
partly that of recognising an old well-remembered style, and
reflecting that here at least was something which might be
"thought to be beyond the reach of change." . . . By the
way, Boole is "one of the profoundest mathematicians of
our time;" but how about "thinkers"?   Certainly his expositions of the principle of a piece of mathematics are
beautiful up to and, I don't doubt, beyond my appreciating.
But that last chapter of the *Laws*, etc., from which you
quote, with Empedocles and Pseudo-Origen and the rest of
them, always seems to me to render a sound as of a largish
internal cavity; and the whole book, taken together with
his R.S.E. paper on testimony and least squares, presents, I
think, too many instances of a particular class of fallacy—I
know I am speaking blasphemies, but there would be a
strike among the postmen if I put in all the necessary
qualifications—too many instances to be got over, not in
absolute number if they were of different kinds, for anybody
may make mistakes, but too many of one kind.   The kind
is *insufficient interpretation, i.e.* letting your equations lead
you by the nose.   The most serious example,—I maintain
it *is* an example,—is his insisting that his theory of logic is
not founded on quantity, so that it furnishes (he holds) an
independent foundation for probabilities, independent of the
usual quantitative foundation.   That this is a fallacy, and
that in particular it is an example of the fallacy of insufficient interpretation, is evident surely when you find that,
even in the higher case of "secondary" propositions, the
elective symbols represent in his own opinion *quantities* of
"time" after all.   With regard to the sentence you quote,
I am always suspicious of any inclination I may feel to find
a question too easy; and, independently of that, your quoting it is itself a staggerer.   But the difficulty I confess
does strike me as a rather artificial one.   There is nothing,
scarcely, in which I think Mill is so right and the Hamilton-
ians so wrong as that question about logic being the laws of
thought.   Hamilton says *as thought*, Mill says *as valid*, and
so does Boole and so do you; but if Mill is right, where is

the difficulty? Why *should* the conditions of thinking correctly be inviolable in the sense of not preventing you from thinking incorrectly, provided they are inviolable in the sense of ensuring that you take the consequences if you do? The laws of projection in geometry are inviolable, but nobody ever thought it a paradox that it is possible for a picture to be *out of drawing* in spite of them, nor is it a paradox that in unfamiliar classes of cases a rigorously accurate piece of perspective *looks* out of drawing. Perhaps you meant, for I suppose the report in *Nature* is incomplete, that it was a difficulty to say in what sense mathematical propositions could be said to be *certain*, considering that one may make mistakes about them. Perhaps something else, which for the above reason or others, is hidden from me.

. . . By the way, I hope it is true that you are to profess experimental physics at Cambridge, or what I hope comes to the same thing, that you are a candidate.

To C. J. Monro, Esq.

Glenlair, Dalbeattie, 15th March 1871.

I have been so busy writing a sermon on Colour, and Tyndalising my imagination up to the lecture point, that along with other business I have had no leisure to write to any one.

I think a good deal may be learned from the *names* of colours, not about colours, of course, but about names; and I think it is remarkable that the rhematic instinct has been so much more active, at least in modern times, on the less refrangible side of primary green ($\lambda = 510 \times 10^{-9}$ inches).

I am not up in ancient colours, but my recollection of the interpretations of the lexicographers is of considerable confusion of hues between red and yellow, and rather more discrimination on the blue side. Qu. If this is true, has the red sensation become better developed since those days? Benson has a new book, Chapman & Hall, 1871, called *Manual of Colour*.

I think it is a great improvement on the Quarto, both in size and quality. It is the size of this paper I write on.

I have not asked you if you wish to go to sermon on Colour, for I do not think the R. I.[1] a good place to go to of nights, even for strong men. I have, however, some tickets to spare.

The peculiarity of our space is that of its three dimensions none is before or after another. As is x, so is y and so is z.

If you have 4 dimensions this becomes a puzzle, for— first, if three of them are in our space, then which three? Also, if we lived in space of m dimensions, but were only capable of thinking n of them, then 1st, Which n? 2d, If so, things would happen requiring the rest to explain them, and so we should either be stultified or made wiser.

I am quite sure that the kind of continuity which has four dimensions all co-equal is not to be discovered by merely generalising Cartesian space equations. (I don't mean by Cartesian space that which Spinoza worked from Extension the one essential property of matter, and Quiet the best glue to stick bodies together). I think it was Jacob Steiner who considered the final cause of space to be the suggestion of new forms of continuity.

I hope you will continue to trail clouds of glory after you, and tropical air, and be as it were a climate to yourself. I am glad to see you occasionally in *Nature*. I shall be in London for a few days next week,— address Athenæum Club.

I think Strutt on sky-blue is very good. It settles Clausius's vesicular theory,

> " for, putting all his words together.
> 'tis 3 blue beans in 1 blue bladder."—*Mat. Prior.*

The Exp. Phys. at Cambridge is not built yet, but we are going to try.

---

[1] Royal Institution.

The desideratum is to set a Don and a Freshman to observe and register (say) the vibrations of a magnet together, or the Don to turn a winch, and the Freshman to observe and govern him.

FROM PROFESSOR TYNDALL.

Monday.

MY DEAR MAXWELL—Why . . . did you run away so rapidly. I wished to shake your hand before parting.—
Yours ever, JOHN TYNDALL.

TO MRS. MAXWELL.

20th March 1871.

There are two parties about the professorship. One wants popular lectures, and the other cares more for experimental work. I think there should be a gradation—popular lectures and rough experiments for the masses; real experiments for real students; and laborious experiments for firstrate men like Trotter and Stuart and Strutt.

FROM C. J. MONRO, Esq.

Hadley, 21st March 1871.

. . . I never observed before that ancient colournomenclature was more discriminate than ours for the more " violently " refracted tints as compared with the less; but I think there must be something in it. But I have always suspected that they referred colour to a positively distinct set of co-ordinates from ours. Gladstone says something of this sort in *Homer;* who put it into his head I can't think; if he made it out for himself I should be very sorry to agree with a man who does not believe in spectrum analysis, and does believe that Leto is the Virgin Mary. Such queer applications of words of colour one does find. You know the " pale " horse of the Apocalypse (vi. 8); well, that is χλωρός, which is usually " green," you know. General Daumas says the Arabs call " vert " what the French call " louvet " in horses; and *louvet,* in Littré, " Se dit, chez le cheval, d'une robe caractérisée par la présence de la nuance jaune et du noir, qui lui donne une certaine ressemblance avec le poil du loup. . . . . Substantivement," he continues, " Le louvet n'est, à proprement parler, qu'un isabelle charbonné." The Arabic for green, and (I have no doubt) the word Daumas speaks of, is akhḍár, kh as ch in Scotch, and the dot marking a modification which, it happens, is imitated by interpolating an L in Spanish and Portuguese; so χλωρός may have been supposed to have something to do with the Semitic word. However, according to dictionaries, " the three greens " in Arabic are " gold, wine, and meat," which beats the green horse. I suppose the Revisionists will leave " pale," and certainly χλωρὸν δέος is the Homeric for a blue funk. But χλωρός, and akhḍár, too, are certainly the colour of chlorophyll, and Daumas's remark is a note on a line in a translation from a poet, which runs " Ces chevaux verts comme le roseau qui croît au bord des fleuves."

I am glad you are going to preach, and I should like to sit under you, but as you assume, it would not do. Thanks all the same.

To Mrs. Maxwell.

Athenæum, 22d March 1871.

I also got a first-rate letter from Monro about colour, and the Arab words for it (I suppose he studied them in Algeria). They call horses of a smutty yellow colour " green." The " pale " horse in Revelation is generally transcribed green elsewhere, the word being applied to grass, etc. But the three green things in the Arabic dictionary are " gold, wine, and meat," which is a very hard saying.

FROM C. J. MONRO, Esq.

Hadley, Barnet, 10th September 1871.

. . . . Of your own things, the *Classification of Quantities* and the *Hills and Dales*, are all I have read to much purpose. Nor them either, you may say, if I go on to ask why you say that " in the pure theory of surfaces there is no method of determining a line of water-shed or water-course, except

as therein is excepted, that is in page 6 ? Why does not
this determine them ? to wit—

$$\left(\frac{dz}{dx}\right)^2 + \left(\frac{dz}{dy}\right)^2 \text{ maximum, } \& \quad \left\{ \begin{array}{l} z \text{ max. for a shed.} \\ z \text{ min. for a course.} \end{array} \right.$$

Or if this does determine them, how does it resolve itself
into "*first* finding," etc. ?

I am glad you like Strutt on sky-blue. You see he
sees his way now to a new theory of double refraction.
Looking at your old letter again, I don't quite see the force
of either of your objections to space of more than three
dimensions. First, you ask if we can think some of the
dimensions and not others, then which ? Surely one might
answer, that depends—depends namely on your circumstances—on circumstances which in your circumstances you
cannot expect to judge of.

" I can easily believe," as Darwin would say, that before
we were tidal ascidians we were a slimy sheet of cells floating on the surface of the sea. Well, in those days, the
missing dimension, and the two forthcoming ones respectively, kept changing with the rotation of the earth,—we *now*
know how, but could not guess then. So, now, the missing
dimension or dimensions, if any, might be determined by
circumstances which we could not tell unless we knew all
about the said dimension or dimensions.

TO PROFESSOR LEWIS CAMPBELL.

Glenlair, Dalbeattie, 19th October 1872.

. . . Lectures begin 24th. Laboratory rising, I hear, but
I have no place to erect my chair, but move about like the
cuckoo, depositing my notions in the chemical lecture-room
1st term ; in the Botanical in Lent, and in Comparative
Anatomy in Easter.

I am continually engaged in stirring up the Clarendon
Press, but they have been tolerably regular for two months.
I find nine sheets in thirteen weeks is their average. Tait
gives me great help in detecting absurdities. I am getting
converted to Quaternions, and have put some in my book, in

a heretical form, however, for as the Greek alphabet was used up, I have used German capitals from 𝕬 to 𝕲 to stand for Vectors, and, of course, ∇ occurs continually.   This letter is called " Nabla,"[1] and the investigation a Nablody.   You will be glad to hear that the theory of gases is being experimented on by Profs. Loschmidt and Stefan of Vienna, and that the conductivity of air and hydrogen are within 2 per cent of the value calculated from my experiments on friction of gases, though the diffusion of one gas into another is " *in erglanzender ubereinstimmung mit $\frac{dp}{dt}$schen Theorie.*"

To Professor W. G. Adams.

Natural Science Tripos,
3d December 1873.

I got Professor Guthrie's circular some time ago.   I do not approve of the plan of a physical society considered as an instrument for the improvement of natural knowledge. If it is to publish papers on physical subjects which would not find their place in the transactions of existing societies, or in scientific journals, I think the progress towards dissolution will be very rapid.   But if there is sufficient liveliness and leisure among persons interested in experiments to maintain a series of stated meetings to show experiments, and talk about them as some of the Ray Club do here, then I wish them all joy; only the manners and customs of London, and the distances at which people live from any convenient centre, are very much against the vitality of such sociability.

To make the meeting a dinner supplies that solid ground to which the formers of societies must trust if they would build for aye.   A dinner has the advantage over mere scientific communications, that it can always be had when certain conditions are satisfied, and that no one can doubt its existence.   On the other hand, it completely excludes any scientific matter which cannot be expressed in the form of conversation with your two chance neighbours, or else by

---

[1] The name of an Assyrian harp of the shape ▽.

a formal speech on your legs; and during its whole continuance it reduces the Society to the form of a closed curve, the elements of which are incapable of changing their relative position.

For the evolution of science by societies the main requisite is the perfect freedom of communication between each member and any one of the others who may act as a reagent.

The gaseous condition is exemplified in the soiree, where the members rush about confusedly, and the only communication is during a collision, which in some instances may be prolonged by button-holing.

The opposite condition, the crystalline, is shown in the lecture, where the members sit in rows, while 'science flows in an uninterrupted stream from a source which we take as the origin. This is radiation of science.

Conduction takes place along the series of members seated round a dinner table, and fixed there for several hours, with flowers in the middle to prevent any cross currents.

The condition most favourable to life is an intermediate plastic or colloïdal condition, where the order of business is (1) Greetings and confused talk; (2) A short communication from one who has something to say and to show; (3) Remarks on the communication addressed to the Chair, introducing matters irrelevant to the communication but interesting to the members; (4) This lets each member see who is interested in his special hobby, and who is likely to help him; and leads to (5) Confused conversation and examination of objects on the table.

I have not indicated how this programme is to be combined with eating. It is more easily carried out in a small town than in London, and more easily in Faraday's young days (see his life by B. Jones) than now. It might answer in some London district where there happen to be several clubbable senior men who could attract the juniors from a distance.

To Professor Lewis Campbell.

Glenlair, Dalbeattie, 3d April 1873.

The roof of the Devonshire Laboratory is being put on,

2 c

and we hope to have some floors in by May, and the contractors cleared out by October. We are busy electing School Boards here. The religious difficulty is unknown here. The chief party is that which insists on keeping down the rates; no other platform will do. All candidates must show the retrenchment ticket.

The Cambridge Philosophical Society have been entertained by Mr. Paley on Solar Myths, Odusseus as the Setting Sun, etc. Your Trachiniæ is rather in that style, but I think Middlemarch is not a mere unconscious myth, as the Odyssey was to its author, but an elaborately conscious one, in which all the characters are intended to be astronomical or meteorological.

Rosamond is evidently the Dawn. By her fascinations she draws up into her embrace the rising sun, represented as the Healer from one point of view, and the Opener of Mysteries from another; his name, Lyd Gate, being compounded of two nouns, both of which signify something which opens, as the eye-lids of the morn, and the gates of day. But as the sun-god ascends, the same clouds which emblazoned his rising, absorb all his beams, and put a stop to the early promise of enlightenment, so that he, the ascending sun, disappears from the heavens. But the Rosa Munda of the dawn (see Vision of Sin) reappears as the Rosa Mundi in the evening, along with her daughters ♀ and ☿, in the chariot of the setting sun, who is also a healer, but not an enlightener.

Dorothea, on the other hand, the goddess of gifts, represents the other half of the revolution. She is at first attracted by and united to the fading glories of the days that are no more, but after passing, as the title of the last book expressly tells us, "from sunset to sunrise," we find her in union with the pioneer of the coming age, the editor.

Her sister Celia, the Hollow One, represents the vault of the midnight sky, and the nothingness of things.

There is no need to refer to Nicolas Bulstrode, who evidently represents the Mithraic mystery, or to the kindly family of Garth, representing the work of nature under the rays of the sun, or to the various clergymen and doctors, who are all planets. The whole thing is, and is intended to be, a solar myth from beginning to end.

To Mrs. Maxwell.

December 1873.

I am always with you in spirit, but there is One who is nearer to you and to me than we ever can be to each other, and it is only through Him and in Him that we can ever really get to know each other. Let us try to realise the great mystery in Ephesians v., and then we shall be in our right position with respect to the world outside, the men and women whom Christ came to save from their sins.

To Professor Lewis Campbell.

11 Scroope Terrace,
Cambridge, 26th February 1874.

Jackson has sent me a MS. of yours about the mechanism of the heavens.[1] After the interpretation of εἰλλομένην, about which Greek appears to meet Greek as to whether it expresses motion or only configuration, the main point seems to be, What is the motion and function of τὸν διὰ παντὸς πόλον?

(a) Is it in one piece with the sphere of the stars? or (β) with that of the sun? or (γ) is it fixed in the earth?

It is evidently a good stout axle, not a mere geometrical line, and it has some stiff work to do.

What is this work?

If the earth is fixed, and the great shaft has its bearings in a hole in the earth, then she (the earth) may, in virtue of her dignity and office, cause the axle to revolve, carrying with it the stars according to a, or the sun according to β. Thus the earth may be the cause of the motion of the Same without moving herself, as a spinster is the cause of the whirling of the spindle, though she does not herself pirouette.

Or we may suppose the earth to act as one who twirls

---

[1] See the *Cambridge Journal of Philology*, vol. v., No. 10, pp. 206, foll.

an expanded umbrella over his head about its stick as
an axis, the holes in the same representing the stars. The
objection to this view (which seems to me to be Jowett's) is
that in stating the relation between the earth and the axis, the
earth is said to be related to the axis (packed or whirling as
the case may be), and not the axis to the earth. Now, I
suppose that without all contradiction the less is related to
the better. Here the earth is like a ball of clay packed
round a graft on the branch of a tree, rather than like a field
in which, by means of a rotatory boring tool, men bore for
water.

But the business of the earth is not so much to keep the
stars in motion as to effect the changes of night and day.
This she may do either by rotating herself from W. to E.,
or by controlling the motion of the sun, by the help of the
great shaft.

Now, if you always observe at the same time of night (a
common practice), you find the eastern stars higher every
day, and the western lower. All have the *same* motion,
which carries them round from E. to W. in a *year*.

Mars, Jupiter, and Saturn, in spite of their wanderings
go on the whole in the same direction, but slower. Venus
and Mercury oscillate about the sun, and the moon goes the
opposite way—from W. to E.

That this way of viewing the matter was really prevalent
at one time is plain, from the expression the rising of such
a star to denote not a time of night but a time of year. It
means either (1) the day when the star rises, just before it is
lost in the brightness of the sun who follows it, or (2) the
day when the star is rising, when it just becomes visible
after sunset.

Virgil, who speaks of stars rising, evidently had no
practical knowledge of what he meant. Plato, if he sometimes gets hazy, is far clearer than Virgil. Grote would
place him far below Mr. Jellinger Symons, who denied the
rotation of the moon, because Grote makes Plato say that
both the heavens and the earth rotate both in the same
direction, and with the same angular velocity.

I think I understand you to make Plato make the earth sit still and preside over the heavenly motions, and so become the artificer of day and night, like a policeman who swings his bull's-eye round to his back. But his words are capable of being used by the movers of the earth, as Milton says,

> If earth, *industrious of herself*, fetch day
> Travelling east.

I hope you will let me know whether I have not misunderstood both you, Plato, and the Truth. I have never thanked you for your Œdipus, etc., which I have enjoyed. But at present I am all day at the Laboratory, which is emerging from chaos, but is not yet cleared of gas-men, who are the laziest and most permanent of all the gods who have been hatched under heaven.[1]

Mrs. Maxwell joins me in kind regards to Mrs. Campbell and yourself.—Your afft. friend.

To W. GARNETT, Esq.

Glenlair, 8th July 1874.

. . . In the MS. he [Cavendish] appears to be familiar with the theory of divided currents, and also of conductors in series, but some reference to his printed paper [on the Torpedo] is required to throw light on what he says. He made a most extensive series of experiments on the conductivity of saline solutions in tubes compared with wires of different metals, and it seems as if more marks were wanted for him if he cut out G. S. Ohm long before constant currents were invented. His measurements of capacity will give us some work at the Cavendish Lab., before we work up to the point where he left it. His only defect is not having Thomson's electrometer. He found out inductive capacity of glass, resin, wax, etc.

---

[1] Alluding to the passage of Plato's *Timæus*, p. 40, which had given rise to the previous discussion.

To Professor Lewis Campbell.

Glenlair, Dalbeattie, 26th September 1874.

Æt. 43.   Yours of the 29th instant is to hand. Whether your devotion to Michael Angelo has urged you to anticipate his day, or whether Time gallops with those who sit to view Necessity, with her weary pund o' tow massed round her rock, being all the remains of the stane o' lint with which she was originally endowed, those who may be set to construe this sentence will be apt to lose much time.

With regard to atoms, I am preparing a hash of them for Baynes of the *Britannica*. The easiest way of showing what atoms can't do is to get some sort of notion of what they can do. If atoms are finite in number, each of them being of a certain weight, then it becomes impossible that the germ from which a man is developed should contain (actually, of course, not potentially, for potentiality is nonsense in materialism unless it is expressed as configuration and motion) gemmules of everything which the man is to inherit, and by which he is differentiated from other animals and men, —his father's temper, his mother's memory, his grandfather's way of blowing his nose, his arboreal ancestor's arrangement of hair on his arms, and his more remote littoral ancestor's devotion to the tide-swaying moon. Francis Galton, whose mission it seems to be to ride other men's hobbies to death, has invented the felicitous expression "structureless germs." Now, if a germ, or anything else, contains in itself a power of development into some distinct thing, and if this power is purely physical, arising from the configuration and motion of parts of the germ, it is nonsense to call it structureless because the microscope does not show the structure; the germ of a rat *must* contain more separable parts and organs than there are drops in the sea. But if we are sure that there are not more than a few million molecules in it, each molecule being composed of component molecules, identical with those of carbon, oxygen, nitrogen, hydrogen, etc., there is no room left for the sort of structure which is required for pangenesis on purely physical principles. Again, suppose that a great many individual atoms take part in a disturbance in my brain, to whom does this signify anything? As for the atoms, they have been in far worse rows before they became naturalised in my brain, but they forget the days before, etc.[1] At any rate the atoms are a very tough lot, and can stand a great deal of knocking about, and it is strange to find a number of them combining to form a man of feeling.

In your letter you apply the word imponderable to a molecule. Don't do that again. It may also be worth knowing that the æther cannot be molecular. If it were, it would be a gas, and a pint of it would have the same properties as regards heat, etc., as a pint of air, except that it would not be so heavy.

Under what form (right or light) can an atom be imagined? Bezonian! speak or die! Now I must go to post with two dogs in the rain.—Your afft. friend.

To the Same.

11 Scroope Terrace,
Cambridge, 4th March 1876.

Aias arrived here about a week ago. I read him with _Æt._ 44. pleasure. He recalled the year 1851, when I got him up. The outline of the play seems very bare and unpromising compared with some others, but this is relieved by other features which are not in the "argument," as _e.g._ the loyalty of the chorus and of Tecmessa to Aias under all circumstances (for the chorus in general veers about, and backs occasionally, according as the wind blows or the cat jumps). This contrasts favourably with the character of Athena, who is but so so, only not so comic as the Atreidæ.

But why do Ulysses and Aias not name each other in the same language? I suppose the last syllable of Odysseus, pronounced Anglicé, is somewhat unpleasant in verse, and Ajax, though familiarised by Pope, has lost the interjectional sound of your hero's name.

---

[1] Tennyson's _In Memoriam._

Two Aberdonians, Chrystal and Mollison, are working at the Cavendish Laboratory. I think Chrystal's work is of a kind not comparable with that done in "a third-class German university," which was the charitable hope of *Nature* as to what we might aspire to in ten years' time. He has worked steadily at the testing of Ohm's Law since October, and Ohm has come out triumphant, though in some experiments the wire was kept bright red-hot by the current.— Your afft. friend.

FROM THE RIGHT REV. C. J. ELLICOTT, D.D., Lord Bishop of Gloucester and Bristol.

Palace, Gloucester, 21st Nov. 1876.

MY DEAR SIR—Will you kindly pardon a great liberty? I have quoted in a forthcoming charge a remarkable expression of yours that atoms are "manufactured articles." Could you in your kindness give me the proper title and reference to the paper and the page? I am now, alas, far from libraries, and have, in matters scientific especially, to ask the aid of others. Will you excuse me asking this further question?

Are you, as a scientific man, able to accept the statement that is often made on the theological side, viz. that the creation of the sun posterior to light involves no serious difficulty,—the creation of light being the establishment of the primal vibrations, generally; the creation of the sun, the primal formation of an origin, whence vibrations would be propagated earthward?

My own mind,—far from a scientific one,—is not clear on this point. I surmise, then, that the scientific mind might not only not be clear as to the explanation, but equitably bound to say that it was no explanation at all. Excuse the trouble I am giving you, for the truth's sake, and believe me, very faithfully yours,

C. J. GLOUCESTER AND BRISTOL.

Maxwell replied as follows by return of post :—

11 Scroope Terrace, Cambridge,
22d Nov. 1876.

MY LORD BISHOP—The comparison of atoms or of molecules to "manufactured articles," was originally made by Sir J. F. D. Herschel in his "Preliminary Discourse on the Study of Natural Philosophy," Art. 28, p. 38 (ed. 1851, Longmans).

I send you by book post several papers in which I have directed attention to certain kinds of equality among all molecules of the same substance, and to the bearing of this fact on speculations as to their origin.

The comparison to "manufactured articles" was criticised (I think in a letter to *Nature*) by Mr. C. J. Monro [*Nature*, x. 481, 15th October 1874], and the latter part of the *Encyc. Brit.*, Article "Atom," is intended to meet this criticism, which points out that in some cases the uniformity among manufactured articles is evidence of want of power in the manufacturer to adapt each article to its special use.

What I thought of was not so much that uniformity of result which is due to uniformity in the process of formation, as a uniformity intended and accomplished by the same wisdom and power of which uniformity, accuracy, symmetry, consistency, and continuity of plan are as important attributes as the contrivance of the special utility of each individual thing.

With respect to your second question, there is a statement printed in most commentaries that the fact of light being created before the sun is in striking agreement with the last results of science (I quote from memory).

I have often wished to ascertain the date of the original appearance of this statement, as this would be the only way of finding what "last result of science" it referred to It is certainly older than the time when any notions of the undulatory theory became prevalent among men of science or commentators.

If it were necessary to provide an interpretation of the text in accordance with the science of 1876 (which may not agree with that of 1896), it would be very tempting to

say that the light of the first day means the all-embracing
æther, the vehicle of radiation, and not actual light, whether
from the sun or from any other source. But I cannot suppose that this was the very idea meant to be conveyed by the
original author of the book to those for whom he was
writing. He tells us of a previous darkness. Both light
and darkness imply a being who can see if there is light,
but not if it is dark, and the words are always understood
so. That light and darkness are terms relative to the
creature only is recognised in Ps. cxxxix. 12.

As a mere matter of conjectural cosmogony, however,
we naturally suppose those things most primeval which we
find least subject to change.

Now the æther or material substance which fills all the
interspace between world and world, without a gap or flaw
of $\frac{1}{100000}$ inch anywhere, and which probably penetrates
through all grosser matters, is the largest, most uniform and
apparently most permanent object we know, and we are
therefore inclined to suppose that it existed before the
formation of the systems of gross matter which now exist
within it, just as we suppose the sea older than the individual fishes in it.

But I should be very sorry if an interpretation founded
on a most conjectural scientific hypothesis were to get
fastened to the text in Genesis, even if by so doing it got
rid of the old statement of the commentators which has long
ceased to be intelligible. The rate of change of scientific
hypothesis is naturally much more rapid than that of
Biblical interpretations, so that if an interpretation is
founded on such an hypothesis, it may help to keep the
hypothesis above ground long after it ought to be buried
and forgotten.

At the same time I think that each individual man
should do all he can to impress his own mind with the
extent, the order, and the unity of the universe, and should
carry these ideas with him as he reads such passages as the
1st Chap. of the Ep. to Colossians (see Lightfoot on Colossians, p. 182), just as enlarged conceptions of the extent

and unity of the world of life may be of service to us in reading Psalm viii.; Heb. ii. 6, etc. Believe me, yours faithfully,   J. CLERK MAXWELL.

FROM THE BISHOP OF GLOUCESTER AND BRISTOL.

Palace, Gloucester,
24th Nov. 1876.

DEAR PROFESSOR CLERK MAXWELL—Allow me not to lose a post in thanking you most warmly for your most kind letter and for the packet of pamphlets,—for which I hardly know how enough to express my gratitude. They are exactly what I needed,—yet I fear I may be taking from your stock more than I ought to take. I have already read a good deal of the *Encyc. Brit.* article on atoms ; so pray, if you are short of copies, don't hesitate to drop me a line. The paper on attraction was also most welcome. I am ashamed to own (for bishops should not enter into these pleasures) that I have of late been speculating a good deal on the physical explanation of gravitation. I seem to feel it must be in the Ether,—and yet how, I see not. In the case of a body near the earth, I can conceive a vast amount of elastic ether behind it, and possibly urging it on, while a small quantity is under it, being excluded by the earth.

I seem also to see how this might be applied to the case of the heavy bodies that fell nearer to the steep side of Schehallion than they ought to have done by calculation ; but then, when I attempt to go farther, I find the theory break down.

It seems to me that we want for several things, *e.g.* light, the conception of an ether-beach all round the visible universe from which waves might be reflexively started, and at which the particles might be more closely packed ; but then again I see not what it is that keeps up the beach.

But I am really ashamed of troubling you, a scientific man, with such wanderings. It will only show that your kindness is thoroughly appreciated.

I cordially agree with you as to the light question. Theologians are a great deal too fond of using up the last

scientific hypothesis they can get hold of. The Christian
Knowledge Society are publishing my charge. When it is
published I shall ask you to do me the favour to accept a
copy. You will then see that the best note in the little
volume is due to your kindness and aid. I remain, with
all good wishes, and sincere thanks, very faithfully yours,

C. J. GLOUCESTER AND BRISTOL.

(*P.S.*)—If you are in London in the spring and near the
Athenæum, do me the kindness of looking in on me, as I
shall be very glad to make your personal acquaintance. I
am commonly in town regularly after Easter.

To PROFESSOR LEWIS CAMPBELL.

Glenlair, Dalbeattie, Christmas 1876.

Æt. 45.   . . . I hope that when this severe weather is past you
will be able to derive benefit from a moderate use of Plato
and Sophocles.

We intended to have gone round by Edinburgh, to pay
Aunt Jane a visit; but we both had such bad colds that we
came home to nurse them, and are now snowed up, and
enjoying the artificial heat of coals, peats, and sticks
judiciously intermingled.

The demonstrator at the Cavendish Laboratory has been
out of sorts all this term, and has had to go home about a
month ago, so we have not been in full force there. I hope
he will be well in February, to absorb the energy of the new
B.A.'s set free from the Tripos and its attendant anxieties.

As we get richer in apparatus, mathematical lectures
give way to experimental, and the black board to the lamp
and scale. I have had a pupil quite innocent of mathematics who has learned to measure focal lengths of lenses,
and has found the electro-motive force from the water-pipes
to the gas-pipes, and from either set of pipes to the lightning-
conductor.

I have been making a mechanical model of an induction
coil, in which the primary and secondary currents are represented by the motion of wheels, and in which I can symbolise

all the effects of putting in more or less of the iron core, or more or less resistance and Leyden jars in either circuit.

I have also been making a clay model of Prof. W. Gibb's thermodynamic surface, representing the relations of the solid liquid and gaseous states, and the different paths by which a body may get from the one to the other.—Your afft. friend,

J. CLERK MAXWELL.

To W. GARNETT, Esq.[1]

Glenlair, 4th January 1877.

By all means take the Groves and coils for your lecture. Are you aware that the electric flash is entirely due to the resinous particles of electricity ? This is well known on the stage, where they blow the particles through a tube over a candle to make stage lightning. The vitreous electricity has nothing to do with it, as you may prove by using pounded glass.

In a letter to Mr. Garnett, dated Glenlair, 9th July 1877, Professor Maxwell gave the following suggestions respecting a projected article on Dynamics, and the letter, like those which follow it, is a good illustration of the help he was constantly rendering to his students and others who asked his advice :—

I think it a pity that the old historical word Dynamics should, for mere considerations of time, be split up into Kinematics, Kinetics, and Statics. With respect to the divisions of the subject, I think they fall thus :—

1. Early attempts at founding the science, ancient Kinematics (mechanical description of curves, etc.) generally correct.

Ancient Statics.—Archimedes.

Modern Dynamics.—Galileo, first founder. Descartes, good up to Kinematics and Statics, failed in Kinetics.

---

[1] See p. 370.

Promoters—
WREN, WALLIS, HUYGHENS, HOOKE.

Laws of collision established, and motion in a circle.

NEWTON.

Three laws of motion. Form suggested by the laws of Descartes. Meaning established by Newton's own copious and complete examples of using them.

Second statement of Newton's third Law.

My notions on the three laws are in "Matter and Motion."

NEWTONIANS.

Cambridge School.	*Popularisers.*	*Scotch School.*
Roger Cotes.	D. Gregory.	Colin Maclaurin.[1]
Robert Smith, etc.	Desaguliers.	James Gregory.
Attwood.	Mme. du Chatelet	J. Playfair.
Whewell.	and Voltaire.	Ivory.

Leibnitz and the Vis Viva Controversy.

Methods of dealing with connected systems.

Example of correct methods by Newton and others before D'Alembert.

D'Alembert's enunciation.—Its historical importance.

Euler. The Bernoullis, etc. Laplace, the flower of this stage of development.

Lagrange and Virtual Velocity.

This is the germ of the method of energy which was fully developed in mathematical form in the *Mecanique Analytique,* but very little appreciated outside the inner circle of mathematicians till the physical theory of energy became generally known.

Mathematical development of higher dynamics. (See Cayley's *Brit. Ass. Report,* 1857 and 1862? specially Hamilton and Jacobi.

Effect of T and T' since 1867.

---

[1] Introduces M $\dfrac{d^2x}{dt^2} = X$ etc. See Maclaurin's *Fluxions.*

Kirchoff's notions in *beginning* of Vorlesungen (not equal to Lagrange, but worth noticing).

I also think that Clausius' equation and definition of "Virial" is important.

The dynamics of other varieties of space than our own requires very brief notice indeed.—Yours truly,

J. CLERK MAXWELL.

To W. GARNETT, Esq.

Glenlair, Dalbeattie, 24th July 1877.

. . . There is a great slur over the word mechanics since a few poets and biologists have misused it. Pratt thought it a fine word.

The result of motion without reference to time I call Displacement. Kinematics must involve the idea of time if it treats of continuous displacements, velocities, and accelerations, though it does not contain within itself materials for comparing different intervals of time. For this we must go to the science which deals with matter; call it Kinetics, Dynamics, or Mechanics.

But I consider that Statics also deserves a place on the same level as Kinematics, as it deals with the equivalence of different systems of forces. But I do not agree with Whewell that Statics is more elementary than Kinematics. . . .

To THE SAME.

Glenlair, 11th August 1877.

Your experiments on electrified paraffin oil are excellent, and may lead to increase of knowledge.

If the fluid dielectric and also the air are perfect insulators, nothing can get electrified, but the equation at the surface, instead of being $P = P_0$, will be

$$P + \frac{1}{8n} (K^2 - 1)\overline{\frac{dV}{dv}}\Big|^2 = P_0$$

(excluding capillary action) where $\dfrac{dV}{dv}$ is the resultant electric force normal to the surface and just outside it. This causes the surface to rise wherever the normal force is great, or close to the electrodes.

.　　　　.　　　　.　　　　.　　　　.

The science of displacements is in Euc. I. 4, etc., and wherever one figure is placed upon another. It belongs to the method of contemplating the relations of two figures which may be supposed to co-exist, though we may also suppose that they are copies of the same figure in different positions.

But just as we assume that distance is a continuous quantity capable of measurement, though all our attempts at measurement are made with instruments made of non-rigid and discontinuous matter, so we may assume that time is a continuously flowing quantity capable of measurement, though we have not yet found out any accurate method of comparing distant intervals of time.

Now Kinematics requires no more than this notion of time, as the common independent variable t. If we suppose that τ is that (unknown) which flows uniformly, then for kinematical purposes it is enough that t is a function of τ; but when we come to Kinetics proper we must have $\dfrac{d^2t}{d\tau^2}$ very small.

Have you read Julius in *Nature*, about the beginning of June? [14th June].

The most constant things we know are the properties of bodies. For instance, water in equilibrium with ice and vapour gives us a good deal.

I. A unit of density (not the orthodox one) $\dfrac{M}{L^3} = D$.

II. A unit of pressure (too small for practical use) $\dfrac{M}{LT^2} = P$.

III. A unit of time (namely, the time of revolution of a satellite just grazing a sphere of water) $= T$.

These three quantities being independent of each other give M, L, and T.

$\dfrac{P}{D}$ gives a (velocity)^2 which could also be got from the $\dfrac{P}{D}$ of the vapour (a different one).

Then this gives also a standard temperature; all that we want is to get pure water.

To THE SAME.

Glenlair, 23d August 1877.

I have been copying Cavendish on the resistance of electrolytes. If there is any one who would try a few of them roughly in the U tube, it might be interesting to compare with Cavendish's results. For weak solutions Kohlrausch may be referred to.

Sea Salt (Chloride of Sodium.)

Experiments in January 1781.

Watered to 1 of Salt.		Resistance.	Resistance × quantity of Salt.
Saturated sol.	3·78	1	
	12	1·91	·602
	30	3·97	·500
	70	8·8	·475
	143	15·75	·416
	1000	93·02	·352
	20,000	18·23	·345

Salt in 20,000 conducts about 7 times better than distilled water.

Salt in 69 of water conducts 1·97 times better at 105° F. than at 58½°.

If Professor Liveing is in Cambridge, could you ask him to put me in the way of finding the best book on chemistry for the year 1777, so as to obtain the equivalents and the names of salts used by Cavendish?

The numbers in the first columns are the quantities which were equivalent to the "acid" in solution of 1 of salt in 29 of water.

3·2   Sal Sylvii (potassium chloride).

2·3   Sal Amm. (ammonium chloride).

14·10 Calc. S.S.A. (?)

2·21 Calcined Glauber's Salt (sodium sulphate).

3·17 Quadrangular Nitre (sodium nitrate).

5·19 Salt D. (?)

The solutions were 3, 10, 12.

I am going to try if this is Troy or Apothecaries' weight.
Saturated solution (1 in 3·78) of common salt has
437,000 the resistance of iron wire.   New distilled water
has more resistance than distilled water kept a year.

All these results and many more were got by comparison
of the strength of shocks taken through Cavendish's body.
I think this series of experiments is the most wonderful of
them all, and well worth verification.

.   .   .   .

Cavendish is the first verifier of Ohm's Law, for he
finds by successive series of experiments that the resistance
is as the following power of the velocity, 1·08, 1·03, ·980,
and concludes that it is as the first power.   All this by the
physiological galvanometer.

.

. . . Can you solve the equation

$$\frac{dz}{dx}\left(\frac{d^2z}{dy^2} - 2\,\frac{d^2z}{dxdy}\right) + \frac{dz}{dy}\left(\frac{d^2z}{dx^2} - 2\,\frac{d^2z}{dxdy}\right) = 0\,?$$

$z = \dfrac{A}{xy}$ is a solution.   Find the general ditto.

To Professor Lewis Campbell.

11 Scroope Terrace,
Cambridge, 5th January 1878.

Æt. 46.   It is more than a month that I have had your letter
lying by me.   I am glad you like Chrystal.   His departure
is a great loss to the laboratory, as it is difficult to find any
one to take up heavy work.   W. D. Niven (brother of the
competitor) is going in for a heavy piece of work on
conduction of heat in gases.   I am no judge of Greek plays,

but I think that your success in choruses is fully equal to
that in dialogue, considering the greater difficulty, not only
in the interpretation, but in guessing the kind of effect,
musical, rhythmical, rhetorical, poetical, and pictorial, which
was aimed at in the delivery of the chorus.

We have all been conversing on the telephone.   Garnett
recognised the voice of a man who called by chance.   But
the phonograph will preserve to posterity the voices of our
best speakers and singers.   See *Nature* of Jan. 3d.

To W. GARNETT, Esq.

Glenlair, 20th September 1878.

. . . . Cavendish would speak of the *pressure* of a voltaic
battery (only he hadn't one), but we require to be educated
up to his mark.

To THE LIBRARIAN OF THE ROYAL SOCIETY.

Glenlair, Dalbeattie, 23d June 1879.

DEAR SIR—Your information about FF.R.S. has been
so useful to me that I now ask about Dr. G. Knight, F.R.S.,
librarian to the British Museum.

(1.) Is his name Gowan, Gowen, Gowin, or Godwin, for
I find all four spellings current ?

(2.) Who is the author of the paper in *Phil. Trans.* for
1776 (near the end of the vol.) describing his great magazines of magnets ?

(3.) Are the magazines [sketch shown] mounted like
great guns still in the possession of the R.S. ?

(4.) Is the portrait of Gowin Knight, by Benjamin
Wilson, F.R.S., among the pictures of the R.S. ?

I have got from the Meteorological Office some Cavendish
MSS. on Magnetism which prompt these enquiries and also
this—

When the R. S. was at Crane Court had it a garden
adjoining ?   Also, where was Crane Court ?

Henry Cavendish and his father Lord Charles worked
together at observations of the variation compass and dipping

needle in the R. S. room and garden.   Are the variation compass and dipping needle still in the R. S. collection ?

Cavendish wrote out directions for using the dipping
needle for Captain Pickersgill, Captain Bayley, Dalrymple.

Dalrymple, I find from Poggendorff, was hydrographer
to the H.E.I.C.   If Cavendish apportioned his instructions
according to the capacity of the recipients, then their capacities would be in descending order, Dalrymple, Pickersgill,
Bayley.   Were any of these F.R.S. ?

Also, was John Walsh, F.R.S., also M.P. ?

Do not answer any of these questions which would
involve trouble, but I have not here any means of answering
them except by the aid of those who are among the records
of the past.   None of the questions are of vital importance,
because I can leave out any statements I have made which
are doubtful.—Yours very truly,

J. CLERK MAXWELL.

Professor Maxwell was frequently invited to join
the Victoria Institute, and in March 1875 he received
a letter from the secretary conveying the special
invitation of the President and Council to join the
Society, "among whose members are his Grace the
Archbishop of Canterbury, and other prelates and
leading ministers, several professors of Oxford and
Cambridge and other universities, and many literary
and scientific men."   The following is all that has
been found of a rough draft of his reply :—

SIR—I do not think it my duty to become a candidate
for admission into the Victoria Institute.   Among the objects
of the Society are some of which I think very highly.   I
think men of science as well as other men need to learn
from Christ, and I think Christians whose minds are scientific
are bound to study science that their view of the glory of

God may be as extensive as their being is capable of. But I think that the results which each man arrives at in his attempts to harmonise his science with his Christianity ought not to be regarded as having any significance except to the man himself, and to him only for a time, and should not receive the stamp of a society. For it is of the nature of science, especially of those branches of science which are spreading into unknown regions to be continually ——[here the MS. ends].

CHAPTER XIII.

ILLNESS AND DEATH—1879—ÆT. 47, 48.

AFTER his recovery from the attack of erysipelas at Glenlair in 1865, Maxwell's health appears to have been fairly good until the spring of 1877. He then began to be troubled with dyspeptic symptoms, especially with a painful choking sensation after taking meat. He consulted no one for about two years. But one day in 1877, on coming into the Laboratory after his luncheon, he dissolved a crystal of carbonate of soda in a small beaker of water, and drank it off. A little while after this he said he had found how to manage so as to avoid pain. The trouble proved obstinate, however, and at last, on the 21st of April 1879, he mentioned it when writing to Dr. Paget about Mrs. Maxwell.

By this time his friends at Cambridge had begun to observe a change in his appearance, and some failure of the old superabundant energy. They missed the elasticity of step, and the well-known sparkle in his eye. During the Easter Term of 1879 he attended the Laboratory daily, but only stayed a very short time. At the end of the term he remarked that he had been unable to do much more than to give his lectures. And before leaving Cambridge for the vacation, he was more than once very seriously unwell.

In June he returned, as usual, to Glenlair. His letters continued to be marked by humorous cheerfulness, and, as was always the case, contained information about everything and everybody except himself. He was still unwearied in his exertions for those to whom his services could be of use. But some casual remarks gave cause for apprehension that he was not gaining strength, and after he had been in Scotland for a few weeks he wrote that " he felt like a child, as for some time he had been allowed no food but milk." By and by the reports were more encouraging, and in September, according to appointment, Mr. and Mrs. Garnett were received at Glenlair. On Maxwell's coming out of the house to welcome them, Mr. Garnett saw a great change in him, and was for the first time seriously alarmed.

In the evening, however, the master of the house conducted family worship as usual for the assembled household. And the days passed much in the same kindly fashion as of old, linking the present to the past in " natural piety." There were the drawings of the oval curves of 1846 ; the family scrap-book, with Mrs. Blackburn's water - colour sketches from the earliest time ; the Glenlair autograph book (see above, p. 16) ; the bagpipes which saved the life of Captain Clerk in the Hooghly ; and a host of other treasures which Maxwell took delight in showing. He led his guests down to the river, and accompanied them a little way along its wooded banks, pointing out where the stepping-stones used to be, where he bathed when a boy, and where the exploit of tub-navigation had been performed. This was the longest walk he had taken for some time. He was unable to drive with them in the afternoon, because he could not bear the shaking of the carriage.

On the 2d of October 1879, in the midst of great weakness and of great pain, he was told by the late Dr. Sanders of Edinburgh, who had been summoned to Glenlair, that he had not a month to live. From that moment he had only one anxiety, the same which had for so long been his chief care—to provide for *her* comfort, whom he now saw that he must leave behind.

He returned to Cambridge; but he was so weak as to be hardly able to walk from the train to a carriage. Under the diligent care of Dr. Paget his most painful symptoms were considerably relieved, and his friends began to entertain fond hopes of his recovery. But his strength gradually failed, and at length it was evident to all that the disease could not be stayed.

During the last few weeks his sufferings were very great, but he seldom mentioned them; and, apart from his anxiety for others, his mind was absolutely calm. The one thought which weighed upon him, and to which he constantly referred, was for the future welfare and comfort of Mrs. Maxwell. During the whole period of their married life (twenty-one years) his ever-present watchfulness and sympathy had supported her even in the smallest domestic concerns ; his knowledge, his constructiveness, his dexterity of hand, had been ever ready to minister to her slightest need,—and now, unable to nurse him as of old, she seemed more than ever dependent on his care. To the last, he regularly gave the orders that were necessary for her comfort, and endeavoured to see that they were carried out.

When too weak to dwell on those scientific inquiries which had been the work of his life, his mind continued active about many of his favourite studies. He remarked one day that he had been wondering why the lines in Shakespeare's *Merchant of Venice*, about the harmony that is in mortal souls (repeating the whole passage) should have been put in the mouth of such a frivolous person as Lorenzo. At another time, when continuous conversation had become impossible, and he had been lying for some time with closed eyes, he looked up and repeated the verse, " Every good gift and every perfect gift is from above," etc., and then added—" Do you know that that is a hexameter ? πᾶσα δόσις ἀγαθὴ καὶ πᾶν δώρημα τέλειον. I wonder who composed it ? " He frequently quoted Richard Baxter's hymn—

> " Lord, it belongs not to my care,
>   Whether I die or live ;
>   To love and serve Thee is my share,
>   And that Thy grace must give," etc.

On the Saturday preceding his death he received the Sacrament of the Lord's Supper from Dr. Guillemard, and it was while Dr. G. was putting on his

surplice that Maxwell repeated to him George Herbert's lines on the priest's vestments, entitled *Aaron*.[1]
Maxwell's mind and memory remained perfectly clear to the very last.

The fortitude with which he bore his sufferings, and the calm self-possession with which he met his end, impressed those most who watched him most

---

[1] AARON.

Holiness on the head,
    Light and perfections on the breast,
Harmonious bells below, raising the dead
    To lead them unto life and rest :
    Thus are true Aarons drest.

Profaneness in my head,
    Defects and darkness in my breast,
A noise of passions ringing me for dead
    Unto a place where is no rest :
    Poor priest, thus am I drest.

Only another head
    I have, another heart and breast,
Another music, making live, not dead,
    Without whom I could have no rest :
    In Him I am well drest.

Christ is my only head,
    My alone only heart and breast,
My only music, striking me e'en dead ;
    That to the old man I may rest,
    And be in him new drest.

So holy in my head,
    Perfect and light in my dear breast,
My doctrine tuned by Christ (who is not dead,
    But lives in me while I do rest),
    Come, people ; Aaron's drest.

narrowly, and had the best reason to know the acuteness of his sufferings.

The end may best be told by those who were nearest to him at the time. I have been favoured with the following communications :—(1) From Dr. Paget; (2) from the Rev. Dr. Guillemard; and (3) from his cousin, Mr. Colin Mackenzie, who acted the part of a brother at the last, as he had done many a time before.

(1.) *Dr. Paget's Statement.*

In April 1879 he began to be troubled with some difficulty in swallowing—the first significant symptom of the disease which was to prove fatal. The summer he spent at Glenlair. At the end of July he consulted Prof. Sanders and Prof. Spence of Edinburgh, and while at Glenlair was attended by Dr. Lorraine of Castle-Douglas. But he grew worse, and at the end of September Prof. Sanders was summoned to him from Edinburgh. He was then suffering from attacks of violent pain, had become dropsical, and his strength was rapidly failing. At Glenlair he was seven miles from Dr. Lorraine. It was therefore decided to remove him to Edinburgh or Cambridge. He chose Cambridge, and arrived there on October 8, accompanied by Mrs. Maxwell, and attended during the journey by Dr. Richard Lorraine.

In Cambridge his more severe sufferings were gradually in great measure relieved, but the disease continued its progress. It was the disease of which his mother had died at the same age.

As he had been in health, so was he in sickness and in face of death. The calmness of his mind was never once disturbed. His sufferings were acute for some days after his return to Cambridge, and, even after their mitigation, were still of a kind to try severely any ordinary patience and fortitude. But they were never spoken of by him in a complaining tone. In the midst of them his thoughts and consideration were rather for others than for himself.

Neither did the approach of death disturb his habitual composure. Before leaving Glenlair he had learnt from Prof. Sanders that he had not more than about a month to live. A few days before his death he asked me (Dr. Paget) how much longer he could last. The inquiry was made with the most perfect calmness. He wished to live until the expected arrival from Edinburgh of his friend and relative Mr. Colin Mackenzie. His only anxiety seemed to be about his wife, whose health had for a few years been delicate, and had recently become worse. He had been to her for some time the most tender and assiduous of nurses. An hour only before his death, when, through extreme bodily weakness, his voice was reduced to a whisper so feeble that it could be heard only when the ear was held close to his mouth, the words whispered to Dr. Paget related not to himself but to Mrs. Maxwell.

His intellect also remained clear and apparently unimpaired to the last. While his bodily strength was ebbing away to death, his mind never once wandered or wavered, but remained clear to the very end. No man ever met death more consciously or more calmly.[1]

On November 5 he gently passed away.

---

[1] Dr. J. W. Lorraine of Castle-Douglas, in a letter addressed to Dr. Paget, and dated 5th October 1879, remarks as follows concerning his patient :—" I must say he is one of the best men I have ever met, and a greater merit than his scientific attainments is his being, so far as human judgment can discern, a most perfect example of a Christian gentleman." This remark, Dr. Paget observes, "is a *very* unusual one in a letter from one physician to another." Dr. Paget also says in writing to Mr. Garnett : "There is a deep interest in the fact of *how* such a man as Maxwell met the trials of sickness, and the approach of death. They are severe tests of amiability and unselfishness, and of the genuineness of religious convictions. It is something to say of a man that his unselfishness and composure remained undisturbed, and it is interesting physiologically and psychologically, that in the very extremity of bodily weakness, when the nourishment

Dr. Paget's report of Maxwell's composure through-out his illness is very strikingly confirmed by his letter to that physician, dated October 3, which is too con-fidential to be inserted here, but consists of a simple unadorned description of the facts of the case, and a request for aid which he knew would be forthcoming. A stranger, in reading that letter, would never divine, and indeed might find it hard to believe, that on *the previous day* (Oct. 2) the writer had been told by medical authority that he had only a month to live. Yet such is the fact. The words "for I am really very helpless," however touching as a description of his condition, are merely the statement of a reason why some one should be got "officially to help" Maxwell himself, but really and chiefly to do for Mrs. Maxwell what *he* had done so long as he had any strength in him. Students of history may perhaps recall Nicias's letter to the Athenians:—"You should also send a general to succeed me, for I have a disease, and cannot remain;"[1] but the words of the unfortunate general, "I claim your indulgence,"[2] though dignified enough, are more than Maxwell would have written.

(2.) The following letter, addressed to me by the Rev. Dr. Guillemard, of the Little Trinity Church, Cambridge, may be left to speak for itself:—

---

of the brain must have become so reduced, the mind remained per-fectly clear." Dr. Lorraine's remark on Maxwell's personal character expresses the feelings of all, nurses included, that were about Maxwell in his last illness.

[1] Thuc. 7, 15.   [2] *Ibid.*

Cambridge, 19th May 1881.

MY DEAR SIR—I shall disappoint you very much in my reminiscences of Maxwell. I never was an INTIMATE FRIEND of his, though we were always on the very best terms, and met not unfrequently, and he was most constant and assiduous in his attendance at church, and interested in all church matters.

But I knew very little about his *inner self* before I was summoned to his dying bed; and he had been brought very low physically, before his return to Cambridge, and was unequal to much *continuous* thought or conversation. He welcomed me warmly whenever I visited him, joined fervently in all acts of prayer, listened with a most intelligent interest to all I read, either out of the Bible (which he knew well-nigh by heart) or out of any of our great devotional writers in prose or poetry; was especially fond of any new hymns, and frequently capped such by reciting from his wonderful memory some parallel passages of his favourite old authors, specially George Herbert.

His faith in the grand cardinal verities was firm, simple, and full; and he avowed it humbly but unhesitatingly, with the deepest gratitude for the revelation of the truth in Jesus. I do not think he had any doubts or difficulties to cloud his clear mind or shake his peace.

He was calmly and serenely resigned to the will of God, and bowed in meek acquiescence before what he believed to be the Word of God.

I never saw a sign of impatience or fretfulness under all his long suffering, or heard an approach to a murmur. His one and only care was for his wife. It was a grand sight to see him day by day girding himself calmly and resolutely for the last struggle, and he passed through it undismayed. I wish I had preserved any of his last words: they have passed away from my shallow memory.—Yours very truly,

W. H. GUILLEMARD.

I am also permitted to insert the following more circumstantial account, which was written by Dr.

Guillemard to the Rev. Isaac Bowman, Vicar of South Creake, Fakenham, Norfolk, on the 9th of December 1879, within five weeks after Maxwell's death :—

He suffered exquisite pain, hardly able to lie still for a minute together, sleepless, and with no appetite for the food which he so required.

He understood his position from the first; knew what it all meant, and calmly girded himself for the awful struggle. He welcomed me at once as visiting him, not only as a friend, but as the Parish Priest come to assist him and to minister to him, and spoke of our relations with a grave, simple cheerfulness. You know the lightheartedness of the man in ordinary times ; and really it abode on him throughout ; he was never downcast or overburdened, and yet he was the humblest and most diffident of men, with the deepest sense of his own unworthiness, of his many short-comings, of his neglected opportunities. " But he loved much, and love had cast out fear." I used to go to him nearly every day of the five or six weeks he was here, to read and pray with him. He preferred the prayers of the Church, and asked for them, and by the wonderful power of his memory knew them all by heart ; but he gladly joined in other devotions, and took special delight in sacred poetry, of which I generally read him two or three short pieces.

He knew all our best writers in that line thoroughly : Milton, Keble, Newman, Wesley, George Herbert—the latter his chief favourite ; and he repeated to me the morning after an unusually bad night, the five stanzas of " Aaron " without a mistake. His knowledge of the Bible was remarkable, and he constantly asked for his most deeply-prized passages. Four days before he was removed from us he received the Holy Communion at my hands, with holy, reverent, fervid devotion, and said what strength it gave him.

I saw him only once again ; he was too weak and restless and exhausted for much intercourse ; but as I rose from my knees he said :—" My dear friend, you have been a true under shepherd to me: read me, before you go, the beautiful prayer out of the Burial Service, ' Suffer me not at my last hour ' "—and his grasp of my hand, as we parted, told me all he felt.

I had known but little of his inner self before his illness; he was singularly reticent; and though we occasionally discussed a text critically, we rarely got upon doctrine, or anything that touched upon the spiritual life. He was a constant regular attendant at church, and seldom, if ever, failed to join in our monthly late celebration of Holy Communion, and he was a generous contributor to all our parish charitable institutions. But his illness drew out the whole heart and soul and spirit of the man: his firm and undoubting faith in the Incarnation and all its results; in the full sufficing of the Atonement; in the work of the Holy Spirit. He had gauged and fathomed all the schemes and systems of philosophy, and had found them utterly empty and unsatisfying—" unworkable " was his own word about them—and he turned with simple faith to the Gospel of the Saviour.

(3.) Mr. Colin Mackenzie, who is by this time well known to the reader, was present at the last. He says :—

A few minutes before his death, Professor Clerk Maxwell was being held up in bed, struggling for breath, when he said slowly and distinctly, " God help me ! God help my wife ! " he then turned to me (Mr. Mackenzie) and said, " Colin, you are strong, lift me up;" He next said, " Lay me down lower, for I am very low myself, and it suits me to lie low." After this he breathed deeply and slowly, and, with a long look at his wife, passed away.

(4.) Another friend who saw him in his last illness, the Rev. Professor.Hort, has summed up his recollections of Maxwell, especially of the graver side of his character, in the following letter :—

FROM PROF. F. J. A. HORT TO PROF. L. CAMPBELL.

Feb. 4, 1882.

It is with extreme regret that I find myself powerless to comply with your request that I should contribute to your Memoir of Professor Maxwell a sketch of his position in reference to theology and religion. A competent and faithful account of Maxwell's inner thoughts during manhood would for several reasons have been of the highest interest. But his habitual reticence as to all that moved him deeply, and my own bad memory, have together left me without the materials needed for a task in itself most attractive. Though the impression of rare greatness which he left upon me in the first days of our acquaintance became stronger and stronger to the end, I have little to offer but a few vague and scattered reminiscences. Such as they are, I am thankful to be allowed to send them.

My earlier recollections of Maxwell are chiefly associated with a small society at Cambridge to which we both belonged, which used to meet on Saturday evenings for the discussion of literary and speculative questions. The aversion to rhetoric which he found traditional among its members was much to his taste, and he always took an animated and interested part in the conversations. Unfortunately his love of speaking in parables, combined with a certain obscurity of intonation, rendered it often difficult to seize his meaning ; but bright and penetrating little sayings, usually whimsical in form, and sometimes accompanied by strange gestures, recurred almost unfailingly at no distant intervals. Whether the tone of his mind was much affected by his participation in our discussions it is difficult to say. During the time that I knew him I can recall no perceptible signs of change other than quiet growth, and suspect that he attained too early and too stable a maturity to receive easily a new direction from any kind of intercourse with his University contemporaries. But it is likely enough that his mind was at least invigorated and consolidated by an influence which others have found reason to count among the strongest and also on the whole most salutary that they have known.

The same may probably be said of the influence of Mr. Maurice's writings, which certainly occupied Maxwell at this time. To what extent he was affected by them I do not know; but the tone in which he used to speak of Mr. Maurice leads me to think that they must have at least given him considerable aid in the adjustment and clearing up of his own beliefs on the highest subjects.

My intercourse with Maxwell dropped when we both left Cambridge. When I returned in 1872, after an absence of fifteen years, he had lately been installed at the new Cavendish Laboratory, and I had the happiness of looking forward to a renewal of friendship with him. I found him, as was natural, a graver man than of old; but as warm of heart and fresh in mind as ever. Owing to accidental circumstances on both sides, we met seldomer than I had hoped, though certainly there was no diminution of cordiality on the part of either. Strangely enough, it was again to the meetings of a small society, in purpose not unlike the former, that I owe most of my impressions of Maxwell in these later years. Though he was often unable to attend its meetings, and could rarely stay for more than an hour, he seemed to find much satisfaction in thus joining in the discussion of speculative questions with a few friends, chiefly middle-aged men, representing among them great diversity of studies, and no less diversity of opinion. The old peculiarities of his manner of speaking remained virtually unchanged. It was still no easy matter to read the course of his thoughts through the humorous veil which they wove for themselves; and still the obscurity would now and then be lit up by some radiant explosion.

Perhaps the most noteworthy of Maxwell's characteristics was his absolute independence of mind, an independence unsullied by conceit or consciousness. Preserved by his simplicity and humility from any fondness for barren paradox, he endeavoured always to see things with his own eyes, without regard to the points of view assumed on one side or another in ordinary controversy: in a word, he was more free from "notionalism" than any one whom I have known.

The testimony of his unshaken faith to Christian truth was, I venture to think, of exceptional value on account of his freedom from the mental dualism often found in distinguished men who are absorbed chiefly in physical inquiries. It would have been alien to his whole nature to seclude any province of his beliefs from the free exercise of whatever faculties he possessed; and in his eyes every subject had its affinities with the rest of the universal truth. His strong sense of the vastness of the world not now accessible to human powers, and of the partial nature of all human modes of apprehension, seemed to enlarge for him the domain of reasonable belief. Thus in later years it was a favourite thought of his that the relation of parts to wholes pervades the invisible no less than the visible world, and that beneath the individuality which accompanies our personal life there lies hidden a deeper community of being as well as of feeling and action. But no one could be less of a dreamer, or less capable of putting either fancies or wishes in the place of sober reality. In mind, as in speech, his veracity was thorough and resolute: he carried into every thought a perfect fidelity to the divine proverb which hung beside yet more sacred verses on the wall of his private room, " The lip of truth shall be established for ever."

During Maxwell's last illness I had the privilege of enjoying two conversations with him; and not long afterwards I put on paper a short and desultory record of some of his words. These notes contain nothing that might not with propriety be brought under other eyes, and therefore I venture to quote them here. Most of what passed on these two occasions presented nothing worthy of remark, unless it be the cheerful naturalness with which Maxwell spoke on all the varied topics that happened to come up before us. His thoughts had evidently been mainly taking a retrospective direction; and every interest of life seemed to be hallowed and brightened by the probable nearness of the Divine summons to a new form of existence.

He told me briefly the story of his illness; how he had been ailing all the year, but had gone northward after

Easter Term without apprehending anything worse than
transient ill-health. He had taken with him Professor
Clifford's *Lectures and Essays* which he had been asked to
review. He had read them with close attention for the
purpose, and had then, at some time in the summer, prepared
to write his criticism. It was a difficult and delicate task,
he said, for " there were many things in the book that wanted
trouncing, and yet the trouncing had to be done with
extreme care and gentleness, Clifford was such a nice fellow."
As soon as he tried to begin, his brain refused its office, and
he found himself incapable of composition ; and then he
knew that his illness had become serious.

Something, I forget what, led the conversation to the
perilousness of strong religious excitement in early youth, on
account of the spiritual exhaustion and permanent religious
insensibility that are apt to follow the dying-out of the
original fervour, and that derive a plausible justification
from the premature and fallacious experience. He spoke
with thankfulness of his own escape from a similar danger.
" The ferment," he said, " about the Free Church movement
had one very bad effect. Quite young people were carried
away by it ; and when the natural reaction came, they
ceased to think about religious matters at all, and became
unable to receive fresh impressions. My father was so
much afraid of this, that he placed me where I should be
under the influence of Dean Ramsay, knowing him to be a
good and sensible man.

" My father was an advocate. This added much to his
usefulness in the country. He was always fond of inventing
plans for country houses ; his note-books shew that he did
this as early as when he was twelve years old. He wanted
to build his house on a scale suited to what he thought he
would require as sheriff, and had so built a small part of it
when he died. We afterwards completed it, as far as possible according to his idea, but on a much smaller scale.
He had wished me to be an advocate ; but I never attended
law classes, as by that time it had already become apparent
that my tastes lay in another direction. Moreover, he looked

up greatly to James Forbes, and desired that I should
be like him. My father died before my marriage, and
before I had been actually elected Professor at Aberdeen.
He had been greatly interested in the sending in of offers of
application. He much wished me to have a Scottish Professorship, that I might have the long vacation free for living
at home.

" My interest is always in things rather than in persons.
I cannot help thinking about the immediate circumstances
which have brought a thing to pass, rather than about any
will setting them in motion. What is done by what is
called myself is, I feel, done by something greater than myself in me. My interest in things has always made me
care much more for theology than for anthropology ; states
of the will only puzzle me. I cannot ascribe so much to a
depraved will as some people do, though I do to a certain
extent believe in it. Much wrong-doing seems to be no
more than not doing the right thing ; and that finite beings
should fail in that does not seem to need the supposition of
a depraved will." On my saying that, though the immediate cause of the miseries of the world is oftener folly than
wickedness, yet men's folly can frequently be traced back to
past misdoing on their part, he warmly assented, and then
added in a different strain : " They were foolish because they
did not ask for wisdom,—not, of course, absolute wisdom,
but the wisdom needed for the moment.

" I have been thinking how very gently I have been
always dealt with. I have never had a violent shove in all
my life.

" The only desire which I can have is like David to serve
my own generation by the will of God, and then fall asleep."

The unexampled impression which his death produced at Cambridge was due to other causes besides
his scientific eminence. Those who lived so near to
him, though they saw little of him, could not fail to
have some feeling of the man. Some rumours of his

wonderful peacefulness in suffering had gone far
enough to touch many hearts. And it was a deep
and widely-spread emotion which found a voice that
Sunday in St. Mary's Church, through the mouth of
one who had known him when both were scholars of
Trinity—the Rev. Dr. Butler, the distinguished head-
master of Harrow School :—

It is a solemn thing—even the least thoughtful is
touched by it—when a great intellect passes away into the
silence, and we see it no more. Such a loss, such a void, is
present, I feel certain, to many here to-day. It is not often,
even in this great home of thought and knowledge, that so
bright a light is extinguished as that which is now mourned
by many illustrious mourners, here chiefly, but also far
beyond this place. I shall be believed when I say in all
simplicity that I wish it had fallen to some more competent
tongue to put into words those feelings of reverent affection
which are, I am persuaded, uppermost in many hearts on
this Sunday. My poor words shall be few, but believe me
they come from the heart. You know, brethren, with what
an eager pride we follow the fortunes of those whom we
have loved and reverenced in our undergraduate days. We
may see them but seldom, few letters may pass between us,
but their names are never common names. They never
become to us only what other men are. When I came up
to Trinity twenty-eight years ago, James Clerk Maxwell was
just beginning his second year. His position among us—I
speak in the presence of many who remember that time—was
unique. He was the one acknowledged man of genius
among the undergraduates. We understood even then that,
though barely of age, he was in his own line of inquiry not
a beginner, but a master. His name was already a familiar
name to men of science. If he lived, it was certain that he
was one of that small but sacred band to whom it would be
given to enlarge the bounds of human knowledge. It was
a position which might have turned the head of a smaller

man; but the friend of whom we were all so proud, and who seemed, as it were, to link us thus early with the great outside world of the pioneers of knowledge, had one of those rich and lavish natures which no prosperity can impoverish, and which make faith in goodness easy for others. I have often thought that those who never knew the grand old Adam Sedgwick and the then young and ever youthful Clerk Maxwell, had yet to learn the largeness and fulness of the moulds in which some choice natures are framed. Of the scientific greatness of our friend we were most of us unable to judge; but any one could see and admire the boy-like glee, the joyous invention, the wide reading, the eager thirst for truth, the subtle thought, the perfect temper, the unfailing reverence, the singular absence of any taint of the breath of worldliness in any of its thousand forms.

Brethren, you may know such men now among your college friends, though there can be but few in any year, or indeed in any century, that possess the rare genius of the man whom we deplore. If it be so, then, if you will accept the counsel of a stranger, thank God for His gift. Believe me when I tell you that a few such blessings will come to you in later life. There are blessings that come once in a lifetime. One of these is the reverence with which we look up to greatness and goodness in a college friend—above us, beyond us, far out of our mental or moral grasp, but still one of us, near to us, our own. You know in part, at least, how in this case the promise of youth was more than fulfilled, and how the man who, but a fortnight ago, was the ornament of the University, and—shall I be wrong in saying it?—almost the discoverer of a new world of knowledge, was even more loved than he was admired, retaining after twenty years of fame that mirth, that simplicity, that child-like delight in all that is fresh and wonderful, which we rejoice to think of as some of the surest accompaniment of true scientific genius.

You know, also, that he was a devout as well as thoughtful Christian. I do not note this in the triumphant spirit of a controversialist. I will not for a moment assume that there is any natural opposition between scientific genius and simple Christian faith. I will not compare him with others who have had the genius without the faith. Christianity, though she thankfully welcomes and deeply prizes them, does not need now, any more than when St. Paul first preached the Cross at Corinth, the speculations of the subtle or the wisdom of the wise. If I wished to show men, especially young men, the living force of the Gospel, I would take them not so much to a learned and devout Christian man, to whom all stores of knowledge were familiar, but to some country village, where for fifty years there had been devout traditions and devout practice. There they would see the gospel lived out; truths, which other men spoke of, seen and known; a spirit not of this world visibly, hourly present; citizenship in heaven daily assumed, and daily realised. Such characters I believe to be the most convincing preachers to those who ask whether Revelation is a fable, and God an unknowable. Yes, in most cases,—not, I admit in all,—simple faith, even peradventure more than devout genius, is mighty for removing doubts and implanting fresh conviction. But having said this, we may well give thanks to God that our friend was what he was, a firm Christian believer, and that his powerful mind, after ranging at will through the illimitable spaces of Creation, and almost handling what he called "the foundation stones of the material universe," found its true rest and happiness in the love and the mercy of Him whom the humblest Christian calls his Father. Of such a man it may be truly said that he had his citizenship in heaven, and that he looked for, as a Saviour, the Lord Jesus Christ, through whom the unnumbered worlds were made, and in the likeness of whose image our new and spiritual body will be fashioned.

There was a preliminary funeral ceremony in Trinity College Chapel, where the first part of the Burial Service was read, in the presence of all the leading members of the University. The body was then taken home to Glenlair, and buried in Parton Churchyard, the funeral being attended by numbers of his countrymen from far and near.

In these reminiscences I have purposely abstained as much as possible from comment. But in concluding this portion of the present work, I may be permitted to record a very few general observations or impressions.

The leading note of Maxwell's character is a grand simplicity. But in attempting to analyse it we find a complex of qualities which exist separately in smaller men. Extraordinary gentleness is combined with keen penetration, wonderful activity with a no less wonderful repose, personal humility and modesty with intellectual scorn. His deep reserve in common intercourse was commensurate with the fulness of his occasional outpourings to those he loved. His tenderness for all living things was deep and instinctive ; from earliest childhood he could not hurt a fly. Not less instinctive was the sense of equality amongst all human beings, which underlay the plainness of his address. But, on the other hand, his respect for the actual order of the world and for the wisdom of the past, was at least as steadfast as his faith in progress. While fearless in speculation, he was strongly conservative in practice.

In his intellectual faculties there was also a balance of powers which are often opposed. His imagination was in the highest sense concrete, grasping the actual reality, and not only the relations of things. No one was ever more impatient of mere abstractions.[1] Yet few have had so firm a hold upon ideas. Once more, while he was continually striving to reduce to greater definiteness men's conceptions of leading physical laws, he seemed habitually to live in a sort of mystical communion with the infinite.

His aunt, Mrs. Wedderburn, who had had the care of him during so much of his early life, said on the occasion of his marriage, "James has lived hitherto at the gate of heaven."

Mr. Colin Mackenzie has repeated to me two sayings of his during those last days, which may be repeated here—" Old chap, I have read up many queer religions : there is nothing like the old thing after all;" and—"I have looked into most philosophical systems, and I have seen that none will work without a God."

Maxwell's humour, which with many passed for mere eccentricity, and to some was the characteristic by which he was chiefly known, at least in earlier life, may be passed over lightly here. With strangers it was sometimes the veil of a sensitiveness which but for this would have made him the victim of his immediate surroundings. In confidential intercourse it was a perpetual fund of delight, the vehicle of his exuberant fancy, as it glanced in all directions from the immediate topic of discourse.

It was his way of acknowledging " the grotesque view "[2] of everything. Like other humourists whom

---

[1] He was particularly indignant at the confusion by some would-be philosophers of *facts* with *laws*.　　[2] P. 262.

I have known, he was never tired of a joke which had once tickled him; only, if retained in employment, it must always be tricked out with some new livery, and have some fresh turn given to it. As late as the summer of 1879, in writing to Professor Baynes about an article on Chemistry for the *Encyclopædia Britannica*, he repeated in some new way the well-worn jest about "an Analyser or a Charlatan." Even on his deathbed at Cambridge, in familiar converse with his cousin and friend, Mr. Colin Mackenzie, he still used the old quaint familiar speech :—" No, not that phial ! the little red-headed chap !"

Nor is it necessary to dwell on the rare freshness of feeling which he carried into middle life. The reader of his correspondence at any period must feel involuntarily that he had "the dew of his youth."

In thinking of him in college days, I used often to associate him in my own mind with Socrates. There is one point in the resemblance which I had not then realised, the "Socratic strength" of Antisthenes —his extraordinary power of moral and physical endurance.—Once at Cambridge, when his wife was lying ill in her room, and a terrier, who had already shown " a wild trick of his ancestors," was watching beside the bed, Maxwell happened to go in for the purpose of moving her. The dog sprang at him and fastened on his nose. In order not to disturb Mrs. Maxwell, he went out quietly, holding his arm beneath the creature, which was still hanging to his face.

What had struck me, I suppose, in making the above comparison, was the eager spirit of inquiry

beneath the ironical shell. I might have added the
union of speculation with mysticism, and of conservatism with progressive thought. But in one essential point, the dialectical cross-questioning method,
the analogy fails. For Maxwell had not spent his
youth in the Athenian agora, but in the solitudes of
Galloway, where he had interrogated Nature more
than Man.

In his conversation he might rather be compared
with the earlier Greek thinkers, " who," says Plato,
(Soph. 243 B) " went on their several ways, without
caring whether they took us with them, or left us
behind." The necessity of utterance was often
stronger with him than the endeavour to make himself understood, and he would pour out his ideas
in simple affectionateness to those who could not
follow them. His thoughts were often tentative,
but his expression of them was always dogmatic,
even in the negative formula "No one knows what is
meant by" so and so.

His indirect, allusive way of speaking was not,
however, wilfully assumed, but was the result of
idiosyncrasy and early habit; and it disappeared
utterly in the presence of any great occasion—a great
joy, a great sorrow, or a great duty. Then his speech
resolved itself into statements of fact, brief and
unemotional, but absolutely simple and direct. And,
latterly at all events, such were generally the characteristics of his style in writing. I have been told by
Mr. Huddlestone, a late Fellow of King's College,
Cambridge, that when consulted about a lightning

conductor for King's College Chapel (a building which he greatly loved) Maxwell called and made a verbal explanation which was unintelligible, but in going away he fortunately left a written statement, and this was perfectly clear.

The Galloway boy was in many ways the father of the Cambridge man; and even the "ploys" of his childhood contained a germ of his life-work. Indeed, it may be said that with him, despite the popular adage, "Work when you work," etc., play was always passing into work and work into play. In twirling his magic discs, his mind was already busied about the cause of optical phenomena. He plied the devil-on-two-sticks with the same eager industry, and with the same simple enjoyment, with which he afterwards spun his dynamical top. And amidst his profoundest investigations, whether about the Rings of Saturn or the Lines of Force, or the molecular structure of material things, the playful spirit of his boyhood was ever ready to break forth. Meanwhile, alike beneath the grave and sparkling mood, a spirit of deep PIETY pervaded all he did, whether in the most private relations of life, or in his position as an appointed teacher and investigator, or in his philosophic contemplation of the universe. There is no attribute from which the thought of him is more inseparable.

He had keen sympathy with ideal aspirations, together with an occasional sense of their fruitlessness. "It's no use thinking of the chap ye might have been." When, in their early married life, Mrs. Maxwell was oppressed with a sense of failure in her

first attempts at Cottage-visiting, he made her sit
down while he read to her Milton's sonnet ending
with the line, "They also serve, who only stand and
wait."

He appears in early days to have been conscious
of some superficial weaknesses, of a certain excitability
of temperament, leading to "preconscious states" and
preventing him from at once setting himself right in
new surroundings; also of the equal danger of shrinking
into himself, and "mystifying" those about him. This
difficulty, and many others "within and without,"
he overcame. But it would be too much to say of
him that "his affections never swayed more than his
reason," or that he obtained as firm a foothold in
practical life as he had by birthright in the region of
scientific thought. This great mass of mind was so
delicately hung as to be guided sometimes by a silken
thread. Few men, if any, would venture to argue or
remonstrate with Maxwell when he had decided on a
course of action in the council-chamber of his own
breast. But he could not consciously hurt any
creature, nor permit the possibility of causing pain
to those he loved. Nor was his power over others
always adequate to the keenness of his perceptions.
For while his penetration often reached the secrets of
the heart, his generosity sometimes overlooked the
most obvious characteristics—especially in the shape
of mean or vulgar motives.

His liberality, in every sense of the word, was
absolute. People have been disposed to criticise
the plainness of his entertainments, without knowing

CHAP. XIII.] CONCLUSION. 431

that while this was a matter of taste, the difference between the plain and the luxurious table was uniformly dispensed in charity. He has also been supposed by some who think that science should disown religion, to have been intolerant as well as orthodox. The contrary was true. And in particular, the mutual admiration and regard for one another of two such men as Maxwell and Clifford, notwithstanding their profound divergence of opinion on subjects of human interest, deserves to be quoted as an honourable exception to the narrow exclusiveness which has been too prevalent alike in the Christian and the anti-Christian world.[1]

He never sought for fame, but with sacred devotion continued in mature life the labours which had been his spontaneous delight in boyhood. Yet, considering the high region in which he worked, he received a large measure of recognition even in his lifetime. The Rumford Medal, conferred in 1860, was the first of a long list of honours, which up to his last year continued to accumulate from all parts of the civilised world.[2] And some of those who had an eye

---

[1] About the year 1860 I remember discussing with him J. Macleod Campbell's book on the Atonement, which had lately appeared. He made some criticisms, which I have forgotten, but I remember the emphatic tone in which he said, "We want light."

[2] In 1870 Maxwell received the honorary degree of LL.D. in the University of Edinburgh; on 11th November 1874 he was elected Foreign Honorary Member of the American Academy of Arts and Sciences of Boston; on 15th October 1875, Member of the American Philosophical Society of Philadelphia; on 4th December 1875, Corresponding Member of the Royal Society of Sciences of Göttingen; on 21st June 1876 he received the honorary degree of D.C.L. at Oxford;

for genius, though their intellectual interests lay in different spheres from his, could not forbear their testimony. Out of many such expressions it is enough to have selected one. Mr. Frederick Pollock in his work on Spinoza, having occasion to refer to Maxwell's views on matter and space, adds the following note (p. 115) :—

Clerk Maxwell was living when these lines were written : I cannot let them pass through the press without adding a word of tribute to a man of profound and original genius, too early lost to England and to Science.

Great as was the range and depth of Maxwell's powers, that which is still more remarkable is the unity of his nature and of his life. This unity came not from circumstance, for there were breaks in his outward career, but from the native strength of the spirit that was in him. In the eyes of those who knew him best, the whole man gained in beauty year by year. As son, friend, lover, husband ; in science, in society, in religion ; whether buried in retirement or immersed in business—he is absolutely single-hearted. This is true of his mental as well as of his emotional being, for indeed they were inseparably blended. And the fixity of his devotion both to persons and

on 5th December 1876 he was elected Honorary Member of the New York Academy of Science ; on 27th April 1877, Member of the Royal Academy of Science of Amsterdam ; on 18th August 1877, Foreign Corresponding Member in the Mathematico-Natural-Science Class of the Imperial Academy of Sciences of Vienna ; and in the spring of 1878 he received the Volta Medal and degree of Doctor of Physical Science *honoris causâ* in the University of Pavia.

ideas was compatible with all but universal sympathies and the most fearless openness of thought. There are no "water-tight compartments,"[1] there is no "tabooed ground;"[2] in spite of much natural reserve, he never really lost his predilection for "a thorough draft."[3] That marvellous interpenetration of scientific industry, philosophic insight, poetic feeling and imagination, and overflowing humour, was closely related to the profound *sincerity* which, after all is said, is the truest sign alike of his genius and of his inmost nature, and is most apt to make his life instructive beyond the limits of the scientific world. He would not wish to be set up as an authority on subjects (such as historical criticism) which, however interesting to him, he had not had leisure to study exhaustively. But our age has much to learn from his example. And in his life, regarded as a whole, there is a depth of goodness which can be but faintly indicated in his biography.

---

[1] P. 205. [2] P. 179.
[3] See the Occasional Poem on St. David's Day.

2 F

CHAPTER XIV.

LAST ESSAYS AT CAMBRIDGE.

AFTER Maxwell's return to Cambridge in 1871, several of those who had been "Apostles" together in 1853-7, revived the habit of meeting together for the discussion of speculative questions. This club of elder men (ἀνδρῶν πρεσβυτέρων ἑταιρία), which included such men as Dr. Lightfoot, now Bishop of Durham, and Professors Hort and Westcott, was christened Erănus (see p. 366), and three of Maxwell's contributions, dated by himself, have been preserved. It seems advisable to print these entire,—although not even chips from his workshop, but rather sparks from the whetstone of his mind,—since what he thought worthy of detaining the attention of such listeners in those ripe years cannot fail to be of interest to many readers.

I.

Does the progress of Physical Science tend to give any advantage to the opinion of Necessity (or Determinism) over that of the Contingency of Events and the Freedom of the Will ?

Æt. 41.   *11th February 1873.*

The general character and tendency of human thought is a topic the interest of which is not confined to professional philosophers. Though every one of us must, each for himself, accept some sort of a philosophy, good or bad, and though the whole virtue of this philosophy depends on it being our own, yet none of us thinks it out entirely for himself. It is essential to our comfort that we should know whether we are going with the general stream of human thought or against it, and if it should turn out that the general stream flows in a direction different from the current of our private thought, though we may endeavour to explain it as the result of a wide-spread aberration of intellect, we would be more satisfied if we could obtain some evidence that it is not ourselves who are going astray.

In such an enquiry we need some fiducial point or standard of reference, by which we may ascertain the direction in which we are drifting. The books written by men of former ages who thought about the same questions would be of great use, if it were not that we are apt to derive a wrong impression from them if we approach them by a course of reading unknown to those for whom they were written.

There are certain questions, however, which form the *pièces de résistance* of philosophy, on which men of all ages have exhausted their arguments, and which are perfectly certain to furnish matter of debate to generations to come, and which may therefore serve to show how we are drifting. At a certain epoch of our adolescence those of us who are good for anything begin to get anxious about these questions, and unless the cares of this world utterly choke our metaphysical anxieties, we become developed into advocates of necessity or of free-will. What it is which determines for us which side we shall take must for the purpose of this essay be regarded as contingent. According to Mr. F. Galton, it is derived from structureless elements in our parents, which were probably never developed in their earthly existence, and which may have been handed down to them, still in the latent state, through untold generations. Much might be said in favour of such a congenital bias towards a particular scheme of philosophy; at the same time we must acknowledge that much of a man's mental history depends upon events occurring after his birth in time, and that he is on the whole more

likely to espouse doctrines which harmonise with the particular set of ideas to which he is induced, by the process of
education, to confine his attention. What will be the
probable effect if these ideas happen mainly to be those of
modern physical science?

The intimate connexion between physical and metaphysical science is indicated even by their names. What
are the chief requisites of a physical laboratory? Facilities
for measuring space, time, and mass. What is the occupation of a metaphysician? Speculating on the modes of
difference of co-existent things, on invariable sequences,
and on the existence of matter.

He is nothing but a physicist disarmed of all his
weapons,—a disembodied spirit trying to measure distances in
terms of his own cubit, to form a chronology in which intervals of time are measured by the number of thoughts which
they include, and to evolve a standard pound out of his own
self-consciousness. Taking metaphysicians singly, we find
again that as is their physics, so is their metaphysics.
Descartes, with his perfect insight into geometrical truth,
and his wonderful ingenuity in the imagination of mechanical
contrivances, was far behind the other great men of his time
with respect to the conception of matter as a receptacle of
momentum and energy. His doctrine of the collision of
bodies is ludicrously absurd. He admits, indeed, that the
facts are against him, but explains them as the result either
of the want of perfect hardness in the bodies, or of the
action of the surrounding air. His inability to form that
notion which we now call force is exemplified in his explanation of the hardness of bodies as the result of the
quiescence of their parts.

"Neque profecto ullum glutinum possumus excogitare,
quod particulas durorum corporum firmius inter se conjungat,
quàm ipsarum quies." *Princip., Pars* II. LV.

Descartes, in fact, was a firm believer that matter has
but one essential property, namely extension, and his influence in preserving this pernicious heresy in existence
extends even to very recent times. Spinoza's idea of matter,

as he receives it from the authorities, is exactly that of Descartes; and if he has added to it another essential function, namely thought, the new ingredient does not interfere with the old, and certainly does not bring the matter of Descartes into closer resemblance with that of Newton.

The influence of the physical ideas of Newton on philosophical thought deserves a careful study. It may be traced in a very direct way through Maclaurin and the Stewarts to the Scotch School, the members of which had all listened to the popular expositions of the Newtonian Philosophy in their respective colleges. In England, Boyle and Locke reflect Newtonian ideas with tolerable distinctness, though both have ideas of their own. Berkeley, on the other hand, though he is a master of the language of his time, is quite impervious to its ideas. Samuel Clarke is perhaps one of the best examples of the influence of Newton; while Roger Cotes, in spite of his clever exposition of Newton's doctrines, must be condemned as one of the earliest heretics bred in the bosom of Newtonianism.

It is absolutely manifest from these and other instances that any development of physical science is likely to produce some modification of the methods and ideas of philosophers, provided that the physical ideas are expounded in such a way that the philosophers can understand them.

The principal developments of physical ideas in modern times have been—

1st. The idea of matter as the receptacle of momentum and energy. This we may attribute to Galileo and some of his contemporaries. This idea is fully expressed by Newton, under the form of Laws of Motion.

2d. The discussion of the relation between the fact of gravitation and the maxim that matter cannot act where it is not.

3d. The discoveries in Physical Optics, at the beginning of this century. These have produced much less effect outside the scientific world than might be expected. There are two reasons for this. In the first place it is difficult, especially in these days of the separation of technical from popular

knowledge, to expound physical optics to persons not pro-
fessedly mathematicians. The second reason is, that it is
extremely easy to show such persons the phenomena, which
are very beautiful in themselves, and this is often accepted
as instruction in physical optics.

4th. The development of the doctrine of the Conservation
of Energy. This has produced a far greater effect on the
thinking world outside that of technical thermodynamics.

As the doctrine of the conservation of matter gave a
definiteness to statements regarding the immateriality of
the soul, so the doctrine of the conservation of energy, when
applied to living beings, leads to the conclusion that the
soul of an animal is not, like the mainspring of a watch, the
motive power of the body, but that its function is rather
that of a steersman of a vessel,—not to produce, but to regulate and direct the animal powers.

5th. The discoveries in Electricity and Magnetism labour
under the same disadvantages as those in Light.   It is difficult to present the ideas in an adequate manner to laymen,
and it is easy to show them wonderful experiments.

6th. On the other hand, recent developments of Molecular Science seem likely to have a powerful effect on the
world of thought.   The doctrine that visible bodies apparently at rest are made up of parts, each of which is moving
with the velocity of a cannon ball, and yet never departing
to a visible extent from its mean place, is sufficiently startling
to attract the attention of an unprofessional man.

But I think the most important effect of molecular
science on our way of thinking will be that it forces on our
attention the distinction between two kinds of knowledge,
which we may call for convenience the Dynamical and
Statistical.

The statistical method of investigating social questions
has Laplace for its most scientific and Buckle for its most
popular expounder.   Persons are grouped according to some
characteristic, and the number of persons forming the group
is set down under that characteristic.   This is the raw
material from which the statist endeavours to deduce general

theorems in sociology. Other students of human nature proceed on a different plan. They observe individual men, ascertain their history, analyse their motives, and compare their expectation of what they will do with their actual conduct. This may be called the dynamical method of study as applied to man. However imperfect the dynamical study of man may be in practice, it evidently is the only perfect method in principle, and its shortcomings arise from the limitation of our powers rather than from a faulty method of procedure. If we betake ourselves to the statistical method, we do so confessing that we are unable to follow the details of each individual case, and expecting that the effects of widespread causes, though very different in each individual, will produce an average result on the whole nation, from a study of which we may estimate the character and propensities of an imaginary being called the Mean Man.

Now, if the molecular theory of the constitution of bodies is true, all our knowledge of matter is of the statistical kind. A constituent molecule of a body has properties very different from those of the body to which it belongs. Besides its immutability and other recondite properties, it has a velocity which is different from that which we attribute to the body as a whole.

The smallest portion of a body which we can discern consists of a vast number of such molecules, and all that we can learn about this group of molecules is statistical information. We can determine the motion of the centre of gravity of the group, but not that of any one of its members for the time being, and these members themselves are continually passing from one group to another in a manner confessedly beyond our power of tracing them.

Hence those uniformities which we observe in our experiments with quantities of matter containing millions of millions of molecules are uniformities of the same kind as those explained by Laplace and wondered at by Buckle, arising from the slumping together of multitudes of cases, each of which is by no means uniform with the others.

The discussion of statistical matter is within the province

of human reason, and valid consequences may be deduced
from it by legitimate methods; but there are certain peculiarities in the very form of the results which indicate that
they belong to a different department of knowledge from the
domain of exact science. They are not symmetrical functions of the time. It makes all the difference in the world
whether we suppose the inquiry to be historical or prophetical—whether our object is to deduce the past state or the
future state of things from the known present state. In
astronomy, the two problems differ only in the sign of t, the
time; in the theory of the diffusion of matter, heat, or
motion, the prophetical problem is always capable of solution; but the historical one, except in singular cases, is in-
soluble. There may be other cases in which the past, but
not the future, may be deducible from the present. Perhaps
the process by which we remember past events, by submitting
our memory to analysis, may be a case of this kind.

Much light may be thrown on some of these questions by
the consideration of stability and instability. When the state
of things is such that an infinitely small variation of the present state will alter only by an infinitely small quantity the
state at some future time, the condition of the system, whether
at rest or in motion, is said to be stable; but when an infinitely small variation in the present state may bring about
a finite difference in the state of the system in a finite time,
the condition of the system is said to be unstable.

It is manifest that the existence of unstable conditions
renders impossible the prediction of future events, if our
knowledge of the present state is only approximate, and not
accurate.

It has been well pointed out by Professor Balfour Stewart
that physical stability is the characteristic of those systems
from the contemplation of which determinists draw their
arguments, and physical stability that of those living bodies,
and moral instability that of those developable souls, which
furnish to consciousness the conviction of free will.

Having thus pointed out some of the relations of physical

science to the question, we are the better prepared to inquire what is meant by determination and what by free will.

No one, I suppose, would assign to free will a more than infinitesimal range. No leopard can change his spots, nor can any one by merely wishing it, or, as some say, *willing* it, introduce discontinuity into his course of existence. Our free will at the best is like that of Lucretius's atoms,— which at quite uncertain times and places deviate in an uncertain manner from their course. In the course of this our mortal life we more or less frequently find ourselves on a physical or moral watershed, where an imperceptible deviation is sufficient to determine into which of two valleys we shall descend. The doctrine of free will asserts that in some such cases the Ego alone is the determining cause. The doctrine of Determinism asserts that in every case, without exception, the result is determined by the previous conditions of the subject, whether bodily or mental, and that Ego is mistaken in supposing himself in any way the cause of the actual result, as both what he is pleased to call decisions and the resultant action are corresponding events due to the same fixed laws. Now, when we speak of causes and effects, we always imply some person who knows the causes and deduces the effects. Who is this person? Is he a man, or is he the Deity?

If he is man,—that is to say, a person who can make observations with a certain finite degree of accuracy,—we have seen that it is only in certain cases that he can predict results with even approximate correctness.

If he is the Deity, I object to any argument founded on a supposed acquaintance with the conditions of Divine foreknowledge.

The subject of the essay is the relation to determinism, not of theology, metaphysics, or mathematics, but of physical science,—the science which depends for its material on the observation and measurement of visible things, but which aims at the development of doctrines whose consistency with each other shall be apparent to our reason.

It is a metaphysical doctrine that from the same antecedents follow the same consequents. No one can gainsay
this. But it is not of much use in a world like this, in
which the same antecedents never again concur, and nothing
ever happens twice. Indeed, for aught we know, one of the
antecedents might be the precise date and place of the event,
in which case experience would go for nothing. The metaphysical axiom would be of use only to a being possessed of
the knowledge of contingent events, *scientia simplicis intelli-
gentiæ*,—a degree of knowledge to which mere omniscience of
all facts, *scientia visionis*, is but ignorance.

The physical axiom which has a somewhat similar
aspect is "That from like antecedents follow like consequents." But here we have passed from sameness to likeness,
from absolute accuracy to a more or less rough approximation.
There are certain classes of phenomena, as I have said, in
which a small error in the data only introduces a small error
in the result. Such are, among others, the larger phenomena
of the Solar System, and those in which the more elementary
laws in Dynamics contribute the greater part of the result.
The course of events in these cases is stable.

There are other classes of phenomena which are more
complicated, and in which cases of instability may occur, the
number of such cases increasing, in an exceedingly rapid
manner, as the number of variables increases. Thus, to take
a case from a branch of science which comes next to
astronomy itself as a manifestation of order: In the refraction of light, the direction of the refracted ray depends on
that of the incident ray, so that in general, if the one direction be slightly altered, the other also will be slightly altered.
In doubly refracting media there are two refracting rays, but
it is true of each of them that like causes produce like effects.
But if the direction of the ray within a biaxal crystal is
nearly but not exactly coincident with that of the ray-axis of
the crystal, a small change in direction will produce a great
change in the direction of the emergent ray. Of course, this
arises from a singularity in the properties of the ray-axis, and
there are only two ray-axes among the infinite number of

possible directions of lines in the crystal; but it is to be expected that in phenomena of higher complexity there will be a far greater number of singularities, near which the axiom about like causes producing like effects ceases to be true. Thus the conditions under which gun-cotton explodes are far from being well known; but the aim of chemists is not so much to predict the time at which gun-cotton will go off of itself, as to find a kind of gun-cotton which, when placed in certain circumstances, has never yet exploded, and this even when slight irregularities both in the manufacture and in the storage are taken account of by trying numerous and long continued experiments.

In all such cases there is one common circumstance,—the system has a quantity of potential energy, which is capable of being transformed into motion, but which cannot begin to be so transformed till the system has reached a certain configuration, to attain which requires an expenditure of work, which in certain cases may be infinitesimally small, and in general bears no definite proportion to the energy developed in consequence thereof. For example, the rock loosed by frost and balanced on a singular point of the mountain-side, the little spark which kindles the great forest, the little word which sets the world a fighting, the little scruple which prevents a man from doing his will, the little spore which blights all the potatoes, the little gemmule which makes us philosophers or idiots. Every existence above a certain rank has its singular points: the higher the rank, the more of them. At these points, influences whose physical magnitude is too small to be taken account of by a finite being, may produce results of the greatest importance. All great results produced by human endeavour depend on taking advantage of these singular states when they occur.

> There is a tide in the affairs of men
> Which, taken at the flood, leads on to fortune.

The man of tact says "the right word at the right time," and, "a word spoken in due season how good is it!" The man of no tact is like vinegar upon nitre when he sings his songs to a heavy heart. The ill-timed admonition hardens the heart, and the good resolution, taken when it is sure to be broken, becomes macadamised into pavement for the abyss.

It appears then that in our own nature there are more singular points,—where prediction, except from absolutely perfect data, and guided by the omniscience of contingency, becomes impossible,—than there are in any lower organisation. But singular points are by their very nature isolated, and form no appreciable fraction of the continuous course of our existence. Hence predictions of human conduct may be made in many cases. First, with respect to those who have no character at all, especially when considered in crowds, after the statistical method. Second, with respect to individuals of confirmed character, with respect to actions of the kind for which their character is confirmed.

If, therefore, those cultivators of physical science from whom the intelligent public deduce their conception of the physicist, and whose style is recognised as marking with a scientific stamp the doctrines they promulgate, are led in pursuit of the arcana of science to the study of the singularities and instabilities, rather than the continuities and stabilities of things, the promotion of natural knowledge may tend to remove that prejudice in favour of determinism which seems to arise from assuming that the physical science of the future is a mere magnified image of that of the past.

II.

ON MODIFIED ASPECTS OF PAIN.

31st October 1876.

Æt. 45.   We often make sensation in general the subject of discussion, but it does not appear that we think much about particular sensations. If we did, we should have had more names for them. Most of the words which seem at first sight to be names of sensations are really used as names of objects which we suppose to be associated with the sensation, and to be indicated by it. All such words as hot and

cold, flat and sharp, green, bright, bitter, frouzy, and so on,
though they may sometimes excite in a very sympathetic mind
a faint image of some actual sensation, call up much more
vividly the idea of some external object or phenomenon.
Language, in fact, has become far more an instrument for conveying information and for recording facts than for awaken-
ing sympathy; and even thought, in articulately speaking
men, is occupied rather in methodising our perceptions than
in chewing the cud of our sensations.

Of the few words which we have left to distinguish
states of feeling, most of the better sort, such as pleasure,
joy, happiness, are remarkably vague, so that the only
available part of the vocabulary of sensations consists of the
names of pains.

Such words as toothache, headache, heartache, certainly
fulfil the condition of suggesting, at least to non-medical
persons, a state of feeling rather than an objective phenomenon.

Now, whatever we may think, each for ourselves, at the
time when we feel a pain or an ache, about consciousness or
subjectivity, we are compelled, when we have to speak about
it, or even to think about it, to view it from the outside, and
to adopt the objective method of treatment.

A pain, then, in a sentient being other than ourselves is
a condition which that being ordinarily endeavours to put a
stop to, and the recurrence of which it tries to prevent;
just as a pleasurable sensation is one which it endeavours to
prolong, and afterwards to reproduce.

The psychologist has therefore to study pleasures and
pains in their effects on a sentient being, just as the
naturalist studies forces in their effects on the motion of
a material system. The motion of the material system
would go on in a determinate manner though no forces were
in action. The actions of a living being, when not in a state
of conscious effort, and often, indeed, even when it supposes
itself to be exercising what it would call its will, go on in a
manner determined by habits established by long custom,

and transmitted from generation to generation, though
varying slightly from individual to individual.

According to the simplest form of the Evolution theory,
those individuals, tribes, or species, which have habits unfavourable to their success in the battle of life, die out, and
help to improve the average quality of the remainder. This
good old rule, however, this simple plan, though it might
suffice for beings of great fecundity, would involve a great
waste of life among higher and less prolific forms. Hence
if any of these should come to have tastes and feelings, and
if these should happen to be such as to repel them from pernicious courses and attract them into salutary ones, this
would give them an advantage in the struggle for existence.
Of course, there would be an equal chance that the tastes
and feelings, when first developed, would be in favour of
courses leading to destruction; but the individuals possessed
by these tastes and feelings would be all the quicker exter-
minated for the good—that is to say, the more vigorous
continuation—of the species.

The susceptibility to pleasure and pain must therefore
be regarded as permitting a certain mitigation of the first
covenant of the evolutionists, "This do and live;" the
original severity of which was all the greater since the word
"this" could be interpreted only by observing what actions
were not followed by death.

By pursuing what we feel to be pleasant, and avoiding
what we feel to be painful, we are following a course which
was probably followed by many of our ancestors, and of
which we know at least this much—that it did not bring
them to such speedy destruction as to prevent them from leaving us as their descendants; and this is more than we can say
for any absolutely new rule of life struck out by ourselves.

But it must be confessed that the monitions of pleasure
and pain are not held in such high esteem by all men, as the
view which we have just taken would seem to warrant.
For though here and there we may find an isolated teacher
who has inculcated the cultivation of all forms of pleasurable
sensation, and the elimination of whatever is disagreeable,

yet there has never been a state of society in which it has
not been reckoned more honourable

To scorn delights and live laborious days,

than to run after pleasure and to shrink from pain.

And this brings me to the main point for our consideration. Why is it that in all times and countries the endurance of pain has been looked upon with great respect, and
has been considered necessary, salutary, honourable, or meri-
torious ? There is no difficulty, of course, about those cases
in which a person undertakes some enterprise, good in itself,
and in pursuance of this enterprise meets with various forms
of suffering. These are only instances of the general maxim
to despise evils of a lower order when they stand in the
way of some good of a higher order.

The ability to endure suffering while engaged in a good
work being recognised as a species of excellence, it is only
natural that this kind of ability should become the subject
of special cultivation. The aspirant to this form of virtue,
therefore, voluntarily submits himself to suffering, not for
the sake of any visible benefit to himself and others to be
obtained by enduring that particular pain, but for the sake
of discipline.

The aim of discipline is that he himself, and still more
that others, may have confidence that he is able to endure
suffering in a good cause, because he has already endured
suffering merely to justify this confidence.

Practices of this kind have in some countries been developed
to an extravagant degree, and may have given rise to various
abnormal sentiments and maxims ; but in the principle of
educational discipline there is nothing but the soundest
wisdom, for the only way in which an individual can
acquire confidence that he can perform a given act when he
wishes to do so is by previous practice ; and when it is
necessary for the public good that there should be a body of
men, of each of whom we can be sure that he will act in a particular manner when the occasion for it arrives, it is absolutely
necessary that they should be drilled to an exercise as like as

possible to the action which is expected of them. But in all
this the inconvenience or suffering which has to be endured
is by no means regarded as good in itself. It is treated
as a necessary concomitant of the act to be performed, just
as muscular effort, which may sometimes arise to the intensity of a pain, is a necessary concomitant of an athletic
exercise in which we may find pleasure.

Indeed the pleasurable or painful character of an elementary
sensation or thought cannot be determined without taking
into account all its surroundings. A sensation or a thought
when separated from its surroundings may be disagreeable
or painful; but when it occurs as a part of an act of perception,
we may not recognize it as painful.

There may even be a pleasurable complex emotion into
which the painful elementary sensation enters as an essential
ingredient, so that if on any occasion the painful sensation
is deficient, we feel that the pleasurable emotion is thereby
marred and rendered incomplete.

There is another class of cases, however, in which the pain
itself is the essential element. I mean penal suffering. The
aim of punishment, when not vindictive, is to prevent the
repetition of certain acts by associating with them penal
consequences.

The painful consequences of an act may be associated
with it by the way of natural consequence, or through the
medium of some external punishing authority, or by the
determination of the actor to punish himself.

When painful effects follow an act by way of natural
consequence, they are not generally called punishment,
because we do not attribute to nature any intention of punishing the particular act. But there can be no better instances
of the conditions of efficacious punishment than those in
which an act is so immediately and so certainly followed by
a painful sensation, that the sensation becomes inseparably
associated with the act, as in the case in which we touch
red-hot iron. In such cases the act is hardly ever repeated.

When the connection between the act and the sensation is not so immediate, but has to be traced through the action of a voluntary punisher, as when an act of a boy or a dog is followed by an unpleasant skin-sensation inflicted by the master, the efficacy of the punishment still depends in a great degree on its promptness and certainty, for these are the conditions under which a permanent association can be effected between the act and the sensation.

Let us now consider the feelings with which punishment may be regarded by the recipient.

There may be beings in whom the feeling is simple—pure repugnance. There is a beast called the Tasmanian Devil, which is said to fight against any odds, however overpowering, as long as any of him can stick together. There is no hope of taming or subduing such a beast by force, and it is probable that he will soon become extinct.

But in the case of less indomitable beings, punishment may soon assume a less hateful aspect.

For even if it is not efficacious in changing the habits, as soon as it is recognised as a certain consequence of transgression, the execution of the punishment may be welcomed as a relief from the expectation of it, and the culprit may have a satisfaction in getting it over, as an honest man has when he pays his debts.

If, however, he is really cured of the habit, he may look on the punishment as an operation by which he has not only paid in full for his past transgression, but has been free from the danger of falling into future transgressions.

Lastly, if his moral perceptions have been so far improved that he recognises that the action for which he was punished was really bad, he will sympathise more with the punisher than with his former self, and will admit not only the justice of his punishment as being according to law, but the justice of the law according to which he was punished.

We come next to those cases in which there is no external punisher, but in which grief or sorrow is awakened within ourselves on account of what we have done.

2 G

Here, again, the effect may be very different according to the mode in which the sensation of grief is applied, and the result is rendered complicated on account of the identity between the punisher and the punished.

The legitimate application of the emotion of grief is of course to the wrong act, so as to associate the act and all that belongs to it with this painful emotion, and so diminish the tendency to repeat the act.

But the association may not be strong enough, or the emotion of grief may altogether miss its mark, and may concentrate attention on itself, and so become transformed into self-pity,—a very complex emotion, in which the sweet is so mingled with the bitter, that its ultimate effect on our conduct becomes very uncertain ; the most probable result, however, being that on future occasions what should have been contrition passes still more easily into self-pity, and the whole performance assumes the character of a graceful play of feelings which we enjoy rather than suffer.

But though such perversions of feeling may occur, there is no doubt that men do make successful use of self-inflicted discipline as a means of influencing their own conduct. The lower degrees of such discipline are put in practice whenever we have to change our habits in the most minute particular.   Without it we could not improve our pronunciation or our handwriting.

The great difficulty, however, in providing for the punishment being applied in the right quarter, when the punisher is identical with the punished person, has led to the adoption of various imperfect solutions of the problem of penance, in some of which the discipline is inflicted by the arms on some other part of the body, and in others hunger is brought into play by abstaining from food. But in no sacrament is the intention of the administrator of such vital importance as in self-penance. Spenser tells us how the Red-Cross Knight underwent discipline under the directions of Patience.

But yet the cause and root of all his ill,
Inward corruption and infected sin,

> Not purged nor healed, behind remained still,
> 　And fest'ring sore, did ranckle yet within,
> Close creeping 'twixt the marrow and the skin ;
> 　Which to extirpe, he laid him privily
> 　Down in a darksome lowly place far in,
> Whereas he meant his corrosives to apply,
> And with streight diet tame his stubborn malady.
>
> In ashes and sackcloth he did array
> 　His daintie corse, proud humours to abate
> And dieted with fasting every day
> 　The swelling of his wounds to mitigate ;
> And made him pray both early and eke late.
> And ever as superfluous flesh did rot,
> 　Amendment ready still at hand did wayt,
> To pluck it out with pincers fiery whott
> That soon in him was left no one corrupted iott.

Here we see the advantage of allegory. In real life the poor man, instead of having Patience and Amendment as his good friends standing by his side, would have to conjure them up inside of him, and to apply the pincers with his own hand. But setting allegory aside, we can find here no exaltation of pain in itself. The pain is a necessary concomitant of the extirpation of evil habits from a conscious being. Amendment is the aim and end of the discipline, and the sole reason that he has to heat his pincers fiery whott is that the evil which he has to extirpe has crept close betwixt the marrow and the skin, and requires the actual cautery to prevent its creeping further.

Again, in the Collect for the first Sunday in Lent, we pray for " grace to use such abstinence," that, our flesh being subdued to the spirit, we may ever obey " the godly motions " of Christ. Abstinence is to be used only as a help towards the subjugation of our lower nature, and the opening up of our higher nature to divine influence.

I have said nothing of the aspect of penal suffering as a satisfaction either of justice from the point of view of a ruler, or of vengeance from the point of view of an aggrieved person.

Nor have I attempted to discuss the process by which the suffering, real or apparent, of one may be a direct source of pleasure to another who has cultivated the spirit of cruelty. For I cannot think that pure cruelty, as distinct from combativeness, anger and love of power, is anything but a morbid growth of feeling, subject, of course, to the general laws of the growth of feeling, whatever they may be, but no more to be regarded as a subject for our discussion here than those forms of perverted feeling which induce dogs to gnaw off their own toes, or devotees to hold their arms above their heads till they can no longer bring them down again.

III.

(PSYCHOPHYSIK.)

5th February 1878.

Whence came we? whither are we going? and what should we do now? are three questions of some celebrity. We have come from somewhere between this and Orion; we are going at — kilomètres per second towards Hercules; and we must therefore observe stars in a direction at right angles to our path;—are the answers suggested twenty-five years ago. It seems to me that a change has come over the questions, so that they now read, What used I to believe about myself? what is it likely I shall have to believe about myself? and what should I believe about myself now?

I used to believe myself to be the Conscious Ego. I am told I shall have soon to believe myself to be a congeries of plastidule souls, and that I must at once study psychophysik in order to obtain a true knowledge of myself.

I propose, therefore, to talk of the Conscious Ego, of Plastidule souls, and of Psychophysik.

(1.) What is your name? is a still more celebrated question. The suggested answer, N. or M., recalls to the mathematician ideas and operations of the most heterogeneous kind. Let us consider some of them. The instructors of my youth would have expected me to answer—My name is the Conscious Ego, one and indivisible, the Subject, in relation to whom all other beings, material, human, or divine, are mere Objects. Whether the being of such Objects can be maintained or upheld apart from my continuous perception of them is the great question of Metaphysic.

Though nothing can rise to the dignity even of an Object, except in so far as it is perceived by me, I regard certain Objects as nearer in rank to me, the Subject, than others, because it is through them that other Objects are perceived. Indeed, I often catch myself, when thinking about my body or my mind, supposing that I am thinking about myself.

There are other objects within the sphere of our perceptions which resemble our bodies, though they are not ours. The actions of these objects are so far like our own that we not only attribute to these objects the power of thinking, but also the consciousness of knowing, feeling, desiring and willing. In short, we suppose that each of these objects, when he asserts himself to be the Conscious Ego, means what we do when we make a similar statement.

The late Professor Ferrier, in his " Metaphysic," makes great use of this Alter Ego, and for his own purposes he treats him as a true Ego, whereas in Metaphysic he can never be more than an Object. This, however, is only a confusion of persons, not an actual division of the substance of the Ego. Our business to-night lies in the abysmal depths of Personality, and relates to the Unity of the Ego.

A great deal of what has been written on this subject relates to the continuity of the Ego in space and time, or in what corresponds in metaphysic to the space and time of physic. And first of Space. Has the Ego anything corresponding to Extension ? Has he parts ? and, if so, are these parts separable in fact or even in idea ?

It has been maintained that he has no parts ; that his state of consciousness at any instant is an inseparable whole, comparable, in respect of extension, to a mathematical point. According to this view, when I think I see an extensive prospect all at once before me, I am in reality either actually rolling my eye in an unconscious frenzy, or without any bodily motion I,—that is, the perceptive Ego,—am attending

first to one minimum visibile, and then to another, so that
what is presented to me is like the idea which a blind man
forms of the shape of an object by stroking it with the end
of his stick. The evidence relates only to the position of
points in the line traced by the end of his stick, but he fills
in the rest of the surface in accordance with his notions of
continuity and probability.

There is no department of psychophysik which has been
so successfully studied as that which relates to vision. When
we keep not only our eyes, but our attention, fixed on one
small object, the field of conscious vision seems to contract, till
only that object remains visible, and even it seems about to
disappear. It generally happens, however, that the feeling of
uneasiness which grows upon us causes at last a slight displacement of the eye, when suddenly a large extent of the
field starts into visibility, and the edges of objects, especially
those normal to the line of displacement, become obtrusively
prominent. This experiment seems at first sight to indicate
that the central spot of the retina has some exclusive privilege in the economy of vision. What it really shows is that
changes in the mode of excitation are essential to perfect
vision, and that vision cannot be maintained under an absolute sameness of excitation. On the other hand, by means
of the instantaneous light of a single electric spark, we may
read a whole sentence of print. Here we know that though
the illumination lasts only for a few millionths of a second,
the image on the retina lasts for a time amply sufficient for
an expert reader to go over it letter by letter, and even to
detect misprints. This experiment suggests certain speculations about memory, a faculty which is often supposed to
be essential to the continuity of the Ego in time.

When men wish to have things remembered, they set up
monuments, and write inscriptions and books,—they draw
pictures and take photographs,—in order that these material
things may help them, in time to come, to call up the
thought of that which they were intended to commemorate.

In our own bodies we have records of past events. Old

wounds may remind us of impressions made years ago, and ocular spectra remind us of impressions made seconds ago. Even in quiet meditation we sometimes find the ideas of visible objects accompanied with sensations hardly less vivid than those produced by real objects, and the memory of spoken words passes in a continuous manner, as the condition of our nerves becomes more exalted, first into a silent straining of the organs of speech, and then into an audible voice.

Beginners in music may practise on a dumb piano, and there is a silent process by which we may improve our pronunciation of foreign words.

We can thus trace a continuous series of the instruments of memory, beginning at the tables of stone and going on to the tables of the heart; and we are tempted to ask whether all memorials are not of the same kind—a physical impression on a material system.

The last American invention of the past year is Edison's Talking Phonograph. This instrument has an ear of its own, into which you may say your lesson, and a mouth of its own, which at any future time is ready to repeat that lesson.

The memory of this machine consists of tinfoil thin enough to be impressionable by the metal style which is set in motion by the voice, and yet thick enough to be retentive of these impressions, and at a proper time to communicate a corresponding motion to the style of the talking part of the machine.

Such is the heart of this instrument, by which it gets its speeches. Are our own hearts essentially different? We know what damage can be done to our memory by physical disturbances. We find ourselves quite unable to recall what we had often perused and reperused on the pages of memory. The page is lost out of our consciousness. Time goes on, and some day we find that the page with all that it contained has been restored to its place. Where was that page when it was out of our consciousness? Not surely in the Ego, unless there be an unconscious Ego. It must be out of the Ego, and therefore an Object. If this be so, it is of no great consequence to the dignity of the Ego, whether this particular object is purely spiritual or has a material substratum, just as history is history, whether or not it has a material substratum of paper and ink.

Memory is sometimes spoken of as if it were essential to individuality. When I wish to convince myself that I am the same person as a certain baby, I may do so by remembering as my own certain acts done by that baby. It may happen, however, that I cannot ascertain whether my present memory of these acts is due entirely to the direct impression made by these acts on me, or whether it is not mainly due to the frequent repetition of the story of these acts, as told to me afterwards by older persons.

But even if I were to find my memory to be all wrong, and that I am not that baby but a changeling, this would not touch my conviction that I who now am, am one.

The phenomenon of double consciousness, though not common, seems to be well established. A person has a double series of alternate states. In one the memory, education, accomplishments, and temper, are quite different from what they are in the other.

Instances have even been adduced of a man believing himself to be two or even three different individuals at the same time. But when we come to examine any particular case, we find that the man has nothing more than an erroneous opinion that he is entitled to the position and rights of the King, or of some other person, as well as his own, so that even in his own imagination he is no more than a person who holds several different offices at the same time.

(2.) The theory of Plastidule souls has been hinted at by several persons of whom Dr. Tyndall has spoken loudest. A much clearer utterance is that of Professor von Nägeli, of Munich, in an address delivered at the Munich meeting of the German Association in 1877.

Du Bois Reymond, in an address to the same body at Leipzig in 1872, had asserted—

"That in the first trace of pleasure which was felt by one of the simplest beings in the beginning of animal life upon our earth, an insuperable limit was marked; while upwards from this to the most elevated mental activity, and downwards from the vital force of the organic to the simple physical force, he nowhere finds another limit."

To this Professor Nägeli replies—

"Experience shows that from the clearest consciousness of the thinker downwards, through the more imperfect consciousness of the child, to the unconsciousness of the embryo, and to the insensibility of the human ovum,—or through the more imperfect consciousness of undeveloped human races and of higher animals to the unconsciousness of lower animals and of sensitive plants, and to the insensibility of all other plants,—there exists a continuous gradation without definable limit, and that the same gradation continues from the life of the animal ovum and the vegetable cell downwards, through organised elementary and more or less lifeless forms (parts of the cell), to crystals and chemical molecules."

Professor von Nägeli accepts the fact of sensation, appetency, and thought in the higher forms of life; but instead of trying to resolve it into a mechanical process, he levels up the discontinuity of the chain of being by attributing sensation not only to all organisms, but also to all cells, molecules, and atoms.   This is what he says—

"Now, if the molecules possess anything which is ever so distantly related to sensation, and we cannot doubt it, since each one feels the presence, the certain condition, the peculiar forces of the other, and, accordingly, has the inclination to move, and under circumstances really begins to move—becomes alive as it were; moreover, since such molecules are the elements which cause pleasure and pain; if, therefore, the molecules feel something that is related to sensation, then this must be pleasure, if they can respond to attraction and repulsion, i.e. follow their inclination or disinclination; it must be displeasure if they are forced to execute some opposite movement, and it must be neither pleasure nor displeasure if they remain at rest."

Professor von Nägeli is what Professor Huxley would call a mere biologist, or he would have known that the molecules,

like the planets, move along like blessed gods. They cannot
be disturbed from the path of their choice by the action of
any forces, for they have a constant and perpetual will to
render to every force precisely the amount of deflexion which
is due to it. They must therefore enjoy a perpetuity of the
highest and most unmixed pleasure, even when, as Professor
Nägeli says, they are the cause of pain to us.

To attribute life, sensation, and thought to objects in
which these attributes are not established by sufficient
evidence is nothing more than the good old figure of
Personification.

If certain bodies, like the sun and stars, move in a
regular manner which we can predict, we may, if it pleases
us, suppose that their nature is like that of the just man
whom nothing can turn from the path of rectitude. If the
motion of other bodies is less simple, so that we cannot
account for it, we may suppose their nature to be tainted
with that capriciousness which we observe in our fellow-men,
and of which we are occasionally conscious in ourselves.

But the study of nature has always tended to show that
what we formerly attributed to the caprice of bodies is only
an instance of a regularity which is unbroken, but which
cannot be traced by us till we acquire the requisite skill.

But granting that the mental powers of atoms may be,
for anything we know, of the very highest order, what step
have we made towards linking our own mental powers with
those of lower orders of being?

(3.) Let us suppose that a thinking man is built up of a
number of thinking atoms. Have the thoughts of the man
any relation to the thoughts of atoms or of one or more of
them? Those who try to account by means of atoms for
mental processes do so not by the thoughts of the atoms,
but by their motions.

Hobbes, in the frontispiece of his *Leviathan*, shows us
a monster like the wicker images of our British antecessors
stuffed with men, and the whole method of his book is

founded on an analogy between the body politic and the individual man.

Herbert Spencer has pushed the analogy both upwards and downwards as far as it will go, and further than it can go on all fours. He shows us how a society is an organism, and how an organism is a society,—how the lower forms of societies and organisms consist of a multitude of homogeneous parts, the functions of which are imperfectly differentiated, so that each can at a pinch undertake the office of any other ; whereas the parts of higher forms of organisms and societies are exceedingly heterogeneous, and discharge more perfectly differentiated functions. Hence to the lower forms a breaking up may be a multiplication of the species, whereas to the higher forms it is death.

In a society, as in an organism, both the working and the thinking will be better done if undertaken by different members, provided that the thinking members can guide the working members, while the working members support the thinking members,—the workers retaining just enough intelligence to enable them to receive the guidance, and the thinkers retaining just enough working power to enable them to appropriate the pabulum presented to them by the workers.

Hence in the more highly developed systems the guiding powers may be concentrated into a smaller portion of the whole system, and may exercise a more undisputed power of guiding the rest, till in the highest organism we arrive at what is called Personal Government, and the organism may bear without abuse the grand old name of Individual. This result is brought about by all the members except one bartering their right of guiding themselves for the privilege of being guided, and so delegating to the one ruling member the functions of government. When the human society has lapsed into the condition of personal government, the consciousness of the head of the state may be expressed by him in the phrase, " L'état c'est moi ; " but though the other members of the society may delegate to the head all their political powers, they cannot delegate their

sensations or any other fact of consciousness, for these
are the incommunicable attributes of that Ego to whom they
belong.

I have now to confess that up to the present moment I
have remained in ignorance of how I came to be, or, in the
Spencerian language, how consciousness must arise. I was
dimly aware that somewhere in the vast System of Philosophy
this question had been settled, because the Evolutionists are
all so calm about it; but in a hasty search for it I never
suspected in how quiet and unostentatious a manner the
origin of myself would be accounted for. I am indebted to
Mr. Kirkman for pointing it out in his *Philosophy without
Assumptions.* Here it is with Mr. Kirkman's comment.
Principles of Psychology, § 179, p. 403 :—" These separate
impressions are received by the senses—by different parts of
the body. If they go no further than the places at which
they are received, they are useless. Or if only some of them
are brought into relation with one another, they are useless.
That an effectual adjustment may be made, they must all be
brought into relation with one another. But this implies
some centre of communication common to them all, through
which they severally pass ; and, as they cannot pass through
it simultaneously, they must pass through it in succession.
So that as the external phenomena responded to become
greater in number and more complicated in kind, the variety
and rapidity of the changes to which this common centre
of communication is subject must increase — there must
result an unbroken series of these changes — there must
arise a consciousness."

On this Kirkman remarks: " He knew he could do it, and
he did it !"—What was the evolution of light to this ? The
next Longinus will put that in before γενέσθω φῶς καὶ
ἐγένετο φῶς.

The opinions about my origin are as various as those
about my nature. Canon Liddon tells me that I was
created out of nothing in the year 1831, though I cannot

make out from what he says on what day of that year the event took place, or why my parents and not some one else found me under a gooseberry bush ; or, indeed, why I should have any part or lot in family matters, from Adam's first sin down to my father's last name.

Mr. Francis Galton tells me that I am developed from structureless germs contributed not only by my two parents but by their remotest ancestors centuries ago. My existence, therefore, does not begin abruptly, but tails off as an exponential function of t does for negative values of that variable. My local existence, however, though at present confined within the periphractic region of my skin, was in former times discontinuous as regards space, being carried about by two, four, or more, distinct human beings.

Dr. Julius Müller tells me that by an analysis of my conscience I shall come to a very different result, namely, that I existed (if, indeed, the fact can be expressed by a past tense and not a pure aorist), in an extra-temporal state, and that in that state I freely determined myself to choose evil rather than good.

He does not say I can remember this transaction. The conviction I am to acquire of it is not to be an experimental or empirical consciousness, but a speculative or philosophical knowledge.

Since, according to Müller's theory, this extra-temporal decision is perfectly free, and since it would be difficult to predicate freedom of a choice which is invariably on one side, he is obliged to assert that in the extra-temporal state some of our species must have chosen the better part. But he has also to maintain that all of us who are born into this world by ordinary generation have already chosen the worse part.

Hence, though he does not say so, he makes the extra-temporal fall a condition of our being born into this world. Whether those of us who make the better choice are born into some other world, or whether, so long as they remain unfallen, they continue in the extra-temporal state—a state certainly not higher but rather more undeveloped than that of time—Müller does not say.

Dr. John Henry Newman has shown us how the doctrine of post-baptismal sin became developed into the discovery of Purgatory with all its geographical details. It would seem as if the doctrine of original sin, in the hands of speculative theologians, might open up to our view a far more transcendental region, compared with which the stairs and terraces and fires of purgatory are as familiar as those of our own hearths and homes.

As to my present state, Du Bois Reymond tells me that not only my bodily but a large part of my mental functions are performed by the motion of atoms under fixed laws, and his result is that the finite mind, as it has developed itself through the animal world up to man, is a double one,—on the one side the acting, inventing, unconscious, material mind, which puts the muscles into motion and determines the world's history : this is nothing else but the mechanics of atoms, and is subject to the causal law ; and on the other side the inactive, contemplative, remembering, fancying, conscious, immaterial mind, which feels pleasure and pain, love and hate : this one lies outside of the mechanics of matter and cares nothing for cause and effect.

Dr. Drysdale tells me that not only my thinking powers but my feelings are functions of the material organism, and that I myself am such a function. He admits that I am not material—no function can be material, for matter is a substance, and a function is not a being at all. Dr. Drysdale, as a Christian materialist, follows his master Fletcher, who says—

"As often as it shall be said that mind or the faculty of thinking is a property of living matter,—that it is born with the body, is developed with the body, decays with the body, and dies with the body,—it is to be understood of the mind only, not the soul. The soul is something not material indeed, but substantial—a divine gift to the highest alone of God's creatures, responsible for all the actions of mind, but as totally distinct from it as one thing can be from another, or rather as something is from nothing."

Dr. Drysdale, however, in order to save the dynamical

theory of life and mind, says that this soul or spirit must either, if now existing, be a passive spectator of the action of the living being connected with it, or else that he has no existence during this present life, but is to be constituted by a divine act after the death of the living being.

In either case, I cannot identify this soul with myself, for I know that I exist now, and that I act, and that what I do may be right or wrong, and that whether right or wrong, it is my act, which I cannot repudiate.

In this search for information about myself from eminent thinkers of different types, I seem to have learnt one lesson, that all science and philosophy and every form of human speech is about objects capable of being perceived by the speaker and the hearer ; and that when our thought pretends to deal with the Subject it is really only dealing with an Object under a false name. The only proposition about the Subject, namely, " I am," cannot be used in the same sense by any two of us, and, therefore, it can never become science at all.

END OF PART I.

PART II.

CONTRIBUTIONS TO SCIENCE.

" He was one of those who took more delight in the contemplation
of truth than in the praise of having discovered it."
                                        PLAYFAIR'S *Memoirs of Dr. Hutton.*

AT the close of the memoir published in the *Proceedings
of the Royal Society,* Mr. W. D. Niven, referring to Maxwell's
scientific work, says :—

It is seldom that the faculties of invention and exposition,
the attachment to physical science, and capability of developing
it mathematically, have been found existing in one mind to the
same degree. It would, however, require powers somewhat akin
to Maxwell's own, to describe the more delicate features of the
works resulting from this combination, every one of which is
stamped with the subtle but unmistakable impress of genius.

It will probably be many years before an approximate
estimate can be formed of the value of Maxwell's work. In
the following pages no attempt has been made to give more
than a brief account of a few of his principal contributions
to science. The chief subjects referred to have, for convenience, been arranged in the following order :—

1. Experiments on Colour Vision and other contributions to Optics.

2. Investigations respecting Elastic Solids.

3. Pure Geometry.

4. Mechanics.

5. Saturn's Rings.

6. Faraday's Lines of Force, and Maxwell's Theory of the Electro-magnetic Field, including the Electro-magnetic Theory of Light and other investigations in Electricity.

7. Molecular Physics.

1. The subject of colour-vision attracted Clerk Maxwell's attention at an early period. In dealing with phenomena of this class, we must remember that it is necessary to distinguish between the sensation itself and its physical cause, or between the subjective and objective aspects of the same phenomenon. Thus a pure musical *tone* consists of a regular succession of similar vibrations, and two tones may differ in the extent of these vibrations, and in the number which take place in a second. Corresponding to these physical differences, are experienced differences in the *intensity* or *loudness*, and the *pitch* of the note. Light, like sound, consists objectively of certain periodic disturbances or vibrations of a medium, but differs from sound in the character of the motion, the nature of the medium which transmits it, and the number of vibrations which take place in a second. The simplest kind of light consists, like a pure tone in sound, of a regular succession of similar vibrations, the extent of which determines the intensity of the light, while their rapidity corresponds to the *colour* sensation produced. The constitution of ordinary white light is much more complicated. Newton allowed a beam of sunlight to pass through a prism, and then to fall upon a screen, when, instead of a white patch of light, he obtained a spectrum, or continuous band of colour, varying from crimson through scarlet, orange, yellow, green, blue, to violet. Rays corresponding to this infinite variety of colour must, therefore, exist together in white solar light, and these rays differ physically from one another in the rapidity of the vibrations of which they consist; the deep crimson corresponding to less than 400,000,000,000,000 vibrations per second, and the extreme violet to more than 700,000,000,000,000, the length of a wave in air being, in the first case, about $\frac{1}{39,000}$ of an inch, and, in the second case, about $\frac{1}{68,000}$ of an inch.

But white light is not necessarily of so complex a nature as sunlight; thus it may consist of a mixture of red, green, and blue lights simply, or of yellow and blue, or greenish-yellow and violet, or of other mixtures of two or more kinds of light, each in itself homogeneous.   When this is the case, the constitution of the light is revealed when it is allowed to pass through a prism, for it is then separated into its simple constituents which, on emerging from the prism, pursue different paths.   Similarly, various kinds of homogeneous light may be *matched* by means of a mixture of lights of other colours appropriately chosen.   Thus orange light of a particular hue may be the homogeneous light of the spectrum, or it may be a mixture of red and yellow or of red and green lights, which is *chromatically* identical with the homogeneous orange, but *optically* different, inasmuch as the mixture can be resolved into its constituents by means of a prism, while the homogeneous orange light may be passed through any number of prisms and yet retain its perfect homogeneity.

These results were accounted for by Thomas Young (1801), on the supposition that there are three separate sensations which are excited to different degrees—the proportion in which each is excited depending on the nature of the light. He was of opinion that these three sensations were red, green, and violet, and that all other hues were compound *colours*, though they might correspond to a simple *kind of light*. " The *quality* of any colour depends, according to this theory, on the *ratio* of the intensities of the three sensations which it excites, and its *brightness* depends on the *sum* of these three intensities."

Young's colour diagram was a triangle, at the angular points of which he placed the three colours corresponding, according to his theory, with the primary sensations; along the sides of the triangle were placed the colours formed by mixing these two and two together, while within the triangle were to be found the colours resulting from mixtures of all three in different proportions.   The position of each colour was determined by finding the centre of gravity, of two or three heavy particles placed at the angular points of the triangle, the weight of each particle being proportional to the quantity of the corresponding light employed in the mixture.

Although the light of the spectrum which corresponds to any particular spectral colour is simple in its constitution, consisting of a definite number of waves per second, and ordinary white light is a combination of an infinite number of such rays, varying in rapidity of vibration from the extreme red to the extreme violet, yet *the mechanism of vision* appears to be of a threefold character, so organised that light of each particular wave-length affects in different proportions the three colour senses, the hue depending on the relative amounts to which they are severally excited, as explained by Young. An analogy may help to make this distinction clearer; thus, it has been customary to speak of the heating power, the illuminating power, and the photographic power, of any particular kind of light; and in the solar spectrum formed by a *glass* prism the greatest heating power is possessed by the ultra red rays, the greatest illuminating power (judged by normal eyes) by the greenish yellow rays, and the greatest photographic power (in relation to silver salts) by the ultra-violet rays; but we do not thereby imply that there are three separate classes of rays in solar light, consisting respectively of heating rays, illuminating rays, and chemical rays : we simply mean that any particular homogeneous ray possesses each of these three powers to a certain extent. Similarly any ray of the spectrum may be capable of stimulating each of three independent colour sensations, but it does not follow that it consists physically of three different kinds of light.

In the *Philosophical Magazine* for 1852 there is an account of an experiment by Helmholtz, which consisted in mixing the colours of the spectrum by forming two spectra from slits at right angles to each other. It is obvious that if the breadth of each spectrum be equal to its length, every pair of colours in the whole spectrum will in this way be superposed. In Helmholtz's experiment it appeared that a mixture of yellow and indigo produced white.

Maxwell's first published experiments on the mixture of coloured lights, executed while he was a B.A. at Cambridge, had reference to the mixtures of coloured lights obtained from coloured papers. One of the best accounts of these experiments will be found in the paper read before the Royal Society of Edinburgh by Clerk Maxwell on 19th March 1855. The instrument first employed was a top, constructed by Maxwell himself; but he afterwards had tops

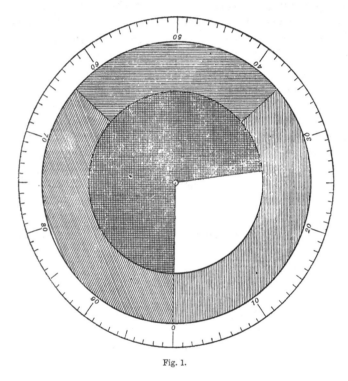

Fig. 1.

made by Mr. Bryson of Edinburgh. The coloured papers were cut into circular discs, with a small hole in the centre to admit the spindle of the top, and were slit along a radius from the circumference to the central hole. By this means two or more discs could be placed together, and by turning one relatively to the others, more or less of any particular disc could be made visible from above. The top consisted of a disc of metal, covered with white paper and mounted on a suitable spindle, the circumference of which was divided into 100 equal parts. Two sizes of discs were employed; the larger discs having been adjusted as required, were placed on the graduated plate, and the smaller discs above them. The plate then presented an appearance similar to that sketched in the adjoining figure. The employment of the apparatus depends on the fact, observed by Hartley the psychologist, that visual impressions remain for some time on the retina even after the source of light has been removed. Thus, suppose that the larger discs are respectively red, green, and blue, and suppose that we look at the middle of the red sector and then spin the top. If the spinning is sufficiently rapid the impression produced by the red light will not have diminished sensibly before the green paper takes the place of the red, and the green light produces its effect on the same portion of the retina, and this again is followed by blue before either the red or the green sensations have sensibly diminished, and then the whole process is repeated many times in a second. The impressions due to the three coloured lights are therefore blended on the retina, and it can be shown experimentally that the tint observed is the same as when the three kinds of light in the same relative proportions are allowed to fall simultaneously upon the eye. The colour top, therefore, enables us to mix together, in any definite proportions, which can be changed at will, the lights from differently coloured papers, and to observe the effect produced by the mixing.

Suppose that the colours of the larger discs are red, green, and blue, and that these are arranged so that the coloured lights may be mixed in the proportions necessary to form white light. Then the colour of the ring exhibited by the larger discs when the top revolves can differ from that of white paper only in respect of illumination. If, therefore, we can diminish the apparent brightness of the central circle at will, we may produce a complete match. This may be effected by combining a black disc, which emits very little light indeed, with the white disc, so that the amount of light received from the central circle is approximately proportional to the angle of the white sector. On spinning the top, the central circle appears of a neutral gray, which is the same as a dull white, and the darkness of the gray increases with the amount of the black sector introduced. If, now, the outer circle appears yellow compared with the central gray when the top is spinning, it is found that, by increasing the amount of blue and diminishing the red and green, the circles may be matched in hue. Similarly, if it appears too green we must diminish the green and increase the red and blue, while if it appears purple we must increase the amount of green. After a little experience it is easy to recognise the character of the change required.

Now, suppose that we have three and only three colour sensations, and that, for the sake of argument, these correspond respectively to red, green, and blue. Then if two colours contain the same amount of red, the same amount of green, and the same amount of blue, they must be in every respect identical. There are, therefore, three conditions to satisfy in order that two colours may match. But in the case of the top, two other conditions must be also fulfilled, for the portions of the smaller discs employed must exactly fill up the circle, and the same must be true of the larger discs. Hence there are in all five conditions to be fulfilled in order to make a colour match, and in general five discs will have to be employed and matched, two against three or three against two. If there had been four primary colour sensations, we should have required in general to employ six discs to ensure a match, and so on for any other number. Maxwell showed that with a normal eye a match could always be obtained with five discs, but that with a colour-blind person a match could always be obtained with four discs, thus demonstrating that the normal eye possesses three independent colour sensations, while the eye of the colour-blind possesses only two. This conclusion will appear more evident from the experiments with the colour box, which will be described presently.

The result of each experiment was first expressed by an equation in which the colours in the outer circle, with their respective quantities as coefficients, appeared on the left of the equation, and those in the inner circle on the right side. Thus, employing V for vermilion, U for ultramarine, and EG for emerald green, SW for snow-white, and Bk for black, the equation

$$\cdot37V + \cdot27U + \cdot36EG = \cdot28SW + \cdot72Bk$$

means that a sector of vermilion occupying 37 divisions of the outer circle, combined with a sector of ultramarine occupying 27 divisions, and a sector of emerald green occupying 36 divisions, produced, on spinning, a colour which matched that obtained by a sector of white paper occupying 28 divisons, and a sector of black occupying 72 divisions. The sum of the coefficients on each side of the equation is of course 1.

The mode of conducting an experiment is best described in Maxwell's own words :—

As an example of the method of experimenting, let us endeavour to form a neutral gray by a combination of vermilion, ultramarine, and emerald green. The most perfect results are obtained by two persons acting in concert, when the operator arranges the colours and spins the top, leaving the eye of the observer free from the distracting effect of the bright colours of the papers when at rest.

After placing discs of these three colours on the circular plate of the top, and smaller discs of white and black above them, the operator must spin the top and demand the opinion of the observer respecting the relation of the outer ring to the inner circle. He will be told that the outer circle is too red, too blue, or too green, as the case may be, and that the inner one is too light or too dark as compared with the outer. The arrangements must then be changed so as to render the outer and inner circles more nearly alike. Sometimes the observer will see the inner circle tinted with the complementary colour of the outer one. In this case the observer must interpret the observation with respect to the outer circle, as the inner circle contains only black and white.

By a little experience the operator will learn how to put his questions, and how to interpret their answers. The observer

Plate 1.

DIAGRAM SHEWING THE RELATIONS OF LIGHT.

should not look at the coloured papers, nor be told the proportions of the colours during the experiments. When these adjustments have been properly made, the resultant tints of the outer and inner circles ought to be perfectly indistinguishable when the top has a sufficient velocity of rotation. The number of divisions occupied by the different colours must then be read off on the edge of the plate, and registered in the form of an equation. Thus in the preceding experiment we have vermilion, ultramarine, and emerald green outside, and black and white inside. The numbers, as given by an experiment on the 6th March 1855, in daylight without sun, are—

$$\cdot 37V + \cdot 27U + \cdot 36EG = \cdot 28SW + \cdot 72Bk.$$

In a similar way matches were obtained in which the resulting tint was a decided colour and not a neutral gray. Thus on the 5th of March 1855 a match was obtained as follows—

$$\cdot 39PC + \cdot 21U + \cdot 40Bk = \cdot 59V + \cdot 41EG,$$

where PC represents pale chrome. The resulting tint in this case is an impure yellow.

Mixtures which appear to make perfect matches by one kind of light are far from matching one another when viewed by a different light. Thus, C representing carmine, the following match was obtained by daylight, viz.—

$$\cdot 44C + \cdot 22U + \cdot 34EG = \cdot 17SW + \cdot 83Bk ;$$

while by gaslight the match was

$$\cdot 47C + \cdot 08U + \cdot 45EG = \cdot 25SW + \cdot 75Bk,$$

which shows that the yellowing effect of the gaslight tells more on the white than on the combination of colours."

Maxwell, following Young, represented the results of his experiments graphically by means of a triangle, at the angular points of which he placed the three colours which he believed to most nearly correspond with the primary sensations. (See Plate I.) The colours he selected for these positions were vermilion, ultramarine, and emerald green, of such strength that, when mixed in equal proportions, they produced a neutral gray which therefore appeared at the centre of the triangle. In the paper published in the *Edinburgh Transactions* the reason alleged for selecting green in preference to yellow is the fact that it was found possible

to produce a distinct yellow from a mixture of emerald green
and vermilion, but impossible to produce a green from a
mixture of blue and yellow. Any colour which can be
produced by a mixture of all the three primary tints will
lie within the triangle; but a colour which must be mixed
with one of the primary tints in order to match a mixture
of the other two will lie without the triangle. Thus the
pale chrome referred to above must be mixed with blue in
order to match a mixture of red and green, and it must
therefore be placed in such a position that the centre of
gravity of the yellow and blue taken in proper proportions
may be on the line joining the red and green, and may
divide this line in the inverse ratio of the amounts of red
and green respectively required. A copy of Maxwell's diagram, showing the chromatic relations of coloured papers,
is given in the accompanying plate (Plate II.) The original
is in the Cavendish Laboratory.

As above mentioned, Maxwell found that in the case of
colour-blind persons only four discs (including black) were
required instead of five to ensure a match. The following
is an example of a colour-blind equation:—

$$\cdot19G + \cdot05B + \cdot76Bk = 1\cdot00R,$$

where G, B, and R represent green, blue, and red respectively. To a normal eye the outer circle represented by the
left-hand side of the equation appears of a dark blue green,
but to the colour blind this matches the full red of the
smaller circle.

By comparing different colour matches, it is possible to
determine the position on the colour diagram of the tint
corresponding to the missing sensation in the colour-blind.
Thus, if two colours appear to a colour-blind person to
match, they can only differ in their effect upon a normal
eye by the degree in which they excite the sensation which
is missing in the colour-blind, and they must therefore lie
upon a line passing through the position of this missing sensation. By determining two such lines, the position of the
pure sensation is located. A line drawn through this point
and the position of white is, to a colour-blind person, a

Plate II

DIAGRAM ILLUSTRATING
THE CHROMATIC RELATIONS OF COLOURED PAPERS.

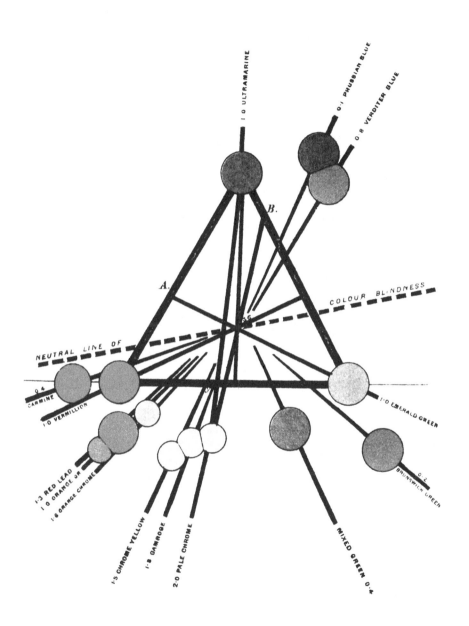

neutral line for its whole length (see Plate II.) The position corresponding to the pure sensation which is missing in the colour-blind is near the red, but outside the triangle, and hence corresponds to a purer red than any in the solar spectrum (see p. 480).

The colour-blind generally class all tints as yellows or blues. According to Maxwell's theory, their sensations correspond to green and blue; but the reason why they regard yellow as brighter than green lies in the fact that although to the normal eye the sensation of yellow is a combination of the sensations of red and green, yet yellows are so much brighter than greens that light from them generally excites the green sensation more powerfully than that from green objects themselves, and hence yellows are more conspicuous than greens to the colour-blind, but only in virtue of the green they contain.

To enable colour-blind persons to distinguish between red and green, Maxwell had a pair of spectacles constructed, one eye-glass of which was red and the other green, so that the object appeared differently to the two eyes. These spectacles cause objects to appear to possess a metallic lustre on account of the different ways in which they are presented to the eyes. To the colour-blind a red object would appear *brighter* when seen through the red glass, while a green object would appear *brighter* through the green glass.

There are colour-blind persons whose vision is very different from that described by Maxwell in his early papers, but to these we shall again refer presently (p. 482).

It would occupy too much space to describe the different methods adopted by Maxwell for the purpose of mixing light of different colours, but his colour box demands a special notice, both on account of the perfection with which it is adapted to the object in view and the extreme beauty of its arrangement. The first two boxes constructed were of inconveniently large dimensions. The form we shall describe is that finally adopted, and the shell, as well as the principal optical apparatus of the box, are still at the

Cavendish Laboratory. The description is taken from Maxwell's paper in the *Phil. Trans.* for 1860.

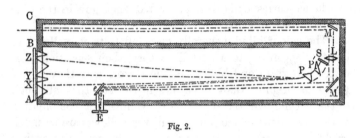

Fig. 2.

Fig. 2 represents the instrument. At A B is placed the apparatus represented in Fig. 3, in which A′ B′ represents a rectangular frame of brass, having a rectangular aperture of

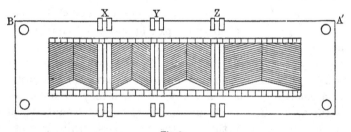

Fig. 3.

6 × 1 inches. On this frame are placed six brass sliders, X Y Z. Each of these carries a knife-edge of brass in the plane of the surface of the frame.

At E (Fig. 2) is a fine vertical slit. At *e*, M′ and M are three plane mirrors, each inclined at 45° to the sides of the box. P P′ are two prisms (angles 45°), and S is a concave silvered glass of radius, 34 inches.

If light enter the box at E, it will be reflected at *e*, and a portion of the pencil, after passing through the prisms, will be reflected from S, and after again traversing the prisms will form a spectrum at A B.

The six movable knife edges above referred to form three slits, X Y Z, which may be so adjusted as to coincide with any three portions of the pure spectrum formed by light from E. The intervals between the sliders are closed with hinged shutters, which allow the sliders to move without letting light pass between them.

The inner edge of the brass frame is graduated to twentieths of an inch, so that the position of any slit can be read off. The breadth of the slit is ascertained by means of a wedge-shaped piece of metal, 6 inches long, and tapering to a point from a breadth of half an inch. This is gently inserted into each slit, and the breadth is determined by the distance to which it enters, the divisions on the wedge corresponding to the 200th of an inch difference in breadth, so that the unit of breadth is ·005 inch. The gauge is balanced on one finger, and inserted into the slit till the pressure just causes it to slide on the finger. This (on the same principle as in Whitworth's Millionth Measurer) ensures uniformity in the pressure in different measurements.

Now suppose light to enter at E, and to be refracted by the two prisms P and P′, and after reflection at S, to again pass through the prisms. A pure spectrum, showing Fraunhofer's lines, is formed at A B, but only that part is allowed to pass which falls on the three slits X Y Z. The rest is stopped by the shutters. Suppose that the portion falling on X belongs to the red part of the spectrum : then, of the white light entering at E, only the red will come through the slit X. If we were to admit red light at X, it would be refracted to E by the principle in optics that the course of any ray may be reversed. If, instead of red light we were to admit white light at X, still only red light would come to E, for all other light would be refracted more or less than the red, and would not reach the slit at E. Applying the eye at the slit E, we should see the prism P uniformly illuminated with red light of the kind corresponding to the part of the spectrum which falls on the slit X when white light is admitted at E.

Let the slit Y correspond to another portion of the spectrum, say the green : then, if white light be admitted at

Y, the prism, as seen by an eye at E, will be uniformly illuminated with green light; and if white light be admitted at X and Y simultaneously, the colour seen at E will be a compound of red and green, the proportions depending on the breadth of the slits and the intensity of the light which enters them. The third slit Z enables us to add a third colour, and thus to combine any three kinds of light in any given proportions, so that an eye at E shall see the face of the prism at P uniformly illuminated with the colour resulting from the combination of the three. The position of these three rays in the spectrum is found by admitting the light at E, and comparing the position of the slits with the position of the principal fixed lines.

At the same time, another portion of the light from the source (generally a sheet of white paper) enters the instrument at BC, is reflected at the mirror M, passes through the lens L, is reflected at the mirror M', passes close to the edge of the prism P, and is reflected along with the coloured light at e, to the eye-slit at E.

In this way the compound colour is compared with a constant white light in optical juxtaposition with it. The mirror M is made of silvered glass, that at M' is made of glass roughened and blackened at the back, to reduce the intensity of the constant light to a convenient value for the experiments. By adjusting the slits properly the two portions of the field may be made equal both in colour and brightness, so that the edge of the prism becomes almost invisible. When light enters at E, the instrument gives a spectrum in which Fraunhofer's lines are very distinct, and the length of the spectrum between Fraunhofer's lines A and H is 3·6 inches. The outside measure of the box is 3 feet 6 inches, by 11 inches, by 4 inches, and it can be carried about, and set up in any position, without readjustment.

In making an observation the centres of the slits X Y Z are placed opposite the divisions of the scale corresponding to the colours to be mixed, the breadth of each slit being varied till the resultant light cannot be distinguished from the white light reflected from M and M'. If the mixture differs from the standard white light only by being too bright or too dull, all the slits X Y Z must be opened or closed in the same proportion. The result of any observation is expressed by an equation in which each particular colour employed is represented by the corresponding division of the scale placed in brackets, while the breadth of the slit is written as the coefficient of this colour. Thus, the equation

$$18\cdot5 \ (24) + 27 \ (44) + 37 \ (68) = W$$

means that the breadth of the slit X was 18·5, as measured by the wedge, while its centre was at the division (24) of the scale; that the breadths of Y and Z were 27 and 37, and their positions (44) and (68); and that the illumination produced by these slits was exactly equal, in the estimation of the observer, to the constant white W.

In the first instance, recorded in the paper above referred to, the observations were made by Professor Maxwell himself (J), and by Mrs. Maxwell (K). Professor Maxwell's complexion was dark and his hair black; Mrs. Maxwell was extremely fair. The observations of each exhibited very small errors from the mean, far less than would have been expected, but there was always a difference apparent between the two observers, and this always in the same direction. A difference of this character was found by Maxwell to be general between dark and fair persons, especially when the light fell on the centre of the retina (the *fovea centralis* or *macula lutea*). To this point reference will again be made.

By considering the errors in the several observations between the standard colours, Maxwell showed that greater accuracy is attainable in the case of red light than in that of green or blue, and that variations in *colour* are more easily detected than variations in *brightness*.

One result of the observations is especially worthy of notice. It follows from a comparison of the equations obtained in the several parts of the spectrum that any tint in the spectrum lying between the divisions (24) and (46) of the scale could be exactly imitated by mixing in proper proportions the red of (24) and the green of (46), while every tint lying between (48) and (64) could be produced by a proper mixture of the green of (48) and the blue of (64). Hence if a colour diagram be constructed as explained on page 473, the whole of the spectrum between the red of (24) and the green of (46) will lie in a straight line, forming one side of the triangle, while that lying between (48) and (64) forms another side (the included angle being not quite closed). It was also found that the violet beyond (68) lay very nearly on the line joining (68) and (24), so that the spectrum exhibited in this diagram "forms two sides of a triangle, with doubtful fragments of the third side." Thus "all the colours of the spectrum may be compounded of those which lie at the angles of this triangle." A diminished copy of one of Maxwell's lecture diagrams, painted with his own hand and deduced (it is believed) from experiments with the colour top, is given in Plate I.[1]

By studying Plate I., it will be seen that a mixture of yellow and blue lights produces varying shades of dirty yellow and pink, as the relative amount of blue is increased, and by properly selecting the yellow and blue a neutral gray (or white) may be obtained.

The pink tint produced by mixing blue and yellow lights had been previously noticed by Helmholtz and others. The well-known fact that an admixture of blue and yellow pigments produces generally a green was explained by Helmholtz on other grounds. Suppose, for instance, that we are provided with properly chosen blue and yellow glasses : let sunlight shine through the blue glass and fall upon a screen, and allow a second beam of sunlight, after reflection at a mirror, to pass through the yellow glass and fall on the same portion of the screen as the blue light. If the two colours

---

[1] The following table gives the position on the scale and wave length in Fraunhofer's measure of the colours selected as standards :—

	Scale	Wave length.
R Scarlet . . .	24	2328
G Green . . .	$46\frac{3}{4}$	1914
B Blue . . .	$64\frac{1}{2}$	1717

are properly selected, the illuminated portion of the screen will appear white or pink. Now place the pieces of glass together, so that the same light passes through both, and the light transmitted will be green. This is explained if we examine with the prism the light transmitted by each glass. That which has passed through the blue glass will show, in addition to blue light, a certain amount of violet and green, while that which has passed through the yellow glass will comprehend, in addition to yellow light, a certain quantity of orange (perhaps red) and green. Now the only light capable of passing through *both* glasses is green, and this, therefore, is the colour observed on looking through the two together. Similarly, when blue and yellow pigments are ground together, the light penetrates a little way into the substance of the pigment, is reflected, and emerges, some of its constituents being absorbed by one pigment and some by the other, and the only light which both pigments transmit is green, which is consequently the colour of the emergent light, and therefore the hue of the mixture.

The set of colour equations given above, and the diagram of colours there described, were obtained from the observations of Mrs. Maxwell (K). The diagram obtained from Professor Maxwell's (J's) own observations was slightly different, as before mentioned.[1] The chief differences were that (J) saw more green in the orange and yellow portions of the spectrum than (K), so that in the diagram obtained from his observations these colours appeared nearer the green than in the other, while the colours between green and blue appeared more blue to (J) than to (K), and were therefore placed higher up in his diagram. Thus, when the instrument was adjusted to suit (K), one of the selected colours being (32), (36), or (40), then (J) saw the mixture too green; but if (48), (52), (56), or (60) were the selected colour when adjusted for (K), it appeared too blue to (J), " showing that there was a real difference in the eyes of

---

[1] Part I. p. 315

these two individuals, producing constant and measurable differences in the apparent colour of objects."

With those colour-blind persons whom Maxwell first examined only two slits were required to produce a colour chromatically identical with white; while the spectrum, a little on the red side of the line F, appeared to be identical with white. " From this point to the more refrangible end the spectrum appears to them ' blue.' The colours on the less refrangible side appear all of the same quality, but of different degrees of brightness; and when any of them are made sufficiently bright they are called ' yellow.' " Thus a colour-blind or dichromic person, in speaking of red, green, orange, and brown, refers to different degrees of brightness or purity of a single colour, and not to different colours. This colour he calls yellow.

Of the three standard colours the red appears to them " yellow," but so feeble that there is not enough in the whole red division of the spectrum to form an equivalent to make up the standard white. The green at E appears a good "yellow," and the blue at ⅔ from F towards G appears a good " blue."

It was for these researches that Maxwell received the Rumford Medal of the Royal Society in 1860.

The only cases of colour-blindness which Maxwell met with in the course of his earlier experiments were those of persons blind to red rays. This, however, is not the only kind of colour-blindness known. Some appear to be deficient in the blue (or violet) sensation. To these persons bright yellow appears white, and the neutral line is a line drawn through the yellow and the point corresponding to the pure blue (or violet) sensation. In a valuable paper published in the *Proceedings of the Royal Society*, vol. xxxi. p. 302, M. Frithiof Holmgren, Professor of Physiology in the University of Upsala, gave an account of a case of red blindness, and a case of violet-blindness in each of which only one eye was really colour-blind, the vision of the other being nearly normal. This peculiarity enabled the observer to compare the experience of colours gained through one eye with that gained by the other, and thus to state in the language of those possessing normal vision, how colours really appear to the colour-blind. His results agree in their general character with Maxwell's, but Holmgren (with many others) makes the third sensation correspond to violet instead of blue, while, according to him, the violet-blind class colours as red and yellow instead of red and green.

Though the largest and most important, Maxwell's theory of compound colours was by no means his only, or even his earliest, contribution to Optics. While a bachelor-scholar at Trinity, at the request of Messrs. Macmillan he wrote a considerable portion of a text-book on geometrical optics. The work was never finished, but the MS. is still extant. The mode of treating the subject, as stated by Professor Maxwell in one of his letters to his father,[1] is decidedly novel, and calculated to bring down a storm of abuse upon the author. He starts by postulating the possibility of obtaining perfect images, and then investigates the laws of reflection and refraction necessary for this.

Perhaps the most remarkable of Maxwell's contributions to Optics was his identification of the velocity of light with the ratio of two quantities, each of which is capable of being measured electrically; but this subject will be again referred to in describing his electrical researches. A paper on a general theory of optical instruments was published by Maxwell in the *Quarterly Journal of Mathematics* in 1858, and another on the same subject in the *Proceedings of the Cambridge Philosophical Society,* 1866. A paper on the best mode of projecting a spectrum on a screen appeared in the *Proceedings of the Royal Society of Edinburgh* in 1869. We may also refer to the papers on "The Focal Lines of a refracted Pencil," in the *Proceedings of the Mathematical Society of London* for 1871-3; on a "Bow seen on the surface of ice," in the *Proceedings of the Royal Society of Edinburgh* for 1872; on "The Spectra of Polarised

---

[1] See Part I. p. 217.

Light," the subject which had interested him so much in 1847 (see Part I., p. 84), in the *Transactions of the Royal Society of Edinburgh* for 1872; and on " Double Refraction in a Viscous Fluid in Motion," in the *Proceedings of the Royal Society* for 1873, and the *Ann. de Phys. and Chimie*, 1874.

Among the many optical contrivances designed by Professor Maxwell, we ought not to omit to mention the real-image stereoscope. This instrument, in which ordinary stereoscopic slides are employed, consists essentially of two convex lenses of short focal length, say 4 inches, placed side by side in a wooden frame at a distance from the pictures, equal to twice their focal length, while the distance between the centres of the lenses is half the distance between the centres of the pictures. The result of this arrangement is that real images of the two pictures of the same size as the pictures themselves, formed one by one lens, and the other by the other, are superposed on the axis of the instrument at a distance in front of the lenses equal to twice their focal length, that is, as far in front of the lenses as the pictures are behind them. The double, or stereographic image so formed is viewed through a large lens at a suitable distance in front of it, the observer standing at a distance of three or four feet. Many persons [1] who cannot appreciate the ordinary box stereoscopes obtain very satisfactory effects with this instrument.

Professor Maxwell prepared a large number of stereoscope slides of geometrical figures. Among them may be mentioned the surface of centres of an ellipsoid, lines of curvature on an ellipsoid, and on elliptic and hyperbolic paraboloids, the parabolic cyclide, the horned cyclide, and the spindle cyclide, a twisted cubic with three asymptotes, a Gordian knot, etc. The wood blocks from which these slides were printed are now in the Cavendish Laboratory.

Another extremely pretty optical toy of his construction, at present in the possession of Mrs. Maxwell, is a Zoetrope,

---

[1] Including the present Lord Bishop of Durham.

or " Wheel of Life." In the ordinary instrument, on looking
through the slits in the revolving cylinder the figures are
seen moving on the opposite side of the cylinder. Maxwell
inserted concave lenses in place of the slits, the lenses being
of such focal length that the virtual image of the object at
the opposite extremity of the diameter of the cylinder was
formed on the axis of the cylinder, and consequently appeared
stationary as the cylinder revolved.[1]

In ordinary light the vibrations take place in all directions at right angles to that in which the light is being
transmitted. A beam of ordinary light is therefore symmetrical on all sides. But it is possible in various ways to
confine all the vibrations to one plane. Thus, if a ray be
reflected from the surface of polished glass at a particular
angle, all the vibrations take place in one plane, and it is
generally believed that this plane is parallel to the surface
of the glass, and therefore perpendicular to the plane of
incidence. Since all the vibrations are now taking place in
one plane, the beam of light has acquired, as it were, sides
or *poles*, and is said to be plane *polarised*. There are other
contrivances by which the movements may be made all to
take place in regular succession in a circle about the direction of the ray as axis, and the light is then said to be
circularly polarised. This may be effected by passing a
beam of plane polarised light through a sheet of mica (or
some other crystals) of a particular thickness.

The velocity with which waves are transmitted through
a substance depends not only on the density but on the
elasticity of the substance *in the direction in which the
vibrations take place*. Now, many crystals have different
elasticities in different directions, and a similar condition
may be induced by mechanical means in other bodies. In
such substances the velocity of light depends on the direction in which the vibrations take place ; and, generally, if
a beam of ordinary light fall upon such a crystal it will be
separated into two, one of which will consist of vibrations in

---

[1] See Part I. p. 37.

the direction in which the elasticity is greatest (of all the directions which it can select), and the other in that in which the elasticity is least. These rays, being transmitted with different velocities, will be differently refracted, and part company in the crystal; and since in each ray the vibrations are in one direction only, each ray will be plane polarised, the vibrations taking place in the two rays in planes perpendicular to one another.

A Nicol's prism consists of a long prism of Iceland spar cut into two along a diagonal plane, the segments being cemented together again by Canada balsam. The effect of the balsam is to reflect one of the polarised rays out at the side of the prism while it allows the other to pass through. Consequently, if common light fall on a Nicol's prism, only half of it will pass through, and this will be plane-polarised in a particular plane. If another prism be placed in the path of the ray and be situated similarly to the first, the light will pass through it, but if it be turned around its axis through a right angle, the light falling on it corresponds to that which is reflected out of the prism by the balsam, and no light passes through the second prism. Under these circumstances the Nicols are said to be *crossed*. Such an arrangement of two Nicol's prisms or any equivalent arrangement which may be made with glass reflectors or crystals of tourmaline, constitutes a polariscope; the first Nicol is called the polariser, the second the analyser.

Fresnel first pointed out that by mechanical stress it is possible to impart to glass a property analogous to that of doubly refracting crystals, making its elasticity (for light) different in different directions. Sir David Brewster showed that a piece of glass heated and then suddenly cooled, *i.e.* unannealed, possesses similar properties. If it be placed between the polariser and analyser of a polariscope and examined by white light, a gorgeous display of colour is observed. The reason is that the plane-polarised light from the polariser on entering the glass is generally split into two rays vibrating in planes at right angles to one another, and these pass through the glass with different velocities.

Plate III.

CHROMATIC EFFECTS OF
A PENTAGON OF UNANNEALED GLASS
IN POLARIZED LIGHT.

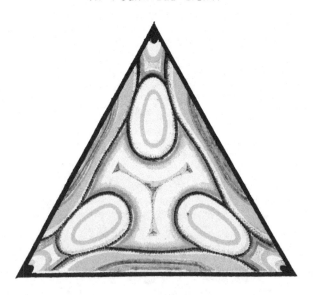

ISOCHROMATIC LINES
IN A
TRIANGLE OF UNANNEALED GLASS.

When they reach the analyser only that component of each set of vibrations can pass through which takes place in the plane corresponding to transmission through the analyser. Then it may happen that the two rays of light of one particular colour are so vibrating that the waves *interfere* and destroy one another, so that this colour is completely absent in the transmitted light while other colours are partially destroyed. The result is that the combination of colours which passes through produces a particular tint, and this will be different according to the obliquity of the ray, the thickness of the glass, and the amount of strain it has experienced, for on the latter depends the difference of velocity of the two rays, and consequently the particular colour destroyed in the case of glass of a given thickness.

Maxwell's first serious experiments on light appear to have had their origin in his visit to Mr. Wm. Nicol in April 1847. After this visit he constructed a polariscope of cardboard, employing blackened glass mirrors as polariser and analyser. This was supplied with lenses for use when a conical beam of polarised light was required. The lenses were mounted in cardboard frames. A very similar instrument constructed by him shortly afterwards, but of wood instead of cardboard, together with the lenses mounted as before on cards, is still preserved in the Cavendish Laboratory. Maxwell, after returning from Mr. Nicol's, prepared some samples of unannealed glass by heating pieces of thick plate glass to redness and allowing them to cool rapidly. By means of a *camera lucida* adapted to the cardboard polariscope he observed and faithfully copied in water-colours some of the chromatic effects exhibited by these plates of glass, showing the manner in which the glass was strained by the rapid cooling. Some of these figures are shown in Plate III., and will be again referred to.[1] As mentioned in the previous part

---

[1] Maxwell had a great faculty for designing, and would frequently amuse himself by making curious patterns for wool-work. His designs are remarkable for the harmony of the colouring. Sometimes they

of this work, of these water-colour drawings he sent some
to Nicol, who presented him in return with a pair of Nicol
prisms of his own construction, which are now, with the
original colour-top, polariscope, specimens of unannealed
glass, etc., in the Cavendish Laboratory. His experiments
on the passage of light through solids exposed to strain
suggested to Maxwell the employment of polarised light as
an analyser of the strains in the different portions of an
elastic solid when exposed to mechanical stress, and led to
the production of the paper on "Elastic Solids" read before
the Royal Society of Edinburgh in February 1850; but
before giving an account of this paper we must mention
another investigation connected with the physiology of
vision.

There are some persons who, when they look at a point
in the sky distant 90° from the sun, at once observe two
conspicuous yellow brushes with their axis in a plane passing
through the sun, while the space between exhibits the complementary violet colour. This phenomenon was first
noticed by Haidinger in 1844, and is known as Haidinger's
Brushes. The appearance is only transitory, and disappears
in a very short time if the eye be kept directed to the same
point in the heavens. The same appearance is produced
whenever " plane-polarised " light enters the eye; as, for example, when light is reflected from a polished surface of glass
at a particular angle. But there are some persons, on the other
hand, who are apparently incapable of seeing Haidinger's
Brushes, or only see them with difficulty. Generally it is
persons of a dark complexion who see Haidinger's Brushes
more readily than others, and in such persons the *yellow spot*
appears to be peculiarly insensitive to blue light. A paper
on this subject was read before the mathematical and physical
section of the British Association by Professor Maxwell in
1866 (*Report*, Part II.) He found that on looking through

---

represented natural objects. A kettle-holder which used to hang by
the fireside was a representation of a square of unannealed glass when
placed between two crossed Nicol prisms.

a solution of chrome alum, the centre of the field of view appears distinctly pink and much paler than the rest of the solution on account of this peculiarity of the *foramen centrale.* The effect decreases if the observer continue to look. Maxwell's attention was first called to this peculiarity of his own eye by noticing a black band in the blue portion of the spectrum whenever he directed his eye straight to that portion. The band never appeared in any other part of the spectrum, but followed the eye up and down the spectrum in the blue, vanishing as soon as the optic axis passed out of the blue into other colours.

The following account of Maxwell's explanation of Haidinger's Brushes, and the personal reminiscences which accompany it, have been kindly contributed by Professor William Swan :—

My earliest noteworthy recollection of Clerk Maxwell dates from 1850, when the British Association was in Edinburgh, and its " Section A " used to meet in the natural philosophy class-room of the University. That year communications were made by Sir David Brewster and Professor Stokes on the remarkable pheno-menon of vision discovered by Haidinger in 1844, and since known as " Haidinger's Brushes." One day when a paper had been read, which, if I remember rightly, was Brewster's, a young man rose to speak. This was Clerk Maxwell. His utterance then, most likely, would be somewhat spasmodic in character, as it continued to be in later times, his words coming in sudden gushes with notable pauses between ; and I can well remember the half-puzzled, half-anxious, and perhaps somewhat incredulous air, with which the president and officers of the section, along with the more conspicuous members who had chosen " the chief seats " facing the general audience, at first gazed on the raw-looking young man who, in broken accents, was addressing them. For a time I was disposed to set down his apparent embarrass-ment to bashfulness ; and such, I daresay, was the general impression. Bashful, very likely, in some degree he was. But, at all events, he manfully stuck to his text ; nor did he sit down before he had gained the respectful attention of his hearers, and had succeeded, as it seemed, in saying all he meant to say.

My only tolerably distinct further recollection is that he handed in a small piece of apparatus which he had made by cementing together on glass sectors of sheet gutta-percha, these being so cut out of the sheet and put together that its fibrous structure, due to rolling in the process of manufacture, should radiate, at least approximately, from a central point. This arrangement, as I understood at the time, in polarised light, reproduced, or rather simulated, Haidinger's phenomenon. But after an interval of thirty years I must speak with caution. Brewster (British Association *Report*, 1850), ascribed Haidinger's phenomenon to "polarising structure existing in the cornea and crystalline lens, as well as in the tissues which lie in front of the sensitive layer of the retina;" while Stokes *proved* that, when the variously coloured rays of the prismatic spectrum were admitted separately into the eye, in the blue rays alone could Haidinger's Brushes be seen. A few years later, in his paper "On the Unequal Sensibility of the Foramen Centrale to Light of Different Colours" (Brit. Assoc. *Report*, 1856), Clerk Maxwell says that, on looking through a prism at a long vertical slit he saw an elongated dark spot running up and down the spectrum, but refusing to pass out of the blue into the other colours. This appearance, he concludes, is due to the "Foramen centrale" of Soemmering; and he adds that when a Nicol's prism is employed the brushes of Haidinger are well seen *in connection with the spot*. The appearance of a dark spot on a blue ground *only*, Maxwell then, or at least afterwards, believed to be due to the yellow pigment of the macula lutea, of which the fovea centralis, as its name imports, is the middle portion, and, as is well known, the place of most distinct vision, having a special selective absorption for the blue rays. His notable discovery that Haidinger's Brushes were only to be seen in connection with the shadow of the yellow spot thus pointed conclusively to the spot itself as the seat of the phenomenon; or, as he himself puts it, makes evident the fact that the brushes are the spot analysed by polarised light. Having thus, to his own satisfaction, localised the polarising structure concerned in the production of Haidinger's phenomenon, he seems to have rested, for I am not aware that he ever wrote again on the subject. No such writing at least appears in the Royal Society catalogue of scientific memoirs. Possibly he hoped that some day he might himself examine the actual structure of the yellow spot; or he may have waited for the result of such examination by other hands than his own. Helmholtz, who cites Maxwell's paper of 1856, attributes the phenomenon of the brushes to a radiating fibrous structure, which, it seems, has actually now been ascertained to exist in the fovea centralis, which he assumes to be feebly polarising, and to possess a special selective absorbing power for the blue rays. (Helmholtz,

Plate IV.

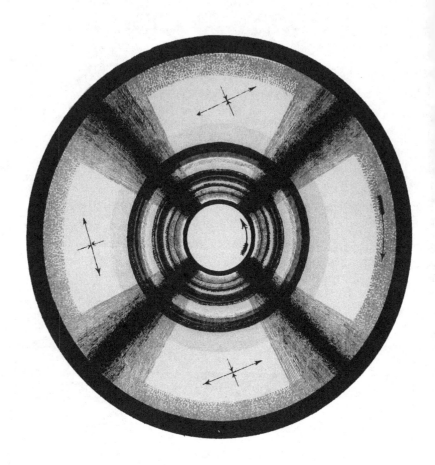

DIAGRAM SHEWING THE
COLOURS EXHIBITED BY A PLATE OF GELATINE
WHEN EXPOSED TO A TORSIONAL SHEAR.

Optique Physiologique, 1867, pp. 548-554.)   All Maxwell's conclusions regarding Haidinger's Brushes, seem thus to be definitely
verified.   WILLIAM SWAN,
               Ardchapel, Helensburgh, *2d April* 1882.

2. In the paper read before the Royal Society of Edinburgh on Feb. 18, 1850, Maxwell describes the mathematical
results of the application of Stokes's theory of elasticity to a
number of cases of the deformation of elastic solids, which
results, when possible, he tested experimentally by subject-
ing the strained solid to analysis by polarised light.   The
first, and perhaps the most interesting, example given in this
paper, refers " to the case of a hollow cylinder, of which the
outer surface is fixed, while the inner surface is made to
turn through a small angle."   The conclusions derived we
give in Maxwell's own words :—

Therefore, if the solid be viewed by polarised light (transmitted parallel to the axis), the difference of retardation of the
oppositely polarised rays at any point in the solid will be inversely proportional to the square of the distance from the axis
of the cylinder, and the planes of polarisation of these rays will
be inclined 45° to the radius at that point.

The general appearance is, therefore, a system of coloured
rings, arranged oppositely to the rings in uniaxal crystals, the
tints ascending in the scale as they approach the centre, and the
distance between the rings decreasing towards the centre.   The
whole system is crossed by two dark bands inclined 45° to the
plane of primitive polarisation, when the plane of the analysing
plate is perpendicular to that of the first polarising plate (see
Plate IV.)

A jelly of isinglass poured, when hot, between two concentric
cylinders, forms, when cold, a convenient solid for this experiment ; and the diameters of the rings may be varied at pleasure
by changing the force of torsion applied to the interior cylinder.

By continuing the force of torsion while the jelly is allowed
to dry, a hard plate of isinglass is obtained, which still acts in
the same way on polarised light, even when the force of torsion
is removed.

It seems that this action cannot be accounted for by supposing the interior parts kept in a state of constraint by the
exterior parts, as in unannealed and heated glass ; for the optical

properties of the plate of isinglass are such as would indicate a strain pressing in every part of the plate in the direction of the original strain, so that the strain on one part of the plate cannot be maintained by an opposite strain on another part.

Two other uncrystallised substances have the power of retaining the polarising structure developed by compression. The first is a mixture of wax and resin, pressed into a thin plate between two plates of glass, as described by Sir David Brewster in the *Philosophical Transactions* for 1815 and 1850.

When a compressed plate of this substance is examined with polarised light, it is observed to have no action on light at a perpendicular incidence; but when inclined it shows the segments of coloured rings. This property does not belong to the plate as a whole, but is possessed by every part of it. It is, therefore, similar to a plate cut from a uniaxal crystal perpendicular to the axis.

I find that its action on light is like that of a *positive* crystal, while that of a plate of isinglass, similarly treated, would be *negative*.

The other substance which possesses similar properties is gutta-percha. This substance in its ordinary state, when cold, is not transparent even in thin films; but if a thin film be drawn out gradually, it may be extended to more than double its length. It then possesses a powerful double refraction, which it retains so strongly that it has been used for polarising light. As one of its refractive indices is nearly the same as that of Canada balsam, while the other is very different, the common surface of the gutta-percha and Canada balsam will transmit one set of rays much more readily than the other, so that a film of extended gutta-percha placed between two layers of Canada balsam, acts like a plate of nitre treated in the same way. That these films are in a state of constraint may be proved by heating them slightly when they recover their original dimensions.

Some pieces of gutta-percha mounted in this way by Professor Maxwell are still preserved in the Cavendish Laboratory; as are also the original plates of isinglass above referred to. The interior cylinder employed for twisting these plates was a cork, and in one of the plates two corks were placed with their circumferences about ⅜ in. apart, and were twisted equally in the same direction. The result of this operation is described as case XIII. in the paper in question, and is determined geometrically by the superposition of the two conditions of strain due to the two twists independently. The isochromatic curves thus obtained are represented in Fig. 4, which is taken from that

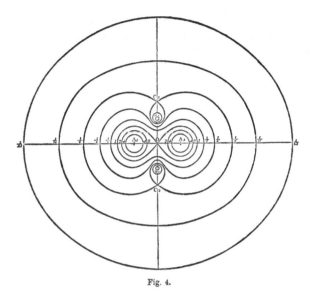

Fig. 4.

in the *Edinburgh Transactions.* The points B_1, B_2, correspond to no retardation. These curves completely agree with those obtained when the plate of isinglass is examined with circularly polarised light. The advantage of employing circularly polarised light lies in the fact that whatever may be the direction of the lines of stress in the isinglass the two component rays into which the light may be supposed separated on entering the medium are of equal intensity, and the colour therefore depends only on the state of strain and not on the position of the plane of primitive polarisation.

The last example of stress in an elastic solid is that of a triangle of unannealed glass, in which "the lines of equal intensity of the action on light are seen without interruption by using circularly polarised light." They are represented in Plate III. In Figs. 5 and 6 A, BBB, DDD, are the neutral points, or points of no action on light, and CCC, EEE, are

the points where the action is greatest; and the intensity of
the action at any other point is determined by its position
with respect to the isochromatic curves.

"The direction of the principal axes of pressure at any point
is found by transmitting plane polarised light, and analysing
it in the plane perpendicular to that of polarisation. The light
is then restored in every part of the triangle, except in those
points at which one of the principal axes is parallel to the plane
of polarisation. A dark band formed of all these points is seen,
which shifts its position as the triangle is turned round in its

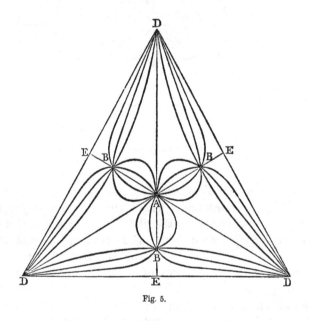

Fig. 5.

own plane. Fig. 5 represents these curves for every fifteenth
degree of inclination. They correspond to the lines of equal variation of the needle in a magnetic chart.

"From these curves others may be formed which shall indicate,
by their own direction, the direction of the principal axes at any
point. These curves of direction of compression and dilatation are
represented in Fig. 6; the curves whose direction corresponds
to that of *compression*, are concave toward the centre of the
triangle, and intersect at right angles the curves of dilatation."

The figures on Plate III. are copies of water-colour sketches made by Professor Maxwell in the very early periods of these investigations, and show the isochromatic lines in unannealed glass in the case of a pentagon and a triangle.

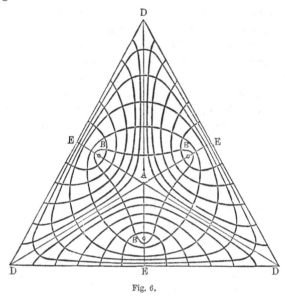

Fig. 6.

The remainder of the paper on elastic solids is taken up with a discussion of a number of examples of great importance to the engineer, such as the flexure of beams, the torsion of cylinders, and the like. To these engineering problems Maxwell again returned many years later, and his paper "On Reciprocal Figures, Frames, and Diagrams of Force," read before the Royal Society of Edinburgh, 7th February 1870, received the award of the Keith Medal. The Theory of Oersted's Piezometer, given in the paper of 1850, is, however, of great scientific value. In it Maxwell points out that the behaviour of a vessel exposed to equal pressures within and without depends only on its cubic compressibility, while the relation between the cubic compressibility and the rigidity in solids, so far from being constant, as Poisson supposed, may be almost any whatever, cork having great rigidity in comparison with its power to resist compression, while caoutchouc, on the other hand, has extremely little.

3. While a bachelor-scholar of Trinity College, on March 13, 1854, Clerk Maxwell read a paper before the Cambridge Philosophical Society "On the Transformation of Surfaces by Bending." This paper, which embodies a great deal of thought, and indicates that its author possessed an extensive acquaintance with the geometrical works of Gauss, Monge, Liouville, and others, is of more interest to the pure mathematician than to the general reader, but in connection with it we may mention a surface which was prepared for Professor Maxwell many years afterwards, and coated by his own hands, as described below, and now preserved in the Cavendish Laboratory. If a heavy uniform string be suspended from two points, and allowed to hang freely, it assumes the form of a curve called a catenary. If a board be cut out in this shape, and the string be stretched around its edge, then cut at the lowest point and one-half of the string unwrapped so as always to be kept stretched, and therefore the free portion of the string a tangent to the board at the point where it leaves it, the extremity of the string will describe a curve called a tractory, because it is the curve traced by a particle lying on a rough horizontal plane, when the end of a string attached to it describes a straight line on the plane, the particle starting from a point outside the straight line. As we go farther along the curve traced by the particle, we of course continually approach the straight line described by the end of the string, but never reach it. This straight line is the "asymptote" of the tractory, and is also the "directrix" of the catenary above referred to. If the tractory be made to revolve about its asymptote it will trace out a surface which has the peculiarity that the *intrinsic curvature* is the same at every point of the surface and is negative, the two principal radii of curvature being opposite in direction and inversely proportional the one to the

other. If a flexible surface be constructed so as to fit a portion of the surface of the solid, thus forming a coat, it will be found to fit equally well at every part of the surface, and it may be inverted or turned through any angle whatever, and it will continue to fit every portion of the solid equally well. By weaving together curved strips of parchment paper and pasting them, Professor Maxwell constructed a coat which fitted every portion of the solid, however it might be placed upon it.

Maxwell's fondness for geometry has already been alluded to more than once, and not unfrequently he would take up geometrical questions of a very generalised character. A simple and very good illustration of some of his investigations in this department is to be found in a paper " On Hills and Dales," published in the *Philosophical Magazine* for December 1870. In this paper, having described the general character of contour-lines, or the lines in which the surface of the earth is cut by level surfaces, Maxwell proceeds to show [1] that the number of summits, or regions of elevation reduced to points, exceeds by one the number of passes, a pass being the point where two regions of elevation unite. Similarly, the number of Bottoms, or regions of depression, exceeds by one the number of Bars, a Bar being the boundary between two regions of depression.   Lines of slope are defined as lines everywhere perpendicular to the contour lines. By following lines of slope we generally reach a Summit or a Bottom, but we *may* reach a Pass or Bar. " Districts whose lines of slope run to the same bottom are called Basins or Dales. Those whose lines of slope come from the same summit may be called, for want of a better name, Hills." " Dales are divided from each other by Watersheds, and Hills by Watercourses." " Lines of Watershed are the only lines of slope which do not reach a bottom, and lines of Watercourse are the only lines of slope which do not reach a summit." These extracts indi-

---

[1] After Listing.   *Censur raümlicher complexe*, etc.   Göttingen Nachrichten, 1861. Also Cayley.

cate Maxwell's anxiety that definite meanings should be assigned to all scientific terms, and that they should be used only in the senses so defined. A further illustration of this trait in his character is found in his paper "On the Mathematical Classification of Physical Quantities," published in *The Proceedings of the London Mathematical Society*, vol. iii. No. 34. We give one other example, taken from *Matter and Motion*, a little work published by the S. P. C. K., and a model of scientific accuracy and philosophic thought. "When the simultaneous values of a quantity for different bodies or places are equal, the quantity is said to be *uniformly* distributed in space." "When the successive values of a quantity for successive instants of time are equal, the quantity is said to be *constant*." Such examples might be multiplied almost indefinitely from Maxwell's published writings. The more elementary the work the greater need did he see for exactness in definitions and accuracy in the use of words; and perhaps no one has done more than Maxwell in giving accuracy to the expression of scientific thought. His influence in this respect has especially been felt by all who have attended the Physical School in Cambridge. On one occasion he remarked, in his half-humorous way, that spheres are *inclusive* figures but circles are *exclusive* :—people are always *trying* to get *into* other circles, but to do so they must get *out of* their own spheres.

4. The keystone as well as the foundation of physical science is dynamics—a subject to which Maxwell was greatly attracted in early days, and to which, even in its most elementary principles, he constantly reverted in after life. We have already referred to his Dynamical Theory of the Electromagnetic field; the sections of his *Treatise on Electricity and Magnetism*, which are devoted to dynamical principles and the general equations of motion, form a most valuable compendium of dynamics; while for those who require less strong meat, his treatment of the subject in the *Theory of Heat* leaves little to be desired; and last, but not least, under the comprehensive title of *Matter and Motion*, we have a treatise on dynamics written for

children in the higher grade schools, so simple that the most casual reader will *think* that he understands it, while it is the admiration of professed mathematicians, and by no means easy reading for those who expect, or have gained, high places in the mathematical tripos : but as the book is well within the reach of every reader, it is unnecessary here to attempt any detailed account of it.

But Maxwell's investigations in dynamics were not confined to paper. Of the problems suggested to him by the devil on two sticks we have no account. During his residence in Cambridge he endeavoured to investigate the process by which a cat is enabled invariably to alight on her feet. The mode of conducting the experiments and the impression they left on the mind of the College will appear from the following extract from a letter written to Mrs. Maxwell, from Trinity, on January 3d, 1870, when Professor Maxwell was examining for the Mathematical Tripos :—

There is a tradition in Trinity that when I was here I discovered a method of throwing a cat so as not to light on its feet, and that I used to throw cats out of windows. I had to explain that the proper object of research was to find how quick the cat would turn round, and that the proper method was to let the cat drop on a table or bed from about two inches, and that even then the cat lights on her feet.

The "Dynamical Top," which was invented by Maxwell to illustrate dynamical propositions, technically so-called, was, in its final form, constructed of brass by Mr. Ramage of Aberdeen. It was this top which Maxwell brought with him to Cambridge when he came up for his M.A. degree in the summer of 1857, and exhibited to a tea-party in his room in the evening. His friends left it spinning, and next morning Maxwell, noticing one of them coming across the court, leapt out of bed, started the top, and retired between the sheets. It is needless to say that the spinning power of the top commanded as great respect as its power of illustrating Poinsot's *Theorie Nouvelle de la Rotation des Corps.*

In the work just referred to, Poinsot shows that a body

supported at its centre of gravity and rotating freely about it will move in the same manner as an ellipsoid whose centre is fixed, but which rolls in such a way as always to rest against a fixed plane (the invariable plane), which is, of course, a tangent plane at the point of contact. The line from the centre to the point of contact is the axis about which the body is at the instant rotating, and this line will, unless the plane touch the ellipsoid at the extremity of one of the principal axes, move both in space and in the body. The curve which the extremity of the axis of rotation describes on the invariable plane is called a herpolhode, while that which it describes on the surface of the ellipsoid is called a polhode. If at any instant the axis of rotation be very near either to the greatest or least axis of the ellipsoid, it will always remain very near that axis, and the polhode will be a small closed curve and the rotation will be stable; but if the axis of rotation be near to the *mean* axis of the ellipsoid, the polhode will be a very large curve, the axis of rotation will in time deviate very widely from its original position in the body, and the motion will be unstable.

Maxwell's top consisted of a brass bell, a long screw passing through the top of the bell and terminating in a steel point " finished without emery and afterwards hardened," serving as the axle. The point rested in an agate cup on the top of a pillar. A heavy nut could be screwed up and down on the axle for coarse adjustments, and the axle could be screwed through the top of the bell, to bring the point to coincide with the centre of gravity of the mass. The moments of inertia could be altered by means of the nut above referred to, and by nine screws with massive milled heads, six of which screwed horizontally into the rim of the bell, while three were arranged symmetrically around the top of the rim, and admitted of a vertical motion. By means of these screws the axle could be made the axis of greatest, least, or mean moment of inertia, or not a principal axis at all. The axle terminated in a point, by which a coloured disc could be fixed upon it. "The best arrangement, for general observations, is to have the disc of card divided into four quadrants, coloured with vermilion, chrome yellow, emerald green, and ultramarine. These are bright colours, and if the vermilion is good they combine into a grayish tint when the revolution is about the axle, and burst into brilliant colours when the axis is disturbed. It is useful to have some concentric circles, drawn with ink, over the colours, and about twelve radii drawn in strong pencil lines. It is easy to distinguish the ink from the pencil lines, as they cross the invariable axis, by their want of lustre. In this way the path of the invariable axis may be identified with great accuracy, and compared with theory." If the top revolve about its axle, the whole disc will appear gray, but when the axle moves, the quadrants in which the " invariable axis" lies will be indicated by a circle of the corresponding colour appearing in full purity. Whenever the invariable line crosses an ink or pencil line, a dark ink dot or lustrous pencil dot will appear at the centre of the circle, and when it passes into another quadrant the circle will contract, then change its colour, and expand again. The use of the nine screws is thus indicated by Professor Maxwell :—" There must be three adjustments to regulate the position of the centre of gravity, three for the magnitude of the moments of inertia, and three for the directions of the principal axes— nine independent adjustments, which may be distributed as we please among the screws of the instrument." A full account of the top and the mode of using it, together with a note on the rotation of the earth, will be found in the *Transactions of the Royal Society of Edinburgh*, vol. xxi., from which the above quotations have been taken.

5. In 1610 Galileo turning his telescope to Saturn, observed a projection on each side of the planet, which led him to conclude that " Saturn consists of three stars in contact with one another." In 1659 Huyghens discovered the true nature of these appendages, which, by their continually varying forms, had puzzled previous astronomers, and concluded that Saturn is girded with a thin flat ring inclined to the ecliptic; its external diameter being about $2\frac{1}{4}$ times that of the

planet itself. In 1665 William Bell observed a dark line running round the northern surface of the ring. In 1675 Cassini noticed the same on the other surface, and concluded that the ring consists of two concentric rings, of which the inner is the brighter. Hadley observed the shadow of the planet thrown upon the surface of the ring, and the shadow of the ring on the planet, and showed that Saturn rotates in the same plane with the rings. In 1714 Moradi observed a want of symmetry in the ring on opposite sides of the planet, and found that when it had disappeared on the eastern side it was still visible on the western, but he did not discover the cause of the phenomena. The observations of Sir William Herschel, published in 1790, corroborated the opinion of Cassini respecting the division. By observing certain spots on the surface of the planet, Herschel found that it rotated on its axis in 10 hours 29 minutes 16·8 seconds in the same direction as the rotation of the earth. In 1789, as the earth passed through the plane of Saturn's ring, Herschel noticed some bright spots on the edge of the ring which were carried almost to the end of the diameter, and appeared to rotate about Saturn in about 10 hours $32\frac{1}{4}$ minutes.

The observations of the present century show that the outer of the two bright rings is permanently divided into two concentric rings by a very narrow gap, and when their plane is inclined to the line of sight at a considerable angle, so that the rings are widely open, a series of dark elliptic curves near the extremities of the major axis of the elliptic projection indicates that each ring is still further broken up into a number of thin concentric rings, the gaps between them being apparently filled by the edges of the inner rings except near the extremity of the major axis of the projections.

In 1850 a dark ring encircling Saturn within the other two rings was discovered independently by three astronomers in England and America. The distance between the outer edge of the dark ring and the inner edge of the adjoining bright ring appeared to vary between narrow limits, the rings sometimes appearing to be in contact. This innermost ring is transparent, so that the edge of the planet can be seen through it, and as it is seen in all positions without distortion it seems that the ring cannot consist of a transparent gas or liquid. The rings appear to be much thicker near the planet than at their outer edges. A comparison of measurements of the diameters of the rings at different times seems to indicate a rapid change in their breadths, the circumference of the outer ring extending outwards, while the inner bright ring appears to be approaching the planet, so that the whole breadth of the ring system is increasing. At least this is the conclusion to which Struve arrived from a comparison of his measurements with those of Huyghens and Herschel.[1] According to Hind the exterior diameter of the outer ring is about 170,000 miles, its interior diameter 150,000; the exterior diameter of the inner bright ring is 147,000 miles, its interior diameter 114,000 miles; and the equatorial diameter of the planet 75,500 miles; but, as above stated, the breadth of the rings appears to be increasing. Sir John Herschel was of opinion that the thickness of the rings did not exceed 100 miles, and Bessel calculated that their mass, as determined by the disturbance they produced in one of Saturn's satellites, did not exceed $\frac{1}{118}$th of the mass of the planet.

If Saturn's rings were solid and at rest, but of course subject to the attraction of the planet, the stress upon them would be such that we can conceive of no material capable of sustaining it. Maxwell remarked that iron would not only be plastic but semi-fluid under such stresses. By allowing the rings to rotate, the attraction of the planet might expend itself in producing the necessary acceleration towards the centre, or, as is generally stated, might be balanced by the centrifugal force, and the stress might be thus relieved to a great extent; but if the rings rotated with the velocity proper to the interior of the inner ring, the rest

---

[1] The evidence of this supposed change in the configuration of the rings is far from conclusive.

would tend to fly off into space, while if they rotated with the velocity proper to the exterior of the outer ring, the inner portions would tend to fall towards the planet; and even if they rotated with the velocity proper to any intermediate portion of the system, the parts beyond would tend to fly outwards, and the parts within would tend to fall towards the planet, unless this tendency were compensated by the attraction of the ring itself. But there is another objection to the hypothesis of the rings being uniform solids, viz. the instability of the motion. If the motion of the ring were slightly disturbed, as is always the case, it would not return to its original orbit, but its deviation would increase until it came in contact with the planet itself.

Laplace was the first to investigate the conditions of stability of Saturn's ring system. He concluded that if the rings are solid, they must consist of a great number of very small concentric rings, each rotating independently with its proper velocity about the planet. The velocity calculated by Laplace for the circumference of the outer ring agreed with that determined, as above mentioned, by Herschel. Laplace avoided the difficulty of the instability of the rings above alluded to by supposing that they were not homogeneous, but that their centres of gravity were at some distance from their geometrical centres. Laplace also showed that, for the attraction of the ring to neutralise its tendency to split up, it is necessary that its density should be $\frac{10}{13}$ that of the planet. In 1851 Professor Pierce showed that the number of rings in the system must be very much greater than Laplace had supposed. This was the position of the question when it was taken up by Maxwell.

On March 23d, 1855, the Examiners announced the subject for the Adams Prize in the following terms :—

The University having accepted a fund raised by several members of St. John's College, for the purpose of founding a prize to be called the Adams Prize, for the best essay on some subject of pure mathematics, astronomy, or other branch of natural philosophy, the prize to be given once in two years, and to be *open to the competition of all persons who have at any time been admitted to a degree in this University* :—

The Examiners give notice, that the following is the subject for the prize to be adjudged in 1857 :—

The Motions of Saturn's Rings.

*** The Problem may be treated on the supposition that the system of Rings is exactly, or very approximately, concentric with Saturn, and symmetrically disposed about the plane of his equator, and different hypotheses may be made respecting the physical constitution of the Rings. It may be supposed (1) that they are rigid; (2) that they are fluid, or in part aeriform; (3) that they consist of masses of matter not mutually coherent. The question will be considered to be answered by ascertaining, on these hypotheses severally, whether the conditions of mechanical stability are satisfied by the mutual attractions and motions of the Planet and the Rings.

It is desirable that an attempt should also be made to determine on which of the above hypotheses the appearances both of the bright rings and the recently discovered dark ring may be most satisfactorily explained; and to indicate any causes to which a change of form, such as is supposed from a comparison of modern with the earlier observations to have taken place, may be attributed.

In his essay, to which the prize was adjudged, and which occupies sixty-eight pages of quarto, Maxwell considered in the first place the hypothesis of Laplace, and showed that to secure stability the irregularity of the ring must be enormously great. Speaking of Laplace's conclusion that the rings must be irregular, Maxwell says :—

We may draw the conclusion more formally as follows :— If the rings were solid and uniform, their motion would be unstable, and they would be destroyed; but they are not destroyed, and their motion is stable, therefore they are either not uniform or not solid.

I have not discovered, either in the works of Laplace or in those of more recent mathematicians, any investigation of the motion of a ring either not uniform or not solid. So that, in the present state of mechanical science, we do not know whether an irregular solid ring, or a fluid or disconnected ring, can revolve permanently about a central body; and the Saturnian system still remains an unregarded witness in heaven to some necessary but as yet unknown development of the laws of the universe.

The following extract shows how Maxwell mentally realised every problem which he considered :—

When we have actually seen that great arch swung over the equator of the planet without any visible connection, we cannot bring our minds to rest. We cannot simply admit that such is the case, and describe it as one of the observed facts in nature, not admitting or requiring explanation. We must either explain its motion on the principles of mechanics, or admit that, in the Saturnian realms, there can be motion regulated by laws which we are unable to explain.

The investigation of the stability of the motion of a solid ring, otherwise uniform but loaded with a heavy particle upon its circumference, led to the conclusion that the mass of the particle must be about four and a half times that of the rest of the ring—

But this load, besides being inconsistent with the observed appearance of the rings, must be far too artificially adjusted to agree with the natural arrangements observed elsewhere, for a very small error in excess or defect would render the ring again unstable.

We are therefore constrained to abandon the theory of the solid ring, and to consider the case of a ring the parts of which are not rigidly connected, as in the case of a ring of independent satellites, or a fluid ring.

There is now no danger of the whole ring, or any part of it, being precipitated on the body of the planet. Every particle of the ring is now to be regarded as an independent satellite of Saturn, disturbed by the attraction of a ring of satellites at the same mean distance from the planet, each of which, however, is subject to slight displacements. The mutual action of the parts of the ring will be so small compared with the attraction of the planet that no part of the ring can ever cease to move round Saturn as a satellite.

But the question now before us is altogether different from that relating to the solid ring. We have now to take account of variations in the form and arrangement of the parts of the ring, as well as its motion as a whole, and we have as yet no security that these motions may not accumulate till the ring entirely loses its original form, and collapses into one or more satellites circulating round Saturn. In fact such a result is one of the leading doctrines of the "nebular theory" of the formation of planetary systems; and we are familiar with the actual breaking up of fluid rings under the action of "capillary force," in the beautiful experiments of M. Plateau.

In this essay I have shown that such a destructive tendency actually exists, but that by the revolution of the ring it is converted into the condition of dynamical stability.

The investigation of the motion of a ring composed of a continuous liquid, led to the result that the waves set up in the ring would cause it to break up into a number of drops which might revolve as satellites about Saturn; but if the condition of stability was not fulfilled, they would coalesce, forming a smaller number of larger drops, until the condition of stability was secured; and the same would be the case if the ring were supposed to consist of a number of concentric narrow rings. The liquid ring, therefore, did not afford a more satisfactory result than the solid one, though it appeared that the internal friction would produce no sensible effect, and Maxwell had recourse to " the dusky-ring, which is something like the state of the air supposing the siege of Sebastopol conducted from a forest of guns 100 miles one way and 30,000 miles the other, and the shot never to stop but go spinning away round a circle radius 170,000 miles."[1] This, however, is not the order in which the investigations appear in the paper before us.

Having dismissed the assumption of the solid rings, Maxwell investigated the effect of disturbance on a ring of small transverse section, the parts of which are not rigidly connected. The disturbances he treated as made up of "harmonic" elements, according to Fourier's method; and, first treating the ring as a number of equal satellites revolving round the planet, he found that such a ring " can always be rendered stable by increasing the mass of the central body and the angular velocity of the ring." If the ring consist of 100 satellites, the mass of the planet must exceed

---

[1] See p. 278.

4352 times that of the ring in order to secure stability for all displacements. " If this condition be not fulfilled, . . . then, although the motion depending upon long undulations may remain stable, the short undulations will increase in amplitude till some of the neighbouring satellites are brought into collision." The satellites of such a ring, when the condition of stability is fulfilled, admit of four vibrations in different periods in ellipses about their mean positions, and these vibrations will be transmitted as waves with different velocities round the ring, so that the form of the ring at any instant resembles " that of a string of beads forming a re-entering curve, nearly circular, but with a small variation of distance from the centre," forming a number of " regular curves of transverse displacement at regular intervals round the circle. Besides these, there are waves of condensation and rarefaction, the effect of longitudinal displacement." Any external disturbance, such as a satellite or wave in another ring, will produce a *forced* wave in the ring, and if the angular velocity of the external disturbing cause around the ring coincide, or nearly coincide, with the velocity of one of the free waves of the ring, the amplitude of the forced wave will increase indefinitely until the ring is destroyed ; but if this condition is not fulfilled, the forced wave will accompany the disturbing cause around the ring as the tide follows the moon.

The effect of one ring upon another is to produce in it a series of *forced* waves travelling with the same angular velocity as the *free* waves in the disturbing ring. Hence in a system of two rings " there will be eight waves in each ring, and the corresponding waves in the two rings will act and react on each other, so that, strictly speaking, every one of the waves will be in some measure a forced wave, although the system of eight waves will be the free motion of the two rings taken together."

The dynamical stability of a ring of satellites is explained by the consideration that when one of the satellites, in consequence of its oscillations, moves with more than its mean velocity, and thus gets in front of its proper position, it will be carried farther away from the planet, and thus not only will its linear velocity be diminished, but, in virtue of its increased distance, its angular velocity will be still farther diminished and, lagging behind, it will fall back into its proper position and, oscillating through it with a velocity less than its mean, will approach the planet and the reverse action will take place.

After considering the manner in which the ring will break up if the condition of stability is not fulfilled, Maxwell takes up the question of "the dusky ring," consisting of innumerable small particles resembling "a shower of rain, hail, or cinders." For the stability of such a ring its *average* density must not exceed $\frac{1}{300}$ of that of the planet, and the density of Saturn being only ·7505, it follows that the average density of the ring cannot greatly exceed that of air at ordinary pressure. Laplace showed that for a ring to rotate as a whole with uniform velocity about Saturn, the density of the planet cannot exceed 1·3 times that of the ring. Hence the particles must move independently, or in a series of concentric rings.

In the case of concentric rings of satellites, their mutual disturbances will destroy one another if the velocity of any one of the four free waves of one coincide, or nearly coincide, with that of any free wave in another; and it is impossible that there should be any great number of rings without this condition frequently recurring.

In the case of a large number of concentric rings, the stability of each pair must be investigated separately, and if in the case of any two, whether concentric rings or not, there are a pair of conspiring waves, those two rings will be agitated more and more till waves of that kind are rendered impossible by the breaking up of those rings into some different arrangement. The presence of the other rings cannot prevent the mutual destruction of any pair which bear such relations to each other.

It appears, therefore, that in a system of many concentric rings there will be continually new cases of mutual interference between different pairs of rings. The forces which excite these disturbances being very small, they will be slow of growth, and it is possible that by the irregularities of each of the rings

the waves may be so broken and confused as to be incapable of
mounting up to the height at which they would begin to destroy
the arrangement of the ring. In this way it may be conceived
to be possible that the gradual disarrangement of the system may
be retarded or indefinitely postponed.

But supposing that these waves mount up so as to produce
collisions among the particles, then we may deduce the result
upon the system from general dynamical principles. There will
be a tendency among the exterior rings to move farther from the
planet, and among the interior rings to approach the planet, and
this either by the extreme interior and exterior rings diverging
from each other, or by intermediate parts of the system moving
away from the mean ring.

The final result, therefore, of the mechanical theory is, that
the only system of rings which can exist is one composed of an
indefinite number of unconnected particles revolving round the
planet with different velocities, according to their respective distances. These particles may be arranged in a series of narrow
rings, or they may move through each other irregularly. In the
first case the destruction of the system will be very slow, in the
second case it will be more rapid, but there may be a tendency
towards an arrangement in narrow rings which may retard the
process.

The transparency of the inner ring, which allows the
planet to be seen through it *without distortion,* and the
observed changes in the configuration of the rings themselves,
all favour this conclusion.

If the changes already suspected should be confirmed by
repeated observations with the same instruments, it will be worth
while to investigate more carefully whether Saturn's rings are
permanent or transitionary elements of the solar system, and
whether in that part of the heavens we see celestial immutability
or terrestrial corruption and generation, and the old order giving
place to new before our own eyes.

The apparatus constructed by Maxwell to illustrate the
motions of the satellites in the rings has been already referred
to.[1] It is an arrangement in which ivory balls are made
to go through the motions belonging to the first or fourth of

---

[1] See p. 295.

the four series of waves above mentioned, except that the balls describe circles instead of ellipses about their mean positions. We give the description as printed in the essay; the apparatus itself is in the Cavendish Laboratory.

The instrument stands on a pillar A (Figs. 7 and 8), in the upper part of which turns the cranked axle CC. On the parallel parts of this axle are placed two wheels, RR and TT, each of which has thirty-six holes, at equal distance, in a circle near its circumference. The two circles are connected by thirty-six small cranks of the form KK, the extremities of which turn in the corresponding holes of the two wheels. The axle of the crank K which passes through the hole in the wheel S is bored

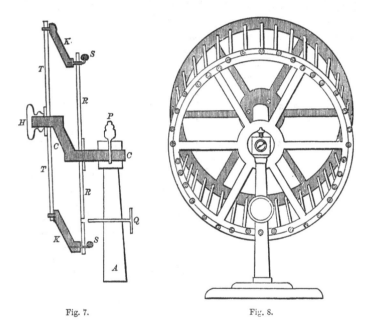

Fig. 7.   Fig. 8.

so as to hold the end of the bent wire which carries the satellite S. This wire may be turned in the hole so as to place the bent part carrying the satellite at any angle with the crank. A pin P, which passes through the top of the pillar, serves to prevent the cranked axle from turning; and a pin Q, passing through the pillar horizontally, may be made to fix the wheel R

by inserting it in a hole in one of the spokes of that wheel.
There is also a handle H, which is in one piece with the wheel
T, and serves to turn the axle.

Now suppose the pin P taken out, so as to allow the
cranked axle to turn, and the pin Q inserted in its hole so as to
prevent the wheel R from revolving; then if the crank C be
turned by means of the handle H, the wheel T will have its
centre carried round in a vertical circle, but will remain parallel
to itself during the whole motion, so that every point in its
plane will describe an equal circle, and all the cranks K will be
made to revolve exactly as the large crank C does.   Each
satellite will therefore revolve in a small circular orbit in the
same time with the handle H, but the position of each satellite
in that orbit may be arranged as we please, according as we
turn the wire which supports it in the end of the crank.

In Fig. 8, which gives a front view of the instrument, the
satellites are so placed that each is turned 60° farther round in
its socket than the one behind it.   As there are thirty-six satellites, this process will bring us back to our starting-point after
six revolutions of the direction of the arm of the satellite; and
therefore, as we have gone round the ring once in the same
direction, the arm of the satellite will have overtaken the radius
of the ring five times.

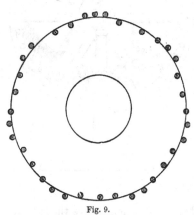

Fig. 9.

Hence there will be five
places where the satellites
are beyond their mean distance from the centre of the
ring, and five where they are
within it, so that we have
here a series of five undulations round the circumference of the ring. In this case
the satellites are crowded
together when nearest to the
centre. . . .

Now suppose the cranked
axle C to be turned, and all
the small cranks K to turn
with it, as before explained,
every satellite will then be carried round on its arm and in the same
direction; but, since the direction of the arms of different satellites is different, their phases of revolution will preserve the same
difference, and the system of satellites will still be arranged in
five undulations, only the undulations will be propagated round

the ring in the direction opposite to that of the revolution of the satellites.

.   .   .   .

If the satellites are arranged as in Fig. 8, where each is more advanced in phase as we go round the ring, the wave will travel in the direction opposite to that of rotation, but if they are arranged as in Fig. 9, where each satellite is less advanced in phase as we go round the ring, the wave will travel in the direction of rotation.

.   .   .   .

We may now show these motions of the satellites among each other, combined with the motion of rotation of the whole ring. For this purpose we put in the pin P, so as to prevent the crank axle from turning, and take out the pin Q, so as to allow the wheel R to turn. If we then turn the wheel T, all the small cranks will remain parallel to the first crank, and the wheel R will revolve at the same rate as T. The arm of each satellite will continue parallel to itself during the motion, so that the satellite will describe a circle whose centre is at a distance from the centre of R, equal to the arm of the satellite, and measured in the same direction. In our theory of real satellites each moves in an ellipse, having the central body in its focus, but this motion in an eccentric circle is sufficiently near for illustration. The motion of the waves relative to the ring is the same as before. The waves of the first kind (Fig. 8) travel faster than the ring itself, and overtake the satellites, those of the fourth kind (Fig. 9) travel slower and are overtaken by them.

This paper was characterised by the late Astronomer Royal as " one of the most remarkable applications of Mathematics to Physics that I have ever seen."

6. But notwithstanding the investigations above referred to, and many other original papers on almost every branch of Physical Science, it is for his researches in Electricity and in Molecular Science that Maxwell stands pre-eminent among the men of science of the present century. After taking his degree in 1854, Maxwell read through Faraday's Experimental researches, a course which he always recommended his students to follow. In Faraday he found a mind essentially of his own type. Thoroughly conversant himself with the Theory of Attractions as developed in Mathematical Treatises,

and with the laws of electrical action as illustrated by Sir
William Thomson in his paper "on the Uniform motion of heat
in homogeneous solid bodies, and its connection with the
Mathematical Theory of Electricity," a paper published in
the *Cambridge Mathematical Journal*, February 1842, and
" on a Mechanical representation of Electric, Magnetic, and
Galvanic Forces," published in the *Cambridge and Dublin
Mathematical Journal*, January 1847, Maxwell saw the
connection between Faraday's point of view and the method
of research adopted by the Mathematicians.   He used to say
that he had not a good nose to smell heresy, but whatever
was good and true Maxwell would detect beneath the mass of
misconception, or even falsehood, which had gathered round
it, and which caused its rejection by nearly every one else
without inquiry.   Faraday's conception of a medium he
adopted as a guide throughout his electrical researches.

Until the sixteenth century all that was known respecting electricity was the one fact that amber when rubbed
possesses the power of attracting light bodies.   This property
was shown (*Physiologia Nova*, 1600) to be possessed by a
variety of substances by Dr. Gilbert of Colchester, who was
Physician to Queen Elizabeth, and who may be regarded as
the founder of the Science of Electricity.   From this time
rapid strides were made in the experimental portion of the
science, and the law according to which the attraction or repulsion between two small bodies charged with electricity varies
with the charges, and the distance between them, was determined by Coulomb with his torsion balance, an instrument
whose value to the experimental investigator can hardly be
over-estimated.   But it is to Cavendish (1771–1781) that we
are mainly indebted for the foundation of the Mathematical
Theory of Electricity, and for the highest experimental
evidence of the law of electrical action.   As the preparation
for the press of *The Electrical Researches of the Honourable
Henry Cavendish* was the last of Maxwell's contributions to
science, the work being published only a few weeks before
his death, we shall again have to refer to Cavendish's investigations, and need only state that his experiments proved

conclusively, and in the best possible manner as far as the instruments at his disposal would allow, that the attraction or repulsion between two small charged bodies varies directly as the product of their charges, and inversely as the square of the distance between them, so that the law of electrical action is the same as Newton's law of gravitation, except that the stress between similarly charged bodies is repulsive, and that between dissimilarly charged bodies attractive. After Cavendish's time comparatively little was added to the theory of statical electricity, if we except the elaborate mathematical investigations of particular problems by Poisson, and the papers of George Green, which until recently were read by few, and appreciated by only two or three, until Faraday took up the subject. Most of Cavendish's work remained unpublished and unknown, and some of his results were independently obtained by Faraday. It is difficult to conceive what would have been the effect on Faraday's mind of perusing Cavendish's "thoughts on electricity," as well as his own accounts of his experiments. Perhaps it is best for the world that Faraday was left to work and think on independent lines ; certainly it has been a boon to Mathematicians and Physicists alike that Maxwell has appeared to expound and develop, if not to perfect, the work of both.

The mathematical theory of attractions had, prior to the time of Faraday, attained a very high degree of development in the hands of Laplace, Lagrange, Poisson, and others, and could be applied to the solution of many very interesting problems in electricity. But Faraday was not satisfied with the hypothesis of *direct action at a distance* between charges of electricity, and held that there must be some mechanism by which electric and electromagnetic actions can be communicated from point to point. Not all the arguments by which he supported this view are conclusive, for the force upon an electrified body and the induced electrification of any conductor will be the same whether we adopt the hypothesis of direct action at a distance or of the transmission of electrical action in lines, straight or curved, through an intervening medium. But any view, whether the arguments in its favour are conclusive or not, is of value if it lead us to inquire more closely into the mechanism by which a phenomenon is brought about; and thus Faraday's conception of lines of force, transmitted through a medium, and exerting tension and pressure wherever they are to be found, are of more value as an instrument of mental research than Weber's Theory of Electro-magnetism, however perfect the latter may be from a mathematical point of view.

The following quotation, from the preface to the *Electricity and Magnetism*, gives Maxwell's views of Faraday in his own words :—[1]

Before I began the study of electricity I resolved to read no mathematics on the subject till I had first read through Faraday's *Experimental Researches on Electricity*. I was aware that there was supposed to be a difference between Faraday's way of conceiving phenomena and that of the mathematicians, so that neither he nor they were satisfied with each other's language. I had also the conviction that this discrepancy did not arise from either party being wrong. I was first convinced of this by Sir William Thomson, to whose advice and assistance, as well as to his published papers, I owe most of what I have learned on the subject.

As I proceeded with the study of Faraday, I perceived that his method of conceiving the phenomena was also a mathematical one, though not exhibited in the conventional form of mathematical symbols. I also found that these methods were capable of being expressed in the ordinary mathematical forms, and these compared with those of the professed mathematicians.

For instance, Faraday, in his mind's eye, saw lines of force traversing all space where the mathematicians saw centres of force attracting at a distance ; Faraday saw a medium where they saw nothing but distance ; Faraday sought the seat of the phenomena in real actions going on in the medium, they were satisfied that they had found it in a power of action at a distance impressed on the electric fluids.

Suppose a small positively electrified body to start from a point close to a positively electrified surface, and suppose

---

[1] See also Maxwell's article on "Faraday" in *Ency. Brit.*, 9th edit.

it to move always in the direction in which it is urged by
the force acting on it, it will, of course, be repelled by the
surface, and will move away along some path straight or
curved, and will continue to move indefinitely, the force
diminishing as it proceeds, unless it meet with a negatively
electrified surface, which will attract it, and coming into
contact with this surface its career will terminate. The path
traced out by such a small electrified body constitutes
Faraday's *line of force*, which is therefore a line whose direction at any point is that of the resultant force at that point.
Such lines of force always proceed from positively electrified
surfaces, and terminate upon negatively electrified surfaces;
or, failing this, they must proceed to infinity. Lines of force
proceeding from a positively electrified body placed in a room,
unless there be other negatively charged bodies in the neighbourhood, will in general terminate upon the walls, floor, and
ceiling of the room, or upon objects in the room in electrical
communication with these. Faraday thus conceived the
whole of the space in which electrical force acts to be tra-
versed by lines of force which indicate at every point the
direction of the resultant force at that point. But Faraday
went further than this : he conceived the notion of causing
the lines of force to represent also the *intensity* of the force
at every point, so that when the force is great the lines
might be close together, and far apart when the force is
small; and since the force in the neighbourhood of a small
charged body is proportional to the charge, he endeavoured
to accomplish this object by drawing from every positively
electrified surface a number of lines of force proportional to
its charge, and causing a similar number of lines of force to
terminate in every negatively electrified surface. In a paper
entitled "On Faraday's Lines of Force," read before the
Cambridge Philosophical Society on December 10th, 1855,
and February 11th, 1856, Maxwell showed that if a system
of lines could be drawn according to Faraday's method, then,
in virtue of the law of electrical action being that of the
inverse square of the distance, the number of lines of force
passing through a unit area of any surface, drawn perpen-

dicular to the direction of the force, is proportional to the magnitude of the force in the neighbourhood, and that the number of lines passing through the unit area of any other surface is proportional to the component of the force at right angles to that surface. Maxwell therefore imagined the positively electrified surfaces from which the lines started to be divided into areas, each containing one unit of electricity, and lines of force to be drawn through every point in each bounding line. These lines therefore divide the whole of space into "unit tubes," whose boundaries are lines of force, and Maxwell showed that, in virtue of "the law of inverse squares," the force at any point in any direction is inversely proportional to the area of the section of the unit tube of force made by a plane perpendicular to that direction. Maxwell further showed that on the negatively electrified surface upon which these tubes terminate, each tube will enclose one unit of negative electricity, and consequently, if a metallic surface be introduced so as to cut the lines of force, the surface being placed at right angles to the tube, a unit of negative electricity will be induced on each portion of the surface contained within the trace of a tube of force ; and hence, in any isotropic medium, these unit tubes of force are also *unit tubes of induction*. If, therefore, a system of tubes of force be drawn in connection with any electrified system, and in accordance with this plan, the whole of the space in which the force acts will be divided into tubes each originating from a unit of positive electricity and terminating upon a unit of negative electricity, while the direction of the force at any point will be indicated by that of the tube, and the magnitude of the force will be inversely proportional to the area of the cross section of the tube. Now, if the law of force had been any other than that of the inverse square, and tubes had been drawn starting from an electrified surface as above, and such that the area of any section of a tube is inversely proportional to the force across the section, these tubes would either leave spaces between them as they recede from the surface, or would intersect one another ; so that it is only for the law of inverse squares that the system of tubes above described is possible. Faraday pointed out that there is not only a tension exerted along each line of force, but that the several lines exert a repulsion upon one another, and Maxwell showed that a tension along the lines of force, accompanied by an equal pressure in every direction at right angles to these lines, is consistent with the equilibrium of the medium. Taking an illustration from the flow of water in a river, Maxwell pointed out that the stream lines or paths along which particles of water flow, are analogous to lines of electric force, the velocity of the water being analogous to the intensity of the force. If the river be supposed to be divided into tubes, the boundaries of which are lines of flow, and if these tubes be so drawn that unit volume of water passes across a particular section of each tube in a second, then, if the flow be steady, unit volume of water will flow across every section of each tube in a second, since no water enters or leaves the tube except at its ends. Such tubes may be called unit tubes of flow, and if no tributaries enter the river there will be the same number of unit tubes crossing each section of the river. Where the bed widens the section of each tube increases, being always inversely proportional to the velocity of the water, and hence the number of unit tubes of flow which cut any unit of area in a cross section of the river will be proportional to the velocity of the water in the neighbourhood. Such a system of tubes, therefore, will represent both the direction of motion and velocity of the water at every point, and will exactly correspond, *mutatis mutandis*, with a system of unit tubes of electric force.

The following letter was addressed to Maxwell by Faraday on receiving a copy of the paper on " Lines of Force :"—

Albemarle Street, W., 25th March 1857.

MY DEAR SIR—I received your paper, and thank you very much for it. I do not say I venture to thank you for what you have said about " Lines of Force," because I know you have done it for the interests of philosophical truth ; but you must suppose it is work grateful to me, and gives me much encouragement to think on. I was at first almost frightened when I saw such mathematical force made to bear upon the subject, and then

wondered to see that the subject stood it so well. I send by this
post another paper to you; I wonder what you will say to it. I
hope however, that bold as the thoughts may be, you may perhaps find reason to bear with them. I hope this summer to make
some experiments on the time of magnetic action, or rather on
the *time* required for the assumption of the electrotonic state,
round a wire carrying a current, that may help the subject on.
The time must probably be short as the time of light; but the
greatness of the result, if affirmative, makes me not despair.
Perhaps I had better have said nothing about it, for I am often
long in realising my intentions, and a failing memory is against
me.—Ever yours most truly,   M. FARADAY.
    Prof. C. Maxwell.

The paper, read before the Cambridge Philosophical
Society, and published in vol. x. of their *Proceedings*, is
confessedly only a translation of Faraday's ideas into mathematical language, with illustrations and extensions, and it
makes no attempt at explaining the nature of the action in
the dielectric, or the mechanism by which the observed
effects are brought about.   About five years later, in a series
of three papers communicated to the *Philosophical Magazine*
in 1861 and 1862, Professor Maxwell gave a simple sketch
of a system of mechanism, capable of producing not only the
electrostatic effects above alluded to, but also of accounting
for magnetic attraction, the action of electric currents upon
one another, and upon magnets, and electromagnetic induction; but before giving an account of these papers it will be
necessary briefly to mention the principal phenomena, an
explanation of which was required.

The ordinary phenomena of magnetism, including the
attraction between dissimilar and the repulsion between
similar poles, as well as the still more familiar phenomena
of the attraction of soft iron by a magnetic pole, are too well
known to require more than a passing mention.   Coulomb
showed that the law of inverse squares obtained equally for
magnetic repulsions as for electrical, so that the stress between
two magnetic poles is proportional to the product of the
strengths of the poles and inversely proportional to the
square of the distance between them, provided the steel of

which the magnets are composed is sufficiently hard to prevent the actions of the magnets on each other altering the strengths of their poles.

If a sheet of paper be supported horizontally above the poles of a magnet, and iron filings be sprinkled over the paper, each filing becomes magnetised by induction in the direction of the resultant magnetic force at the point where it is situated, and if the paper be gently tapped so as to overcome friction, the mutual attraction of the unlike poles in the filings causes them to adhere together in threads or filaments, the North pole of one filing attaching itself to the South pole of a neighbouring filing, and so on, the points of attachment all lying along a line of force. In this way the filings form a graphic representation of the lines of magnetic force, and it was this experiment which first suggested to Faraday the idea of the physical existence of such lines; and as he found it difficult to conceive of curved lines of force being due to "direct action at a distance" (Exp. Res. 1166), he considered that there must be some medium which is the vehicle both of magnetic and electric forces, and that such forces are propagated from particle to particle of the medium. Faraday also supposed that the same medium might serve as the vehicle for the transmission of light. The investigation of the properties of the medium necessary to account for observed electric and magnetic actions, the explanation of these actions, and the determination of the velocity of light from purely electro-magnetic considerations on the hypothesis of the existence of a such a medium constitute Maxwell's greatest contribution to electrical science.

The action of an electric current upon a magnet was first observed by Œrsted. It is said that he made many attempts in his laboratory to discover an action between a magnet and a wire conveying a current, but in all his attempts he carefully placed the wire at right angles to the magnetic needle, and could detect no effect whatever. On attempting to repeat the experiment in the presence of his class he placed the wire parallel to the needle, and the latter immediately swung round and ultimately came to rest nearly at right angles to the wire. Whenever the North pole (*i.e.* the *North seeking* pole) of a magnet is brought near to a wire conveying a current, the pole tends to go round the wire in a certain direction, while the South (or South seeking) pole of the magnet tends to go round the wire in the opposite direction, and hence if the magnet be free to turn about its centre, the magnet will come to rest at right angles to the wire. Many *memoriœ technicœ* have been given for determining the manner in which a magnet will behave in the neighbourhood of a current. Maxwell's rule was as follows :—Suppose a right-handed screw to be advancing in the direction of the current, and of necessity rotating as it advances, as if it were piercing a solid. The North pole of a magnet will always tend to move round the wire conveying the current in the direction in which such a screw rotates, while the South pole will tend to move in the opposite direction.

We may thus suppose every wire conveying a current to be surrounded by lines of magnetic force which form closed curves around the wire, and the direction of the force is that in which a right-handed screw would rotate if advancing with the current. In the case of a straight wire of infinite length, these curves are of course circles. Since action and reaction are equal and opposite, it follows that whatever be the mechanical force exerted by a current upon a pole of a magnet, the latter will always exert an equal and opposite force upon the wire or other conductor conveying the current. Many experiments have been devised to show this. Maxwell used to illustrate it in a very simple way. Having attached a piece of insulated copper wire to a small round plate of copper, he placed the plate at the bottom of a small beaker. A disc of sheet zinc was then cut of such size as to fit loosely in the beaker, a small " tail " of zinc being left attached to it ; this was bent up and united to the copper wire above the top of the beaker, while the plate of zinc was suspended in a horizontal position an inch or two above the copper plate. The beaker was filled up with dilute sulphuric acid and placed on one pole of an electromagnet, some sawdust or powdered resin being placed in the liquid to show its movements.   On exciting the magnet the liquid rotated in one direction, and on reversing the polarity of the magnet the direction of rotation was reversed.   If the plates be suspended by a string, so that they can readily turn round in the beaker about a vertical axis, the action of the magnet on the current in the vertical wire will cause the plates to turn always in the direction opposite to that of the liquid.

The laws of the mechanical action of conductors conveying currents upon magnets and upon each other were investigated by Ampère in a series of experiments which were at once conclusive and exhaustive.   These experiments were alluded to in the highest terms by Professor Maxwell.   Any account of them would be out of place here, and we only refer to them as furnishing the experimental evidence for the statements which follow.

We have already described the manner in which magnetic lines of force may be supposed to surround a wire conveying a current.   Now let such a wire be bent into a closed curve or ring which need not necessarily be circular.   The lines of force, which themselves form closed curves around the wire, will all pass in the same direction through the ring formed by the wire conveying the current, as if they were strung upon the wire, and hence the North pole of a magnet will tend to pass through the ring in the direction of the lines of force ; and a moment's reflection will show that this direction is that in which a right-handed screw would advance if rotating in the direction of the current in the wire.   Hence, if the North pole of a magnet be brought near to such a small closed circuit, on the one side it will be attracted and tend to pass through the circuit ; on the other side it will be repelled.   The South pole of a magnet will be acted upon in precisely the opposite manner.   Hence if a small magnetic needle be suspended within a coil of wire conveying a current, it will tend to set itself at right angles to the plane of the coil.   Such an arrangement constitutes a galvanometer.

Now suppose that we have a small disc of steel of the same size and shape as the ring formed by the wire, and that this disc is magnetised so that one side is a north pole and the other a south pole. Such a disc will act upon external magnets in the same manner as the current if it be magnetised, so that a right-handed screw rotating with the current would enter at the south face and emerge at the north face. Such a magnetised disc is called a *magnetic shell*, and it will of course be acted upon by a magnet with forces exactly equal and opposite to those with which the magnet is acted upon by it. The magnetic lines of force proceeding from a circuit conveying an electric current are therefore the same as would proceed from the magnetic shell above described, the strength of the magnetisation being properly adjusted ; in other words, the magnetic field around such a circuit is the same as that surrounding the magnetic shell, and hence it follows that two circuits, each conveying electric currents, will act upon one another in the same way as two magnetic shells whose circumferences coincide with the wires, and which are magnetised as above described.

Now if the shells be parallel and magnetised in the same direction, they will have their opposite faces presented towards each other, and will attract one another. If they are magnetised in the opposite directions they will repel one another. Similarly, two parallel circuits will attract one another if the currents be passing in the same direction in both, and will repel one another if they be going in opposite directions. Also two parallel wires, which may be considered as parts of such circuits, will attract one another when the currents in them are going in the same direction, and repel one another if they are going in the opposite directions. Maxwell's rule for determining the manner in which a circuit conveying a current will behave in the presence of other currents or of magnets is a very simple expression of Faraday's results. Defining the *positive* direction through a circuit as that in which a right-handed screw would advance if rotating with the current, he enunciated the rule thus :—

If a wire conveying a current be free to move in a magnetic field it will tend to set itself so that the greatest possible number of lines of magnetic force may pass through the circuit in the positive direction.

Since the magnetic field may be produced either by magnets or by electric currents themselves, as above described, this rule combined with the principle that action and reaction are equal and opposite will serve to determine the character of the action either upon circuits conveying currents or upon magnets in every possible case which may arise, and, in fact, embodies the magnificent results of Ampère's investigations in this subject.

Previously to the experiments of Faraday the *induction* of electric currents was unknown. The principal phenomenon depending upon this action, which had been observed, and of which no satisfactory explanation had been offered, was that of Arago's rotating disc. In this experiment a disc of copper was made to rotate rapidly in its own horizontal plane above a compass needle, when the needle was observed to follow the disc and rotate on its vertical pin. This experiment was subsequently repeated by Sir John Herschel and Mr. Babbage, who employed discs of various substances, and found that it was only when the discs were good conductors of electricity that Arago's result was obtained. Faraday, in the first series of his Experimental Researches, describes an experiment in which a copper disc was made to rotate between the poles of an electro-magnet, while one electrode of a galvanometer was connected with the axis of the disc, and the other with a wire which was held in contact with the edge of the disc, which edge was amalgamated to secure a good connection. On spinning the disc a current was immediately obtained, the direction of which was reversed with that of the rotation. This experiment may be regarded as the starting-point of the dynamo machines of Wilde, Gramme, Siemens, and others, which seem destined to play so important a part in the civilised life of the future.

Faraday also showed that when two circuits are placed near to one another, if a current be started in one circuit there is an instantaneous current produced in the *opposite direction* in the neighbouring circuit, while on stopping the "primary" current a transient current in the same direction as the primary occurs in the other or "secondary" circuit. This experiment was the origin of the now well-known induction coil. Again, when the current was flowing steadily in the primary circuit, if the secondary circuit were brought nearer to it, a current was *induced* in the secondary in the direction opposite to that in the primary, and continued during the approach of the circuits. On removing the secondary circuit a transient current was set up in the same direction as that in the primary.

We cannot here spare space to trace the development of the laws of induced currents. The character of the action may in all cases be inferred from the very concise statement of Lenz, generally quoted as Lenz's law, and which may be thus expressed :—

If a conductor move in a magnetic field, an electromotive force will be induced in the conductor which will tend to produce a current in such direction that the mechanical force upon the conductor tends to oppose its motion.

This law, taken in conjunction with the statements made above respecting the mechanical action in a magnetic field upon a conductor conveying a current, serves to determine the character of the induced current whenever a conductor moves in the neighbourhood of magnets or electric currents. Moreover, the starting of a current in a neighbouring circuit must have the same effect upon the wire as if the conductor were suddenly brought from an infinite distance into the position which it actually occupies. Hence Lenz's law will apply to every case of induced currents.

Maxwell's statement expresses the laws of induced currents quantitatively as well as qualitatively. It is as follows :—

Whenever the number of lines of magnetic force passing through a closed circuit is changed there is an electro-motive force round the circuit represented by the rate of diminution of the number of lines of force which pass through the circuit in the positive direction.

If, then, the number of magnetic lines of force passing through a circuit is diminished, there will be an electro-motive force round the circuit in the direction in which a right-handed screw would rotate if advancing along the lines of force ; a line of force being always supposed to be drawn in the direction in which a north magnetic pole tends to move along it. If the number of lines of force passing through the circuit is increased, the electro-motive force will be in the opposite direction. This law can be deduced from that which expresses the mechanical action upon a circuit conveying a current when placed in a magnetic field together with the principle of the conservation of energy. That it may be numerically true all the quantities involved must be expressed in terms of the electromagnetic system of units.

The telephone is a beautiful example of the application of this law. Every movement of the iron disc in front of the pole of the magnet alters the number of magnetic lines of force passing through the coils of wire surrounding the pole, and hence induces a current in one direction or the other in the coil, which current, increasing or diminishing the strength of the magnetism in the receiving telephone, causes a cor-responding motion in the iron disc of the receiver, which therefore emits sounds similar to those incident upon the receiving instrument.

From what has been stated it will appear that the motion of a conductor will produce a current therein only when the conductor is moving in a *magnetic field*, that is, a portion of space through which magnetic lines of force pass. Faraday supposed that a conductor under these circumstances was thrown into a peculiar condition, which he termed " the electrotonic state," and that a current was induced whenever this state varied. Maxwell showed that this electrotonic state, on the variations of which the induced current in a circuit depends, corresponds to the number of magnetic lines of force which pass through the circuit. Because every change in this quantity involved the action of electromotive force, its relations to electromotive force being the same as those of momentum to force in dynamics, he called the quantity itself *electromagnetic momentum*. Maxwell's conception of the physical nature of this quantity will be described presently.

The determination of the laws of self-induction in electric currents is another of Faraday's many contributions to electrical science. After one of the Friday evening lectures at the Royal Institution, a certain Mr. Jenkin informed Faraday that when he broke the connection of the circuit in his electromagnet by separating two pieces of wire which he held in his hands, he felt a smart shock. Faraday said that this was the only suggestion, out of a very great number, made to him by ordinary members of a popular audience which ever led to any result. On investigating the matter, Faraday found that when a current is flowing in a coil of wire if the battery be removed there is a tendency for the current to continue after the removal of the battery, and that this tendency is increased by increasing the number of turns of wire in the coil, and still more so by inserting soft iron in the centre of the coil. This tendency does not depend so much on the length of the wire as upon the relative positions of its parts, and if the wire be first doubled and then wound into a coil the tendency disappears. If a few Grove's cells send a current through a short straight piece of wire and the circuit be broken a very feeble spark will be seen on breaking, but if a large electromagnet be introduced into the circuit a very much brighter spark will appear on breaking contact, though the current sent by the battery is feebler. Thus, when a current flows in such a coil its behaviour reminds us of that of water flowing in a pipe which, when an obstruction is suddenly introduced so as to stop the flow, exerts an enormous pressure for a short time upon the pipe and obstruction, in virtue of the momentum which the water has acquired ; but that the action is not due to any momentum actually possessed by the moving electricity is shown by the fact that it depends on the configuration of the wire. This property of a coil is called self-induction. If the poles of an electro-magnet be joined by a wire of great resistance as well as by the battery, when the battery is removed a considerable current will flow through the wire. This current Faraday called the *extra*-current. It is more generally referred to as the self-induction current.

A similar action takes place when connection is made between a battery and a coil. The current does not at once acquire its full value, but for a short time goes on steadily increasing; the self-induction of the coil causing it to behave as if the current in it possessed considerable mass, which has in the first instance to be put into motion. All these actions are immediate consequences of the law of induced currents stated on p. 526.

There is a well-known experiment of Faraday in which a specimen of his heavy glass, or borate of lead, was placed between the poles of a powerful electro-magnet and a beam of plane polarised light was passed through the glass in the direction of the magnetic force. Faraday found that when the light passed from the north to the south pole of the magnet the plane of polarisation was turned through an angle in the same direction as a right-handed screw would rotate if piercing a solid and advancing with the light. When the light passed in the opposite direction, the rotation of the plane of polarisation was in the same direction with respect to the magnet, and therefore reversed with respect to the path of the light. In this respect the heavy glass under the influence of the magnet behaved differently from a solution of sugar which always turns the plane of polarisation of the light in the same direction with reference to its direction of transmission. This was the first experiment which showed any relation between light and magnetism, and indicated that the medium which serves as the vehicle of light—the luminiferous ether—must at least be affected by the presence of magnetic force, though the fact that the presence of ponderable matter is necessary to the production of this rotation, and that the direction of the rotation depends on the nature of the matter, renders it doubtful how far magnetic force affects the ether directly.

All transparent solids and liquids exhibit the same action on light in different degrees. If a tube of water with plate

glass ends be placed within a coil of wire through which an
electric current is passing, and plane polarised light be transmitted through the tube, the plane of polarisation will be
turned through an angle in the direction in which the current
circulates, and this angle will be proportional to the current.
Verdet showed that in the case of a transparent (para-)
magnetic substance the rotation is in the opposite direction
to that of the current.

The curious effect of a magnet upon the luminous discharge in a vacuum tube and the recent experiments of
Dr. Kerr, may indicate other relations between light and
electricity and magnetism.

Having thus very briefly referred to the principal phenomena of magnetism and electromagnetism, we may proceed
to give a short explanation of the medium or mechanism by
which Maxwell accounted for these phenomena and their
mutual interdependence.

From the well-known laws of the propagation of light,
Maxwell assumed "as a datum derived from a branch of
science independent of that with which we have to deal, the
existence of a pervading medium, of small but real density,
capable of being set in motion, and of transmitting motion
from one part to another with great, but not infinite, velocity."
Inasmuch as this medium can transmit undulations with
finite velocity, it follows that it possesses a property analogous to mass, so that its motion implies kinetic energy; in
addition to elasticity, in virtue of which its deformation
implies potential energy.

It is well known that if a body rotate about a fixed centre
there will be a tension along any radius drawn in the plane of
rotation.   The form which the earth would assume under the
action of gravity only, if there were no rotation, would be
that of a sphere.   The diurnal rotation tends to cause the
polar axis to contract and the equatorial diameter to increase;
and this action would go on indefinitely were it not that at
a certain early stage it is balanced by the attraction of gravitation, and thus the earth assumes a nearly spherical form, in
which the polar axis is shorter than the equatorial diameter.

Referring again to the case of the earth, it is demonstrable from the fundamental laws and principles of dynamics that if matter were conveyed from the equatorial regions to the poles, and there deposited so as to lengthen the polar axis at the expense of the equatorial diameter, the rate of rotation of the earth would be increased and the length of the day would be diminished; while if the earth became more oblate its velocity of rotation would diminish. In fact, if any body be in rotation, and be unacted upon by external forces, or if the forces acting upon it be such as not to affect its rotation, and if the system be altered in shape by internal stresses or otherwise, so that its *moment of inertia* about the axis of rotation is increased, the angular velocity will be diminished and, in the case of a sphere becoming an oblate spheroid, the velocity at the circumference will also be diminished, while if the moment of inertia be diminished, the reverse effect takes place.

Now Maxwell supposed that any medium which can serve as the vehicle of magnetic force consists of a vast number of very small bodies or *cells* capable of rotation, and which we may consider to be spherical or nearly so when in their normal condition, until we have reason to believe them to be of some other form. When magnetic force is transmitted by the medium, these bodies are supposed to be set in rotation about the lines of magnetic force as axis, and with a velocity depending on the intensity of the force. For the sake of fixing our ideas he supposed the rotation to be in the direction in which a right-handed screw would turn if it advanced in the direction of the force. We thus have the magnetic field filled with "*molecular vortices,*" all rotating in the same direction about the lines of magnetic force as axes. As we have seen, these vortices will tend to contract in the direction of their axes of rotation, and to expand at right angles to this direction, so that if initially they are elastic spheres, they will tend to become oblate spheroids like the earth. This tendency will involve a *tension* in the medium along the lines of force, these being the lines along which contraction tends to take place, and this will be accompanied by an equal pressure in every direction perpendicular to the lines of force, on account of the tendency of the vortices to expand equatorially.

Now suppose that we have a north magnetic pole and a south magnetic pole placed near to one another. Lines of force will proceed from the North pole, generally in curved lines, to the South pole. The space in the neighbourhood of the poles will be filled with molecular vortices, which will be most energetic along the line joining the poles, and the velocities of the vortices will diminish as we pass into weaker portions of the field. The tension along the lines of force, tending to draw the North and South poles together, affords sufficient explanation of the apparent attraction between the poles; the kinetic energy of the molecular vortices accounts for the potential energy of the separated poles, which we thus suppose to be really kinetic energy, though possessed by the *medium* between the apparently attracting bodies and not by the bodies themselves. (Perhaps all examples of so-called potential energy we shall some day find to be really kinetic energy possessed by a medium with the properties of which we have been hitherto unacquainted.) When the poles approach one another, the field which is occupied by the vortices is diminished in extent, and though the velocity of the vortices is increased, the whole energy of the field is diminished, and the difference is expended in work done upon the approaching magnets. If the poles are of equal strength, and can come absolutely to coincide, the field is destroyed, all the vortices come to rest, and the energy possessed by them is all expended in work done on the magnets.

If two like poles, north poles for example, be placed near to one another, the lines of force proceeding from the one, instead of going to the other, will be turned aside, and if the poles be of equal strength, a plane bisecting, at right angles, the line joining the poles, will separate the lines of force due to the one from those due to the other, so that no line will cut the plane (Fig. 10). The lines of force thus passing nearly parallel to one another, the pressure exerted by the molecular vortices in every direction at right angles to the lines of force will cause an apparent repulsion between the poles.

To account for the transmission of rotation *in the same direction* from one molecular vortex to the next, Maxwell

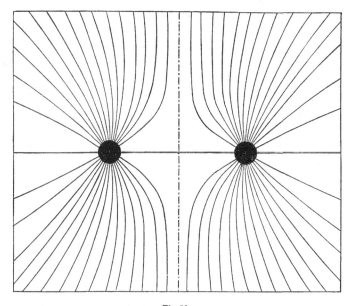

Fig. 10.

supposed that there exists between them a number of extremely minute spherical bodies which roll, without sliding, in contact with the surfaces of the vortices. These bodies serve the same purpose as " idle wheels " in machinery, which, coming between a driver and follower, transmit the motion of the former to the latter unchanged in direction. These minute spherical particles Maxwell supposed to constitute electricity. They roll upon the cells or vortices as if the surfaces in contact were *perfectly* rough, or provided with teeth gearing into one another, and thus, whatever forces may be applied, sliding is impossible. What we ordinarily consider as molecules of matter are supposed to be very large compared with the molecular vortices, and therefore *à fortiori* with the particles of electricity. In an insulator, or dielectric, it is supposed that the electric particles are unable to pass from molecule to molecule of the body, but in a conductor they can do so, the passage, however, being opposed by friction, so that heat is generated and energy dissipated in the transfer.

Now suppose that we have a current of electricity flowing through a conducting wire. Let us confine our attention at first to the central line of particles. These, as they flow, will cause all the cells they touch to rotate about axes perpendicular to the line of flow, so that the stream of particles will be surrounded by rings of vortices. Each ring of vortices will behave like an indiarubber umbrella ring when we pass it over the finger or the stick of an umbrella. Instead of sliding into its place it proceeds by a rolling motion, continually turning itself inside out, as it were, each circular section of the ring or tore rotating about its own centre. Now this motion of the vortices would tend to cause the layer of electric particles outside them to move in the opposite direction to the central stream, and this tendency, to which we shall again refer when we speak of induction, can only be overcome by causing the next ring of cells to rotate in the same direction as the inner ring, when the particles may simply roll round between the coaxial rings of vortices without moving backwards or forwards. But if the layer of particles be compelled to move forwards like the inner stream the layer of vortices surrounding it must rotate more rapidly that the layer within it, and so on, each successive shell of vortices rotating more rapidly until we reach the extreme layer contained within the conducting wire. The shell of vortices which bounds the conductor must by the same mechanism set up molecular vortices in the dielectric, the motion being communicated in ever-widening circles to an unlimited distance. It does not follow that this communication of motion is instantaneous. The cells may consist of elastic material which does not assume its final state of motion as soon as the tangential action of the electric particles is exerted upon it, but begins at first to undergo a deformation, the time taken to set up a given rotation in each depending on its density and elasticity.

Hence electro-magnetic induction, which is the name given to the action we are now discussing, will be propagated through space with a finite velocity, but of this we must say more hereafter.

From what has been said it appears that when a steady (*i.e.* constant) current is flowing in a wire, molecular vortices will be set up in the surrounding dielectric, the axis of rotation of each vortex being perpendicular to the plane passing through the wire and the vortex. The axes about which the vortices turn will therefore form circles surrounding the wire, while the vortices themselves will constitute vortex rings, spinning with very great velocity in the same manner as the indiarubber ring above referred to, or the rings of smoke which are sometimes seen to emerge from a tobacco pipe. But the lines about which the molecular vortices rotate are magnetic lines of force, there being a tension in the medium along these lines, and a pressure everywhere at right angles to them. Hence a straight line carrying an electric current will be surrounded with magnetic lines of force, forming circles with their centres on the axis of the wire, and since the direction of the magnetic force is that in which a right-handed screw would advance if rotating with the vortices, it follows that the direction of the magnetic force around the wire will be that in which a right-

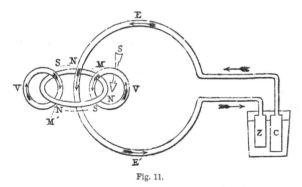

Fig. 11.

handed screw would rotate if advancing with the current. The medium will be subject to tension in circles around the wire, and to pressure in planes passing through the wire, reminding us of the cylinders of an Armstrong gun.

If the wire be bent the same will be true in kind, but the lines will no longer be accurately circles. All the magnetic lines of force pass through a closed circuit in the direction in which a right-handed screw would advance if rotating in the direction of the current. Fig. 11, taken from the paper in the *Philosophical Magazine*, shows the relations between the current, the lines of magnetic force, and the direction of motion of the vortices, the arrows E E′ representing the current, S N indicating the direction of the magnetic force, while the arrows V V′ show the direction of rotation of the vortices.

Now suppose a wire conveying a current to be placed in a magnetic field at right angles to the lines of force. Let S N (Fig. 12) represent the lines of force, A the section of

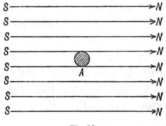

Fig. 12.

the conductor, and let the current be travelling from the reader through the paper. In the space immediately above the wire, the molecular vortices due to the magnetic force originally in the field will be rotating in the direction in which the current in A urges them, while in the space below the conductor the reverse will be the case. Hence the velocity of the vortices above the wire will be increased by the current, while that of the vortices below the wire is diminished. The pressure of the medium at right angles to the lines of force will therefore be greater above the wire than below it, and the wire will be urged downwards at right angles to the lines of force and to its own direction.

Again, suppose two parallel wires to be near together and to convey currents in opposite directions. The strength of a current determines the difference between the velocities of the molecular vortices on opposite sides of it, the electric particles being related to the vortices in the same way as a differential wheel in mechanism; but the vortices on one side of a moving stream of electric particles may be brought to rest if the velocity of those on the other side be doubled, the current remaining the same though the electric particles themselves will now have to spin round, but this makes no difference. Hence, when parallel wires convey currents in opposite directions, the vortices between them being made to spin in the same direction by both currents, will rotate faster than those on the opposite sides of the wires, and pressing as they do with force proportional to the squares of their circumferential velocities, the wires will be pushed apart *as if* they repelled one another.

When two parallel wires convey currents in the same direction, they tend to make the cells in the space between them spin in opposite directions, and the velocities of the molecular vortices there will consequently be less than on the other side of the wires. The pressure of the medium between the wires will therefore be less than in the space beyond, and the wires will be pushed together *as if* they attracted one another.

Now, suppose that a current of electricity commences to flow in a wire. Molecular vortices will be set up in the immediate neighbourhood of the wire, and these vortices acting on the electric particles on the other side of them, remote from the wire, will endeavour to set them in motion in the direction opposite to the current in the wire. But if the medium be a dielectric, the particles cannot be displaced through a sensible distance. They will therefore be made to rotate and start another and larger layer of vortices surrounding the wire, and so the motion will be propagated as above explained. But suppose that at a certain distance there is placed another wire parallel to the first, and forming part of a closed circuit in which no current is flowing. The

particles of electricity in this wire will be acted on in the
same way as those in the dielectric, but meeting with very
little resistance to their motion along the wire, they find it
easier to move through the wire than at once to transmit
the vortex motion to the elastic bodies on the other side of
them. But when a driver and follower are connected by a
differential wheel, if the follower be retarded only by its
own inertia, however small a resistance the differential wheel
may experience to its motion of translation it will at
length cause the follower to turn at the same rate as the
driver, and will itself cease to move. Hence the resistance
of the conductor at length brings the electric particles to rest,
and causes them to communicate the vortex motion to cells
beyond them. Thus when a current is started in a wire
transitory currents in the opposite direction will be induced
in neighbouring conductors, while electric stress will be produced in the dielectric, the elastic cells whose motion constitutes the molecular vortices being at first deformed by the
tangential stress of the electric particles, but both the
induced currents and the stress will have entirely ceased
as soon as all the molecular vortices are in full swing.

Before a current can be maintained in a primary wire, the
molecular vortices in the surrounding field must be properly
started, and this requires the expenditure of work in consequence of the mass of the bodies which constitute the
vortices. It is therefore impossible for a finite electro-
motive force to start a finite current in an indefinitely short
time, in the same way as it is impossible for a finite force
to produce instantaneously a finite velocity in a material
body, and, just as in dynamics we sometimes speak of the
reaction of a body against acceleration as though it were a
force opposing the force applied, so we sometimes speak of
the corresponding action in the case of the current as though
it were a force opposing the battery or other electro-motor,
and speak of it as the electro-motive force of self-induction.
As, however, this depends not on the current in the wire
simply, but on the molecular vortices in the surrounding
medium, it is clear that the self-induction of a wire will

depend on the energy of these vortices, and this must depend on the relations of the several portions of the wire to one another and to the medium, as well as on the density of the medium. The density of the medium Maxwell identified with its *magnetic permeability*. This is greater in (para-)magnetic substances than in air or vacuum; greatest of all in iron. In fact, it is so great in the case of iron, that Maxwell supposed the particles of the iron itself to take part in the vortex motion. Hence the energy of the field, and therefore the self-induction of the wire, is greater the greater the magnetic permeability of the surrounding medium, and the presence of an iron core in a coil immensely increases its self-induction and the energy corresponding to a given current flowing in the coil.

If, after a current has been established in a wire, the circuit be broken or the electro-motive force removed, the molecular vortices refuse to come to rest till they have expended their energy. The only outlet for this energy is a current in the wire, since there is no opportunity of doing work in a non-conducting medium, where there can be no slipping between the elements of the mechanism. The vortices, therefore, keep the electricity moving in the wire after the battery has been removed, until they have expended all their energy in doing work against the resistance of the wire.

But if there be another conductor in the field parallel or slightly inclined to the first, there is another partial outlet for the energy of the system, and a " secondary " current will be set up in the second wire in the same direction as the current in the primary, while that in the primary will be less than it would have been if no secondary circuit had existed. In this way the hypothesis of molecular vortices affords an explanation both of the mutual induction of two circuits and the self-induction of one.

Suppose a wire to be placed in a magnetic field at right angles to the lines of force, and then to be moved so as to cut the lines at right angles, we should expect that in front of the moving wire the lines of force or threads of vortices would be squeezed together transversely, but extended in the direction of their length, somewhat in the same way as elastic strings would be affected by the wire before they broke and allowed it to pass through. Behind the wire the lateral pressure will be relieved, the vortices will contract in the direction of their axes and expand equatorially. But we have seen that the effect of stretching a rotating elastic body in the direction of its axis of rotation, and compressing it at right angles to this direction, increases the velocity of rotation so that the actual velocity of every point on the surface is increased; while the contraction of the body along the axis of rotation diminishes the velocity. Hence as long as the wire is moving across the lines of force the velocity of the vortices in front of the wire will be greater than that of the vortices behind, and the electric particles in the wire, coming, as they do, between two sets of vortices, which are rotating with different velocities, will flow in a stream along the wire. The direction of the current in the wire will be that which would cause the vortices in front to rotate more rapidly than those behind, and therefore to exert a greater pressure on the wire; in other words, there will be a current induced in such direction as to oppose the motion of the wire. We arrive at a similar result if we suppose the lines of force to be cut obliquely instead of orthogonally. Thus Lenz's law is a consequence of the hypothesis of molecular vortices. If we suppose the magnetic force to act from south to north horizontally, the wire to be vertical and to move from west to east, we have magnetic force acting from south to north, mechanical force acting from east to west, and opposing the motion of the wire, and electro-motive force acting in the wire vertically *upwards*.

Suppose that all over a certain area the electricity is pushed forwards through a very small distance along the normal, so that it does not pass from molecule to molecule of the substance, but in each molecule undergoes a displacement from back to front. The electric particles pressing tangentially on the walls of the elastic cells are unable to set them rotating, because each cell is acted upon equally all round in the direction in which the electricity tends to move, and the substance of the cell therefore undergoes a shearing strain which is resisted by its elasticity, and the state of strain of the cells is propagated through the dielectric by means of the displacement of the electric particles which behave like perfectly incompressible bodies. When the force producing the original displacement is removed the cells resume their original form in virtue of their elasticity, the electric particles return to their normal positions, and the energy of the strained elastic cells expends itself in the work done during the electric discharge. Thus the same medium which serves as the vehicle of magnetic force and produces all the phenomena of electromagnetism also serves for the transmission of the force between charges of statical electricity and as a reservoir of the energy due to electrostatic charges. If the dielectric be divided into cells by unit tubes of force and equipotential surfaces drawn for every unit difference of potential, each cell will contain the same amount of energy.[1] The following quotations from the paper in the *Philosophical Magazine* explain the application of the hypothesis of molecular vortices to statical electricity in Maxwell's own words :—

According to our theory the particles which form the partitions between the cells constitute the matter of electricity. The motion of these particles constitutes an electric current; the tangential force with which the particles are pressed by the matter of the cells is electromotive force, and the pressure of the particles on each other corresponds to the tension or potential of the electricity.

.   .   .   .

A conducting body may be compared to a porous membrane which opposes more or less resistance to the passage of a fluid; while a dielectric is like an elastic membrane, which may be impervious to the fluid, but transmits the pressure on the one side to [the fluid] on the other.

.   .   .   .

In a dielectric under induction, we may conceive that the electricity in each molecule is so displaced that one side is

---

[1] See *Elementary Treatise on Electricity* by Professor James Clerk Maxwell, published by the Clarendon Press, 1881.

rendered positively and the other negatively electrical, but that
the electricity remains entirely connected with the molecule, and
does not pass from one molecule to another.

The effect of this action on the whole dielectric is to produce a general displacement of the electricity in a certain direction. This displacement does not amount to a current, because
when it has attained a certain value it remains constant, but it
is the commencement of a current, and its variations constitute
currents in the positive or negative direction, according as the
displacement is increasing or diminishing. . . . When we find
electromotive force producing displacement in a dielectric, and
when we find the dielectric recovering from its state of electric
displacement with an equal electromotive force, we cannot help
regarding the phenomena as those of an elastic body, yielding to
a pressure and recovering its form when the pressure is removed.

Suppose we have a body positively electrified. This
means that a displacement of the electricity in the medium
takes place in all directions around the body and away from
its surface. The cells are thus exposed to a shearing strain,
diminishing as the distance increases, because the surface
over which the displacement takes place being increased the
linear displacement of the electricity is proportionately
diminished, the particles of electricity behaving like a perfectly incompressible fluid. The medium being isotropic the
lines of electric displacement coincide with those of electric
stress, which stress is everywhere proportional to the displacement. The distortion which the cells experience by
the pressure of the electric particles induces an elastic pressure in all directions, at right angles to the direction of displacement, so that there is a pressure in the medium at right
angles to the lines of force.

Now suppose that we have two positively charged
bodies in the field, which we may suppose to possess equal
charges. Each produces a displacement of the medium
outwards from itself, but the electric particles behaving like
an incompressible fluid, it is clear that there can be no lines
of displacement from the one to the other, but that between
the bodies the lines of displacement will be curved so as to
avoid one another in the same way as the stream lines

CONTRIBUTIONS TO SCIENCE. 543

emanating from two pipes, each of which is supplying water to a tank, would be curved round, and would avoid one another. The lines of displacement, and consequently the lines of force which coincide with them, will therefore be bent in exactly the same manner as the magnetic lines of force represented in Fig. 10, p. 533, and the pressure in the medium at right angles to these lines will cause an apparent repulsion of the bodies.

For the same displacement, that is, for the same charges of the little bodies, the repulsion will be proportional to the elasticity of the medium. It is also proportional to the product of the charges, or, since they are equal, to the square of one of them. Suppose then that the medium is exchanged for one of greater elasticity. If we wish to keep the repulsion between the bodies the same, the displacements and therefore the charges must be diminished, the product of these charges, that is, the square of either charge, being made inversely proportional to the elasticity of the medium. The magnitude of each charge must therefore vary inversely as the square root of the elasticity of the medium when the dielectric is changed. Hence, if we define the electrostatic unit of electricity as "that quantity of positive electricity which, acting on an equal quantity at unit distance repels it with unit force," it follows that the unit will vary with the character of the dielectric, being inversely proportional to the square root of its elasticity.

But the attraction or repulsion between two given charges of electricity varies inversely as the specific inductive capacity of the dielectric, so that the electrostatic unit of electricity varies directly as the square root of the specific inductive capacity, and thus the specific inductive capacity is a quantity which varies inversely as the elasticity of the medium.

Suppose we have two parallel wires conveying equal electric currents in the same direction. Other things remaining unchanged, the velocity of the molecular vortices at any point is proportional to the strength of the currents. The attraction between the wires we know to be proportional to the product of the strength of the currents, that is to the square of one of them. The pressure excited by the vortices is, *cæteris paribus*, proportional to their density and the square of their velocity. Suppose we keep the attraction between the wires the same, but change the density of the medium. Then the velocity of the vortices at any point must vary inversely as the square root of the density of the medium. But the velocity of the vortices is proportional to the strength of the currents. Hence the strength of each current must vary inversely as the square root of the density of the medium. If then the electromagnetic unit of current be defined as that current which, flowing in a certain wire, attracts an equal current in another given wire with unit force, the unit of current, and therefore the unit of electricity, which is the amount flowing per second across any section of a wire conveying a unit current, will vary inversely as the square root of the density of the medium.

The ratio of the electromagnetic to the electrostatic unit of electricity will therefore be proportional to the ratio of the square root of the elasticity to the square root of the density of the medium. But this is known to be the velocity with which a transverse vibration is propagated through the medium. Hence the ratio of these units is a concrete velocity, and is proportional to the velocity of propagation of an electromagnetic disturbance, or of the vortex motions above described, through the dielectric. If the units are chosen according to the ordinary system their ratio is not only proportional to but identical with this velocity.

In a paper published in the *Phil. Trans.* for 1868, Professor Maxwell gave an account of an experiment for determining the ratio of the electrostatic and electromagnetic units of electricity where air is the dielectric. The principle of the method lay in balancing the attraction between two electrified discs by the repulsion between two coils of wire in which currents were flowing in opposite directions. One of the discs and one coil was placed at one end of the beam of a torsion balance, the other disc and coil being fixed, but a third coil, conveying the same current as the other two, was placed at the other end of the beam in order to eliminate the magnetic action of the earth and the suspended coil. The apparatus is now in the Cavendish Laboratory. The result of the experiment gave for the ratio of the units a velocity of 288,000,000 metres, or 179,000 statute miles per second. The result obtained by another method by MM. Weber and Kohlrausch is 310,740,000 metres per second. The battery employed for the electrostatic charges was M. Gassiot's battery of 2600 cells, charged with corrosive sublimate. The accuracy of this result depends on that of the B. A. unit of resistance, the velocity being in fact represented by 28·8 Ohms.

Now, according to the undulatory theory, light consists of transverse vibrations of an elastic substance pervading space and all bodies, and the velocity of light as determined by Foucault is 298,000,000 metres per second, or very near the mean of the values obtained by Maxwell, and by Weber and Kohlrausch, for the velocity of propagation of electromagnetic disturbances. If this is found to be always the case, clearly the same medium will serve to account for the phenomena of electrostatics and electromagnetism, and for the propagation of light which must consequently be of the nature of an electromagnetic disturbance.

If an electromagnetic disturbance take place in a perfect insulator we have seen that it must be transmitted to an unlimited distance, for as no slipping can take place between the electric particles and the cells, and as the particles themselves cannot be displaced except by inducing a corresponding elastic stress in the medium, there is no outlet for the energy of the disturbance, which must therefore be communicated from cell to cell without limit. But if the medium be a conductor, that is, if the electric particles can undergo a permanent displacement passing from molecule to molecule against a frictional resistance and without any tendency to return, the energy of the electromagnetic disturbance will be gradually dissipated; for the electric particles, instead of communicating the whole of the motion of one layer of cells to the next, will themselves be set in motion,

and part of the energy will be dissipated as heat instead of
being imparted to the external layer of cells. The disturbance will therefore continually diminish as it is propagated,
until it very soon becomes insensible. The behaviour is the
same as that of a driver and follower connected by a differential
wheel, whose epicyclic motion is retarded by forces of the
nature of friction. Hence electromagnetic disturbances cannot be propagated in conductors of electricity, and we therefore infer that all true conductors are *opaque* to light.

The transparency of electrolytes, such as saline solutions
and the like, offers no difficulty in the face of this conclusion,
as the transference of electricity in them is by a process
entirely different from true conduction and more allied to
the convection of heat, but Maxwell pointed out that the
transparency of gold leaf is much greater than the theory
would indicate. Thus the resistance of a particular piece of
gold leaf was such that it ought to transmit only 10^{-50} of
the light incident upon it, which would be totally imperceptible, while the amount of green light actually transmitted
by it was easily perceived. This result Professor Maxwell
could reconcile with the theory only by supposing " that
there is less loss of energy when the electromotive forces are
reversed with the rapidity of the vibrations of light than
when they act for sensible times, as in our experiments."

We have seen that the velocity of transmission of an
electromagnetic disturbance in any medium is expressed by
the quotient of the square root of the elasticity divided by
the square root of the density of the dielectric. We have
learned that the elasticity is inversely proportional to the
specific inductive capacity of the medium while the density
corresponds with the magnetic permeability. Hence we
infer that the velocity of transmission of an electromagnetic
disturbance varies inversely as the square root of the
specific inductive capacity, and also inversely as the square
root of the magnetic permeability of the dielectric, and this
must be true for the velocity of light if light be an electromagnetic disturbance. Now the magnetic permeability of
most transparent media, such as glass, quartz, sulphur, hydro-

carbons, and the like, does not differ sensibly from that of a vacuum, and hence in these substances the velocity of light must be inversely proportional to the square root of their specific inductive capacity ; or, since the index of refraction of a medium is the ratio of the velocity of light in a vacuum to its velocity in that medium, it follows that the refractive index must be directly proportional to the square root of the specific inductive capacity. As all our measurements of specific inductive capacity refer to the action of electromotive forces which continue for a much longer time than the duration of a luminous vibration, we should expect the last mentioned relation to agree most nearly with experiment the longer the wave length of the light, or, as it is sometimes stated, the specific inductive capacity of a dielectric should be equal to the square of its refractive index for " *light of infinite wave length.*"

The results of the measurements of the specific inductive capacity of certain liquids by Silow, and of gases, sulphur, paraffin, and resin, agree with this theory as well as can be expected. Boltzmann also finds that the specific inductive capacities of crystalline sulphur along its three crystallographic axes are different, these differences coinciding with the differences of the squares of the refractive indices for light transmitted along these three directions.

Dr. Hopkinson (*Phil. Trans.* Part II. 1881) has recently measured the specific inductive capacities of turpentine, benzol, petroleum, ozokerit lubricating oil, castor oil, sperm oil, olive oil, and neats' foot oil. The hydrocarbons give results which are quite in accordance with Maxwell's theory, but the fatty oils, which are compounds of glycerine with fatty acids, have inductive capacities far too great. The same appears to be the case with all the varieties of glass tested by Hopkinson, the specific inductive capacities of which vary from 6·61 in the case of very light flint to 9·896 for " double extra-dense " flint. In the case of solid paraffin, Hopkinson's result agrees very nearly with that of Boltzmann and with Maxwell's theory. In the case of glass, as in that of the fatty oils, the high specific inductive capacity is associated with a complex chemical constitution, glass consisting essentially of metallic silicates, including silicates of the alkaline and alkaline-earthy metals.

The measurement of the specific inductive capacity of glass is attended with great difficulty on account of the phenomenon generally known as residual charge or electric absorption, that is the *apparent* soaking of the electricity into the substance of the glass. This is a subject in which Maxwell took very great interest, and in his work on electricity and magnetism he has given a mechanical illustration of the action on the supposition that it is due to a want of homogeneity in the glass, some parts of which he supposed to conduct electricity better than others, though badly at the best. A form of experiment, very beautiful in its design, was devised by Maxwell for measuring specific inductive capacities, and was carried out by Mr. J. E. H. Gordon, who was able to reverse the electric stress in the glass 12,000 times per second; but this is of course no approximation to the rapid alternations of the " waves " of light. With the apparatus employed, however, the reduction of the observations involves great mathematical difficulties, and the results must therefore be received with caution whether we regard them as supporting the theory or as opposed thereto.

In applying the hypothesis of molecular vortices to the action of a magnetic field on polarised light, Maxwell " found that the only effect which the rotation of the vortices will have on the light will be to make the plane of polarisation rotate in the *same* direction as the vortices, through an angle proportional—

(A) to the thickness of the substance.

(B) to the resolved part of the magnetic force parallel to the ray.

(C) to the index of refraction of the ray.

(D) inversely to the square of the wave length in air.

(E) to the *mean radius* of the vortices.

(F) to the capacity for magnetic induction."

The relation (E) between the amount of rotation and the size of the vortices, shows that different substances may differ in rotating power independently of any observable difference in other respects. We know nothing of the absolute size of the vortices ; and on our hypothesis the optical phenomena are probably the only data for determining their relative size in different substances.

Now, independently of the action of a magnetic field on polarised light, all the phenomena of diamagnetism can be accounted for on the hypothesis that the magnetic permeability of diamagnetic substances is less than that of a vacuum, so that they behave like a paramagnetic substance immersed in a medium more magnetic than itself. But Maxwell has pointed out that " since M. Verdet has discovered that magnetic substances have an effect on light opposite to that of diamagnetic substances, it follows that the molecular rotation must be opposite in the two classes of substances."

We can no longer, therefore, consider diamagnetic bodies as those whose coefficient of magnetic induction is less than that of space empty of gross matter. We must admit the diamagnetic state to be the *opposite* of the paramagnetic ; and that the vortices, or at least the influential majority of them, in diamagnetic substances, revolve in the direction in which positive electricity revolves in the magnetising bobbin, while in paramagnetic substances they revolve in the opposite direction.

Perhaps we cannot conclude this account of the hypothesis of molecular vortices better than by quoting Maxwell's own words :—[1]

I think we have good evidence for the opinion that some phenomenon of rotation is going on in the magnetic field ; that this rotation is performed by a great number of very small portions of matter, each rotating on its own axis, this axis being parallel to the direction of the magnetic force, and that the rotations of these different vortices are made to depend on one another by means of some kind of mechanism connecting them. The attempt which I [have] made to imagine a working model of this mechanism must be taken for no more than it really is, a demonstration that mechanism may be imagined capable of producing a connection mechanically equivalent to

---

[1] *Electricity and Magnetism*, vol. ii. Art. 831 (1st ed.)

the actual connection of the parts of the electro-magnetic field. The problem of determining the mechanism required to establish a given species of connection between the motions of the parts of a system always admits of an infinite number of solutions. Of these some may be more clumsy or more complex than others, but all must satisfy the conditions of mechanism in general.

The following results of the theory, however, are of higher value :—

(1) Magnetic force is the effect of the centrifugal force of the vortices.

(2) Electromagnetic induction of currents is the effect of the forces called into play when the velocity of the vortices is changing.

(3) Electromotive force arises from the stress on the connecting mechanism.

(4) Electric displacement arises from the elastic yielding of the connecting mechanism.

In a paper entitled "A Dynamical Theory of the Electromagnetic Field," read before the Royal Society on December 8, 1864, Maxwell deduced all the above results by purely mechanical reasoning, only assuming the existence of a medium capable of receiving and storing up potential and kinetic energy, and therefore capable of doing work in " recovering from displacement in virtue of its elasticity," while the parts of the medium are connected by " a complicated mechanism capable of a vast variety of motion, but at the same time so connected that the motion of one part depends, according to definite relations, on the motion of other parts, these motions being communicated by forces arising from the relative displacements of the connected parts, in virtue of their elasticity." For the existence of such a medium we have evidence independent of electrical actions. With regard to the mechanism no attempt is made in the paper to give to it any definite constitution. This paper has been regarded as Maxwell's greatest contribution to electrical science, but most of the results obtained in it have been already mentioned.

The following is a good specimen of Maxwell's humorous irony, of which there are many samples in his scientific works. He is discussing certain developments by Bernhard Riemann Lorenzo, of Weber and Neumann's theory of Electro-magnetism, which is based on the assumption that the action between two quantities of electricity is direct action at a distance, and depends not only on the distance between the charges but upon their relative motion.

From the assumption of both these papers we may draw the conclusions — first, that action and reaction are not always equal and opposite ; and second, that apparatus may be constructed to generate any amount of work from its own resources.

. . . . .

I think that these remarkable deductions from the latest developments of Weber and Neumann's theory can only be avoided by recognising the action of a medium in electrical phenomena.

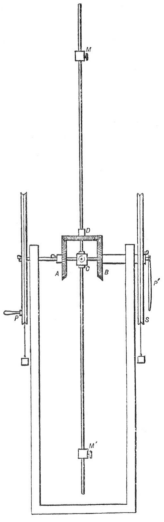

Fig. 13.

While at the Cavendish Laboratory Maxwell constructed a mechanical model which illustrates in a very beautiful manner the principal phenomena of induced currents. As a piece of mechanism it is simply a differential train, such as is often employed as a dynamometer for measuring the power absorbed by a machine. The apparatus is sketched in Fig. 13. The grooved wheel P is keyed to the same shaft as the bevel wheel A, which therefore turns with it, and the rotation of this piece represents the primary

current. A second bevel wheel D turns loosely on the arm
C D, which is one of four arms (of which only two are shown
in the figure) forming a cross, which can turn freely on the
central shaft at C. Sliding weights M M', etc., can be
fixed in any desired position on these arms so as to alter
the moment of inertia of the cross, which is the differential
piece in the mechanism. A third bevel wheel B is keyed
to the same hollow shaft with the wheel S, which is similar
to P, and the rotation of the piece B S represents the current
in the secondary circuit. As the shaft B S is hollow, and
rides loosely on the shaft A C, the wheels A and B can
turn quite independently of one another, except in so far
as they are connected by the wheel D. P' is an index
attached to the interior shaft and turning with P. A loop
of string is hung over each of the wheels P and S, and
carries a small weight. These strings act as friction brakes
to the wheels, and the friction represents the resistance of
the primary and secondary circuits respectively. The
moment of inertia of the loaded cross, or differential piece,
represents the moments of inertia of the cells which constitute the molecular vortices in the dielectric. Its kinetic
energy when rotating represents the energy of the vortices,
and its angular momentum is proportional to the electromagnetic momentum of the system. The moments of inertia
of the other portions of the mechanism are very small compared with that of the loaded cross. The motion of the cross
and the wheel D is impeded by as little friction as possible.

Suppose that the wheel P is made to revolve, representing a current in the primary wire ; the heavy cross will
not at first move, but the wheel D will revolve and communicate the motion to B, which, with S, will rotate in the
direction opposite to that of P, representing a current in the
secondary circuit opposite in direction to that in the
primary. But the motion of S is resisted by the friction
brake, and a finite force must therefore be exerted by D on
B to drive it. The reaction of B, together with the force
exerted by A, will constantly tend to make the cross revolve
in the same direction as P, and the velocity of the cross

being constantly accelerated, it will presently revolve with a velocity half that of P, and then D will roll round B, which, with S, will remain at rest. The piece B S will then continue at rest as long as the rotation of P remains constant corresponding to the cessation of the current in the secondary circuit, while that in the primary remains unchanged, but if P be accelerated, S will revolve in the direction opposite to the motion of P. Now suppose P to be suddenly stopped. The kinetic energy of the cross will cause it to continue to revolve until it has done a corresponding amount of work against resistances, and A being at rest, D will roll upon it and compel B, with S, to revolve in the same direction as the cross, that is, in the same direction in which P formerly revolved, and whatever be the resistance to the motion of S, it will be overcome, and S will revolve till the work done against resistance is equal to the kinetic energy originally possessed by the cross. This corresponds to a current induced in the secondary coil on stopping the current in the primary, which current is in the same direction as the primary current, and continues until the energy of rotation of the molecular vortices has been used up in work done against electrical resistance.

If one operator lay hold of the wheel S, and endeavour to keep it at rest while another applies a steady force to P, the motion of P will be accelerated much less rapidly than if the same force had been applied to it, and S had been free, because P can only move by setting in motion the cross with its great moment of inertia. If the operator who is turning P now suddenly stops it, a great shock will be experienced by the machinery, and the wheel S will slip from the grip of the other operator however firmly he may hold it. The force applied to S may correspond to an air-break in the secondary coil, and this is sufficient to prevent a spark when the battery current is started in the primary, but by suddenly stopping the primary current, as in Ruhmkorff's coil, a disruptive discharge or spark passes through the air between the terminations of the secondary wire. (If the operator who endeavours to keep the wheel

S at rest is inexperienced the effect upon him is very striking).

If a pin be placed in the face of the wheel S, and one end of a spring press against the pin, while the other end is fixed to the frame of the apparatus, we have a representation of a secondary coil in which the circuit is broken, and a Leyden jar inserted with its coatings in connection with the ends of the wire. When the motion of P is changed, S will begin to move, and will deflect the spring, corresponding to a current in the secondary coil charging the Leyden jar. If the spring admit of very great deflection, so that a great amount of work must be done upon it before it slips from the pin, the primary current may have attained its full strength before the slip takes place. This corresponds to the capacity of the Leyden jar being too great to allow of its being charged to a sufficient potential to produce a spark. In this case no spark takes place, but when the force between the wheels D and B diminishes on account of the diminution of the acceleration of P, the spring relieves its strain by forcing the wheel S backwards, and the Leyden jar under corresponding circumstances quietly discharges itself through the secondary coil, reversing the operation by which it was charged. But if the pin slips from the spring, the wheel S will revolve, and the spring will fly back corresponding to a disruptive discharge through the air, and if the acceleration of P continue long enough several such disruptive discharges may take place.

We must, of course, be careful not to endeavour to learn from such a model lessons which it was not designed to teach, and we must remember that the behaviour of the mechanism does not represent the electrical action in *all* respects.

For many years Maxwell rendered valuable service to the British Association, especially in connection with electrical science. Some account of the meetings which he attended will be found in the letters printed in another part of this work, and though during the last few years of his life other engagements prevented his attendance at the annual gatherings, he always showed signs of keen en-

joyment when discussing the " British Asses." In 1862
he was appointed a member of " The Committee on Standards
of Electrical Resistance." In the report issued in 1863, the
Appendix, " On the Elementary Relations between Electrical
Measurements," bears the name of Professor Maxwell in conjunction with that of Professor Fleeming Jenkin, while the
general description of the method employed in the determination of the Ohm or B. A. unit of resistance, together
with the mathematical theory and details of the experiments,
are from Maxwell's pen. In 1863-4 Maxwell was again at
work on the same subject in the laboratory of King's College,
and most of the " spins " [1] were conducted under his own
supervision. In 1869 the results of Maxwell's experiments
on the relation of the electromagnetic to the electrostatic
unit of electricity, described above, were embodied in the
Report to the British Association at the meeting at Dundee,
and this forms the last of the Reports of the Committee.

In 1874 Professor Maxwell was elected a member of
the committee appointed by the British Association for the
purpose of investigating Ohm's law. Most of the work
executed by this committee was carried out by Professor
Chrystal in the Cavendish Laboratory, under the supervision
and at the suggestion of Professor Maxwell. An account of
the investigations will be found in the report presented to
the Association at the Glasgow meeting in 1876.

Before concluding our notice of Maxwell's contributions
to electrical science we must mention the preparation for the
press of *The Electrical Researches of the Honourable Henry
Cavendish*, published in 1879, only a few weeks before the
death of its editor. The amount of labour which Professor
Maxwell bestowed on this work during the last five years of
his life can only be known to those who were constantly in
his company. Nearly all the MS. he transcribed with his
own hand, the greater part being copied after midnight, while
he watched over Mrs. Maxwell during the long illness to
which allusion has elsewhere been made. Every obscure

---

[1] See Part I. p. 316.

passage or allusion was the subject of a long and searching investigation; and many were the letters written to the Librarian of the Royal Society and to scientific and literary friends in different parts of the country, to gain information respecting the meaning of obsolete words and symbols, or the history of individuals. But besides this, and a comparison of Cavendish's results with those obtained by subsequent investigators, Maxwell repeated many of Cavendish's experiments almost in their original form, only employing modern instruments for the purposes of measurement. The introduction and the appendices to the work evidence much labour, patient investigation, and very extensive acquaintance with the literature bearing on the subject. Maxwell was by no means one of the class of "thinkers" who only read their own writings; his acquaintance not only with scientific literature, but with nearly every other class of books was astonishing; and if any question of physics was brought before him, he could generally give an account of nearly all that had been done in the subject. In this respect he resembled the late Professor W. H. Miller, whom Cambridge men used to consult about *everything*.

It would be impossible here to give any adequate account of Maxwell's work in connection with the Cavendish papers, but we may mention one experiment both on account of its intrinsic importance and of the interest which Maxwell took in it. Cavendish describes an experiment (see p. 104 of *The Electrical Researches*) in which a sphere of 12·1 inches in diameter was enclosed between two copper hemispheres 13·3 inches in diameter, and the copper hemispheres, which were supported in frames hinged together, and opened or closed by silk cords, were brought together so as to enclose the sphere, contact being made at the same time between it and the hemispheres. The outer sphere was then charged, and the connection between it and the inner sphere removed. On separating the hemispheres, the inner sphere was found to be discharged, or at least the charge, if any, left upon it was too small to affect Cavendish's pith ball electrometer. Knowing what fraction of the original charge could not fail to be de-

tected by the electrometer, Cavendish deduced from the
results of the experiment that the law of variation of
electrical action with the distance must lie between the
inverse $(2 - \frac{1}{50})$th and $(2 + \frac{1}{50})$th power of the distance,
and inferred that it is the inverse square.

Maxwell placed a copper sphere about 10 inches in
diameter within a hollow copper sphere, made of two hemispheres, about 12 inches in diameter, supporting it upon a
ring of ebonite, so that the spheres might be concentric.
The outer sphere was carefully insulated, and a hole cut in
its surface, which could be accurately filled by a little trap
door.   To this trap door was attached a wire which rested
on the inner sphere when the door was shut down.   The
door was carried on a metal arm which turned about a hinge
attached to the outer sphere, and could be raised or lowered
by a silk thread which passed over a pulley and hung in
front of the operator.   When the trap door was raised a
wire connection could be allowed to fall by relaxing a second
silk cord, so as to make contact with the inner sphere through
the hole in the outer.   The trap door was first closed and
the outer sphere charged by means of a Leyden jar which
was immediately removed (the electric machine being placed
in a distant apartment).   The trap door was then raised, the
outer sphere discharged by connecting it to earth, and the
inner sphere then put in communication with a quadrant
electrometer by means of the wire above mentioned.   Not a
trace of electricity could be detected on the sphere.   To test
the accuracy of the method, a small brass sphere was insulated, and supported at a distance of about 60 centimetres
from the outer sphere, and the experiment was repeated.
When the copper sphere was charged the brass sphere was
connected to earth, and thus a small negative charge was
induced upon it by the positive electricity on the copper
sphere, and the ratio of this charge to that on the copper
could be easily calculated.   The brass sphere was then insulated, the outer copper sphere discharged as before, and the
inner sphere examined with the same result as previously.
The brass ball was then discharged, the outer sphere being

insulated, when it was found that there was sufficient positive
electricity on the outer sphere to deflect the electrometer
through more than 300 times the largest deflection which
could escape notice. Now when the brass ball was insulated,
the negative charge upon it was about $\frac{1}{64}$th of the charge
originally on the copper sphere, and this induced a positive
charge on the sphere when in connection with the earth,
which was about $\frac{1}{486}$ of its original charge. Hence $\frac{1}{486}$ of
the original charge is more than 300 times the largest charge
which could escape observation. From these figures it follows that, in the expression of the law of electrical action,
the true index of the inverse power of the distance must be
either 2 or differ from it by less than $\frac{1}{21600}$. These experiments were carried out in the Cavendish Laboratory by Mr.
MacAlister of St. John's College. In note 19 appended to
the Cavendish papers, Maxwell describes the experiment,
and gives its complete mathematical theory.

The idea that electricity flowing in conductors behaves
like an incompressible fluid is at least as old as Cavendish;
but before he had the opportunity of reading Cavendish's
papers, Maxwell taught that all electric discharges involve a
displacement of electricity in a closed circuit, the electricity
behaving both in conductors and in dielectrics like a perfectly incompressible fluid. Thus, when a Leyden jar is
discharged, a certain quantity of electricity flows from the
inner coating through the knob of the jar, and hence to the
outer coating, but an equal quantity also passes across any
surface which we may imagine drawn within the glass, so as
to include the inner coating and exclude the outer. The
glass, therefore, may be regarded as completing a circuit across
every section of which the same quantity of electricity flows.
Again, if positive electricity be communicated to a conductor
by means of a wire which serves as an *electrode*, an equal
quantity passes out from the surface of the conductor and is
squeezed into the surrounding dielectric, causing a similar
transfer of electricity across every surface drawn in the
dielectric so as to surround the conductor, the extent of the
displacement of course diminishing as we recede from the

conductor on account of the increased area of the surface, but the displacement continuing until some external conductor is reached through which the circuit is completed. Thus, when we say that the charge of a body is increased, we mean that positive electricity is communicated to it through an *electrode*, but we know that an equal amount passes out of the body through its external surface and is squeezed into the dielectric, and it is this dielectric which is the only portion of the system really affected by the charge, and in which the whole *energy* of the charged body resides. This view has very recently been put forth under the title of *The Conservation of Electricity*. According to Maxwell's view of the constitution of dielectrics, the squeezing of a quantity of electricity into a dielectric does not imply a condensation of the electricity, but a *strain* in the dielectric on account of the displacement of the electricity, which cannot move without distorting the " cells " of which the dielectric is supposed to be made up.

7. Ever since men began to think about Nature, philosophers have differed in their views respecting the primary constitution of bodies. When, with all the freshness of a first impulse, though unprovided with the barest means of verification, the human mind first went forth in search of physical truth, two paths at once appeared, leading to opposite poles, and both were trodden by pilgrims full of hope. Some found their satisfaction in contemplating the continuous fulness of the universe. They could imagine no gap in Nature, whose endless variety they were contented to refer to a principle of infinite divisibility and to the inter-play of contiguous elements whose changes were ordered by the mind that was interfused with all. To others, of a more analytical turn, it appeared impossible to account for the most obvious phenomena, except by the hypothesis of atoms moving in a void.

The penetrating intellect of Democritus had early given consistency to an atomic theory, which became the foundation of the system of Epicurus. Yet, widely as this philosophy prevailed at certain periods, the doctrine of a *plenum* was on the whole the more prevalent in the ancient world.

The controversy has been handed on to modern science, or rather it has inevitably reappeared; some, with Descartes and Spinoza, resolving matter into continuous extension; while others, as Bacon would say, cut deeper into Nature, and would divide her if possible into her ultimate constituent parts. Experimental investigations, especially those through which chemistry became an exact science, have greatly favoured atomistic views, yet the tendency to maintain the continuity and plasticity of matter is still apparent, and the bold hypothesis of the vortex atom, due to Sir William Thomson, affords us perhaps the nearest approach to a reconciliation between these two contradictory theories. According to this hypothesis the whole of space is filled with a uniform and perfect fluid, and material atoms are vortices of one form or another which have been created within this fluid.

In order to reach the starting-point of Maxwell's investigations in molecular physics we must go back to Daniel Bernoulli who, in his hydrodynamics, published in 1738, explained the pressure of the air on the hypothesis that air consists of a number of particles moving about in all directions and impinging on any surface exposed to its action. Le Sage of Geneva in 1818 explained the pressure of a gas in strict accordance with the modern Dynamical Theory. The theory of Le Sage was based upon his doctrine of "ultramundane corpuscles," a doctrine borrowed from the older atomists, and employed by Le Sage to explain gravitation. For a concise account of this and of the older atomic theories the reader is referred to Maxwell's article "ATOM" in the ninth edition of the *Encyclopædia Britannica*. In 1847 John Herapath published his *Mathematical Physics*, in which he supposed gases to consist of perfectly hard molecules impinging against one another, and against any surface exposed to their action; and he points out the relation of the motion to the temperature and pressure of a gas and explains gaseous diffusion.

But we are indebted for all the modern developments

of the molecular theory of gases, as well as for its establishment on a sound dynamical basis, mainly to the researches of three men—Professor R. Clausius, Dr. Ludwig Boltzmann of Vienna, and James Clerk Maxwell. Maxwell's principal contributions to the literature of this subject are his papers on " Illustrations of the Dynamical Theory of Gases," presented to the British Association at the meeting in Aberdeen in 1859 ; " On the Viscosity or Internal Friction of Air and other Gases," which constituted the Bakerian lecture read before the Royal Society on February 8th, 1866, and published in the *Phil. Trans.* for that year; " On the Dynamical Theory of Gases," a paper read before the Royal Society on May 31st, 1866, and also published in the *Phil. Trans. ;* the lecture " On Molecules " delivered before the British Association at Bradford in 1873, and which has been referred to in Part I. ; a paper " On the Dynamical Evidence of the Molecular Constitution of Bodies " which was read before the Chemical Society in the Spring of 1875, and published in the June number of its Journal ; and the article " ATOM," in the ninth edition of the *Encyclopædia Britannica.*

According to the molecular theory all bodies are made up of molecules which are more or less free to move relatively to one another. In solids, each molecule can only move through a small distance from its normal position, so that all its movements are essentially of the nature of vibrations. In liquids the molecules are free to move through any distances within the substance of the liquid, but their motion is constantly impeded by neighbouring molecules, from whose interference they are never altogether free, each molecule in its movements resembling an individual endeavouring to work his way through a dense crowd. In a gas each molecule is perfectly free—except on the occasions, comparatively rare, unless the gas be very dense—when it comes into collision with other molecules of gas, or of some other body in contact with the gas. According to this theory, a gas consists of molecules moving in all directions in straight lines, except when they strike one another or some foreign body, when their motion is changed in accordance with the laws of impact of perfectly elastic bodies.

It is probable that the investigations connected with Saturn's rings, which led Maxwell to contemplate the condition of the air in the midst of " a flight of brickbats," first led him to take up the subject of the kinetic theory of gases. At any rate, these investigations led him to the adoption of what he aptly termed " the statistical method," which in his hands became so fruitful in its applications to molecular science. This method consists in the separation of all the things considered into classes, which fulfil certain conditions, and the determination of the number of individuals which at any instant are to be found in each class, without reference to the behaviour of any particular individual. In " the historical method," on the other hand, the life history of each individual is traced, and if classes are considered at all it is only in their relation to the particular individual we are contemplating. The first method deals with the interests of the community at large and disregards the fate of the individual, except as affecting the average condition of the community; the second deals with the interests of the individual alone. It is obvious that when the individuals are to be counted by millions the only way in which they can be successfully treated by a finite mind is by the statistical method, and this was the method which Maxwell adopted.

In the earliest papers of Clausius he treated all the molecules of a particular gas at the same temperature as moving with the same velocity. Maxwell showed that this could not be the case, and that, even if all the molecules were started with the same velocities, their mutual collisions would increase the velocity of some and diminish that of others, until at length the velocities would be distributed " according to the law of errors." This law will be best understood from Maxwell's own illustration. He prepared a diagram illustrating the distribution of bullet marks on a target on the hypothesis that all the shots are *aimed* at the centre of " the bull " by the same marksman, while the probability of an error of any given magnitude occurring is less the greater the error : diminishing very rapidly as the error increases, according to the ordinary hypothesis respecting " errors of observation " developed by Laplace.   If on such a diagram a series of concentric circles be drawn, say an inch apart, about the centre of the target, and the number of shots between each consecutive pair of circles counted, the number will be found to be greatest for a particular pair, but there is nothing to prevent some shots from hitting the target at any distance from the centre.   If the length of the line joining the centre of the target with any bullet mark be taken to represent the velocity of a molecule, the diagram will represent the law of distribution of velocities among the molecules of a gas when they have acted on one another for an indefinite time. They will congregate about a particular value, but there is nothing to prevent some individuals possessing a velocity as great or as small as we please.

The momentum of a particle depends on the product of its mass and its velocity ; its kinetic energy on the product of its mass and the square of its velocity.   One of the first and most important properties of gases which Maxwell showed to be a consequence of the kinetic theory was, that when two groups of molecules are in equilibrium, being separated only by a diaphragm which can transmit the impacts of the particles, the average kinetic energy of the molecules must be the same in each group.   Identifying the kinetic energy of the molecules with the heat of the gas, since we know that for equilibrium the gases must be at the same temperature, it follows that the specific heat varies inversely as the mass of a molecule, which is the dynamical expression of the law of Dulong and Petit, a law which states that the products of the specific heat and combining weight is the same for each element.   The pressure of a gas is proportional to the kinetic energy of unit volume of a gas, and, since the average kinetic energy of the molecules is the same for each gas at the same temperature, it follows that equal volumes of two gases at the same pressure and temperature contain the same number of molecules, and hence the density of a gas at standard temperature and pressure is proportional to the mass of a molecule, *i.e.* to its combining weight, which is Gay Lussac's law of equivalent volumes.

If two gases be in communication with one another, the particles of each are found to penetrate the other until the gases become uniformly mixed, unless some external force act to prevent the uniformity of the mixture. This phenomenon, known as gaseous diffusion, has been studied by Graham, Loschmidt, and others, and is a direct consequence of the dynamical theory. It is plain that the rate of diffusion of one gas into another will be proportional to the *average* velocity of the particles in the case of two gases being separated by a diaphragm perforated by very fine holes, so that each mixture may be considered homogeneous up to the diaphragm. Maxwell showed that in this case, when the pressures on the two sides of the diaphragm are equal, " the volumes diffused will be as the square roots of the specific gravities inversely, which is the law of diffusion established by Graham."

The slowness of gaseous diffusion was for some time regarded as an objection to the kinetic theory of gases. Clausius overcame this difficulty by the introduction of the conception of *the mean free path* of a molecule, and by showing how short this path is. The mean free path is the average distance through which a particle passes between two successive collisions with other particles. The first estimate of the mean free path in air at ordinary pressure and temperature was made by Maxwell from determinations of viscosity or rate of diffusion of momentum in air. Depending on the length of the mean free path are (1) the diffusion of matter; (2) the diffusion of momentum or viscosity; and (3) the diffusion of energy, or the thermal conductivity of the gas. From determinations of the rate at which these three kinds of diffusion proceed, independent determinations of the mean free path in different gases have been made. For hydrogen, at standard pressure and temperature, it is about $\frac{1}{250000}$ of an inch. For other gases and air it is somewhat less. The length of the mean path in air given by Maxwell in 1859 was $\frac{1}{389000}$ of an inch.

Suppose two trains to be moving in opposite directions, side by side on parallel lines, and suppose that as they pass each other a number of passengers from each train jump into the other.   It is clear that each passenger will carry with him the *momentum* he possesses into the other train which is moving in the opposite direction, and will consequently diminish the momentum and therefore the velocity of that train.   If a continuous interchange of passengers, backwards and forwards, between the trains were to take place, the trains would ultimately be brought to rest relatively to one another.   This is an illustration of the *diffusion of momentum,* and was originally given by Balfour Stewart.   Now, suppose two streams of gas to be passing each other tangentially. There will be a continuous interchange of molecules by diffusion between the two streams, and the molecules carrying with them the momentum they possess, the effect of the diffusion will be to tend to bring the two streams to relative rest.   Diffusion of momentum consequently introduces a tangential action between layers of gas which are moving relatively to one another, and therefore causes a resistance to any "continuous change of form, depending on the rate at which that change is effected," that is, it confers upon the gas the property of viscosity.   Thus, the so-called viscosity of gases received from Maxwell its complete explanation in accordance with the kinetic theory.

The viscosity of a gas depending on the rate of diffusion of momentum, and therefore on the rate of diffusion of matter, is proportional, like the latter, to the average velocity of the molecules.   If the gas be very rare, each particle will meet with fewer collisions, and consequently its course will be less interrupted than when the gas is denser, so that Maxwell found that between two strata of gas at a given distance apart the rate of diffusion was the same whatever the density.   He thus "arrived at the startling result that the co-efficient of internal friction is independent of the density of any particular kind of gas." The experimental verification of this result occupied a considerable portion of Maxwell's leisure time in 1865.   The description of the experiments and the statement of the result arrived at form the subject of the Bakerian lecture of 1866. There plates of glass were suspended by a steel piano wire, and caused to oscillate in their own planes between four fixed plates, which were placed at equal distances apart, and so that the moving plates were suspended half way between each pair of fixed plates (see Part I. p. 341). The distance between the plates was varied in different experiments, and the whole system was enclosed in a receiver which could be filled with different gases and the pressure regulated at will. The oscillations of the discs were observed by means of a mirror and scale, and the experiment consisted in determining the rate at which the oscillations died away, from which the viscosity of the gas was calculated. For experiments at high temperatures the receiver was surrounded by a steam jacket. The results arrived at showed that dry air is more viscous than damp air, and that the viscosity of air is greater than that of hydrogen or carbonic acid, nearly in the same ratio as was determined by Graham in his experiments on the transpiration of gases through capillary tubes. But the most remarkable results were (1), that the viscosity of any particular gas is independent of the pressure; and (2) that it is directly proportional to the temperature measured from the absolute zero of the air thermometer.[1] From the last result, Maxwell deduced " that the force between two molecules is proportional inversely to the fifth power of the distance between them." If the molecules only acted by impact, it would follow that the viscosity must be proportional to the square root of the absolute temperature.

In the paper " On the Dynamical Theory of Gases," read before the Royal Society on May 31, 1866, the molecules are dealt with in the most general manner, no hypothesis being made respecting their constitution. The paper discusses (1) the phenomena of diffusion, depending on the average value of the velocities; (2) the phenomena depending on the average square of the velocities, which deter-

---

[1] More recent experiments indicate that this statement should be modified.

mines pressure and viscosity; and (3) the phenomena depending on the average value of the cubes of the velocities, on which depends the diffusion of energy or the conduction of heat in the gas. In this paper Maxwell showed that if a number of gases be enclosed in any space under the action of any external forces, each gas will arrange itself as if the others were absent, and this result is independent of the law of action of the particles on each other. This is the arrangement which Dalton suggested would probably exist in the atmosphere if the effect of winds could be annulled. He also showed that in a vertical column of air in equilibrium under the action of gravity the temperature must be the same throughout. This was contrary to the then current opinion that the temperature must diminish as we ascend, but Maxwell showed independently that it is a consequence of the principle of the dissipation of energy. Boltzmann was, however, the first to show how, in dealing with a collection of molecules, to take account of external forces acting upon them. In discussing the diffusion of energy Maxwell showed that the thermal conductivity of iron at $25°$ C, as determined by Principal Forbes, is 3525 times that of air at $16·6°$ C.

In the paper " On the Dynamical Evidence of the Molecular Constitution of Bodies," published in the *Journal of the Chemical Society*, June 1875, Maxwell introduces Clausius's conception of the " virial " of a system, a quantity of which great use has been made in the discussion of molecular theories since a name has been given to it by Clausius. This quantity is the sum of the products of the attractions between each pair of molecules into the distance between them. Maxwell commences by explaining Clausius's equation, which states that the energy of a quantity of gas is the sum of two quantities, one of which is the virial, and the other is proportional to the product of the pressure and the volume. Now Joule showed that when a gas expands into a vacuum its temperature remains sensibly unaffected, though no energy is communicated to or taken from it, and since the gas obeys Boyle's law the product of its pressure and volume is constant. Hence the energy being unchanged, and the product of pressure and volume remaining the same, it follows that the *virial* must be unaffected by expansion. But the virial must depend on the density if there be any sensible forces between the molecules when at sensible distances, unless the force between two molecules varies inversely as the distance between them, a law which Newton showed to be inadmissible in the case of molecular forces. We must therefore conclude that the force between two molecules is sensibly zero, unless the distance between them be very small compared with the average distance between molecules, and that the virial is therefore sensibly zero in a gas, so that the energy and pressure of a gas depend on the motion of the molecules and not on the forces between them.

The experiments of Regnault indicated that in most gases as the density increases the pressure falls below that indicated by Boyle's law, which shows that when the particles approach very near together the virial is positive, or there is attraction between the molecules. When the pressure is made still greater the gas reaches a condition in which a small increase of density is accompanied by an enormous increase of pressure, so that the virial is negative, and the forces between the particles repulsive. This is the case in the liquid state. Maxwell infers that as the particles of a gas approach each other the forces between them become attractive, attain a maximum, then diminish, and when they are within a certain distance become repulsive, the repulsion increasing so rapidly "that no attainable force can reduce the distance of the particles to zero." The same conclusion as far as the attraction between the particles at small distances is concerned, was indicated by the experiments which Joule conducted, in company with Sir William Thomson, on the expansion of air into a vacuum, when very careful measurements indicated that a slight cooling effect took place.

The greatest difficulty which the kinetic theory of gases has to face is the observed relation between the specific heats of gases at constant pressure and at constant volume. Boltzmann showed that if each molecule possess n " degrees of freedom" the ratio of the kinetic energy of translation to the whole kinetic energy of the system must be equal to the ratio of 3 to n. Now in a rigid body capable of rotating in any manner n is equal to 6, and this makes the total energy equal to twice the energy of translation. This requires that the ratio of the two specific heats should be 1·33 instead of 1·408, and the observed ratio therefore disproves the hypothesis of hard bodies. If we suppose the molecules to be material points, incapable of rotation or vibration n is equal to 3, and the energy of translation is the whole of the kinetic energy possessed by the molecules. This would make the ratio of the specific heats to be 1·66, which is too great for any real gas except mercury vapour, for which the ratio has been shown by Kundt and Warbourg to be nearly 1·66.

The spectroscope shows that the molecules of a gas are capable of executing vibrations in various periods. They must therefore be material systems, and cannot have less than six degrees of freedom. The ratio of the specific heats cannot therefore be greater than 1·33, and this is too small for most gases. Every additional degree of freedom possessed by the molecules makes the ratio less, and requires that the specific heat of the gas should be greater than is observed to be the case.

In a paper "On Boltzmann's Theorem on the Average Distribution of Energy in a System of Material Points," read before the Cambridge Philosophical Society on May 6, 1878, Maxwell showed that whatever be the forces acting upon or between the molecules, provided they be subject to the principle of conservation of energy, the average kinetic energy of any two given portions must be proportional to the number of degrees of freedom of these portions, and hence the total kinetic energy corresponding to an increment of temperature of 1° C is shown to be proportional to the product of the number of degrees of freedom into the absolute temperature.

The actual dimensions of molecules were first estimated by Loschmidt in 1865, then by Stoney in 1868, and by Thomson in 1870. At his lecture "On Molecules," before the British Association at Bradford, Maxwell gave the following

TABLE OF MOLECULAR DATA,

Divided into three ranks, according to the degree of accuracy
with which the Quantities are known.

	Hydrogen.	Oxygen.	Carbonic Oxide.	Carbonic Acid.
Mass of Molecule (Hydrogen=1)	1	16	14	22
Rank				
I. Velocity (of mean square) metres per second at 0° C . .	1859	465	497	396
Mean path, tenth-metres . .	965	560	482	379
Rank				
II. Collisions in a second (millions)	17,750	7646	9489	9720
Rank				
III. Diameter, tenth-metres . .	5·8	7·6	8·3	9·3
Mass, twenty-fifth-grammes .	46	736	644	1012

30th June 1877.

DEAR GARNETT— . . . I have been considering diffusion
of gases, and the method of separating heavy gases from light
ones, and I find it hopeless to do it by gravity, but if a tube
10 cm. long with two bulbs, and the straight part stuffed with
cotton-wool, were filled with equal volumes of H and CO_2, and
spun 100 times round per second for about half an hour, then
the ratio of CO_2 to H by volume would be greater in A than in
B by about $\frac{1}{150}$, which is measurable. I have got a new light
about equilibrium of temperature in two different gases. Let
forces having potentials act on the molecules of two gases, but
differently on each. Let the potential of forces acting on the
gas a be zero in the region A and very large in B, diminishing
continuously in the stratum C. Let the potential for gas b be
zero in B and very great in A, diminishing continuously in C.
Then the region A will contain the gas A nearly pure, and B
the gas B nearly pure, and in the stratum C there will be en-
counters between the two kinds of molecules. By Boltzmann
and Watson the average kinetic energy of a single molecule is
the same throughout the whole vessel. Hence the condition of
thermal equilibrium between two gases (not mixed, but kept
pure though in contact) is that the mean kinetic energy is the
same in each. And it is difficult to see where this method
breaks down when applied to solids.

I find the electric conductivity of air supposed of conducting
spheres to be $\frac{1}{18} \pi^2 s^2 N V$.

Where s = distance of centres at striking.
N Number in cubic centimetre.
V Mean velocity.

Now $\frac{\pi}{4}s^2N = 17,700$ for air, and $V = 48,500$.

But this is in electrostatic measure. In electromagnetic measure the resistance is

$$\frac{4}{\pi}\frac{v^2}{48500000},$$

so that $r = 2.10^{13}$ per cubic centimetre, or about 10^{10} greater than that of copper; but this is far smaller than that of gutta-percha. Hence the insulating power of air is not consistent with its molecules being conducting spheres.

But why should the molecules be conductors ?—Yours very truly,   J. CLERK MAXWELL.

Maxwell's investigations in the Kinetic Theory of Gases led him to a conclusion which is of great value in the theory of energy. The principle of the dissipation of energy, sometimes called the second law of Thermodynamics, states that it is impossible by means of inanimate material agency to obtain work at the expense of heat by cooling a body below the temperature of the coldest body in the neighbourhood. This principle was first distinctly given by Sir William Thomson. Maxwell showed that it obtains only in consequence of the coarseness of our faculties not allowing us to grapple with individual molecules. If we could seize upon individual molecules, and bring them to rest in the same way as we can lay hold of a fly-wheel, and compel it to do useful work until it has been deprived of all its motion of rotation, we could convert the whole of the heat of a body into work, and bring it to the absolute zero of temperature. As it is, we are at the mercy of the molecules, and capable of obtaining from them only so much work as they are willing to give in the most favourable circumstances in which we are able to place them. Maxwell imagined a quantity of gas, all initially at the same pressure and temperature, to be divided into two portions, A and B, by a partition full of little trap doors which might be opened or closed without the expenditure of energy. Each trap door he supposed placed in charge of a "*demon*," that is a creature whose eyes are sharp enough to see the molecules and estimate their velocities, and hands agile enough to open and close the trap doors in time to allow or prevent the passage of any particular molecule which is approaching the partition. The operation depends on the difference of the velocities of the particles in the same mass of gas, and the office of the *demon* is purely selective, so that any mechanism which could be devised to *sort* the molecules in the same way would be equally effective. Suppose each demon to open his trap door when a molecule is approaching the partition from A with a velocity above the average, but to keep it closed when the velocity of the molecule approaching from A is below the average; while in the case of a molecule approaching the partition from B the door is opened if the velocity be small, and closed if it be great. In this way all the slowly moving molecules will gradually be sorted into the compartment A, while the rapidly moving particles will be accumulated in B. Thus the temperature of the gas in B will be raised, and that in A lowered without any loss of energy or any work being done by an external agent. A heat engine may now be employed, using B as the source and A as the condenser, and doing work at the expense of part of the heat of A until equilibrium of temperature between B and A has been produced, when the services of the demons may be again utilised, and the process repeated until the whole of the heat of the gas has been converted into work. This is contrary to the principle of dissipation of energy, which has thus been circumvented by intelligence. No corresponding method of overcoming the principle of conservation of energy can be devised, and this principle is thus shown to rest on an entirely different kind of footing from that of the second law of Thermodynamics.

The following extract from Maxwell's article "ATOM" in the ninth edition of the *Encyclopædia Britannica* is characteristic. Speaking of the teaching of molecular science respecting the size of atoms, he says :—

It forbids the physiologist from imagining that structural details of infinitely small dimensions can furnish an explanation of the infinite variety which exists in the properties and functions of the most minute organisms.

A microscopic germ is, we know, capable of development into a highly organised animal. Another germ, equally microscopic, becomes when developed an animal of a totally different kind. Do all the differences, infinite in number, which distinguish one animal from another arise each from some difference in the structure of the respective germs? Even if we admit this as possible we shall be called upon by the advocates of Pangenesis to admit still greater marvels. For the microscopic germ, according to this theory, is no mere individual, but a representativo body, containing members collected from every rank of the long-drawn ramification of the ancestral tree, the number of these members being amply sufficient not only to furnish the hereditary characteristics of every organ of the body, and every habit of the animal from birth to death, but also to afford a stock of latent gemmules to be passed on in an inactive state from germ to germ, till at last the ancestral peculiarity which it represents is revived in some remote descendant.

Some of the exponents of this theory of heredity have attempted to elude the difficulty of placing a whole world of wonders within a body so small and so devoid of visible structure as a germ, by using the phrase structureless germs. Now, one material system can differ from another only in the configuration and motion which it has at a given instant. To explain differences of function and development of a germ without assuming differences of structure is therefore to admit that the properties of a germ are not those of a purely material system.

A paper " On Stresses in Rarified Gases arising from Inequalities of Temperature," by Professor Maxwell, was read before the Royal Society on April 11, 1878, and published in the *Phil. Trans.* for 1879. The notes and appendix added to the paper in May and June 1879 embodied the results of Maxwell's last investigations in the kinetics of gases. In this paper Maxwell showed that when inequalities of temperature exist in a gas the pressure at a point is not generally the same in all directions, but the maximum and minimum pressures differ by an amount depending on the rate of change of the increase of temperature per unit length in the direction in which this rate is greatest.

The stress thus arising from variation in rate of change of temperature varies inversely as the pressure of the gas, and is therefore most conspicuous in high vacua. If two small bodies are warmer than the air, the line joining them will be a line of maximum pressure, and they will repel each other, while they will attract one another if they are colder than the air. If, however, a ring be placed so as to have the line joining the bodies for its axis and be sufficiently heated the repulsion may be changed into attraction. In the case of a cup, as noticed by Stokes, the variation of the rate of change of temperature is much greater on the convex side than on the concave, where it is nearly uniform, like the electric potential within a hollow vessel, and hence the normal pressure is greater on the convex surface than on the concave, which will account for the motion of the cup radiometer if tangential stresses are neglected. But when the tangential stresses on any portion of gas are considered, it appears, that they, with the normal forces, form a system which is in equilibrium, so that inequality of temperature has no tendency of itself (*i.e.* without the action of gravity, etc.), to produce currents in the gas. Maxwell therefore concludes that the above explanation is insufficient, and that the true cause of the motion is to be found in the character of the tangential action between the solid and the gas, allowing the gas to slide over the surface of the solid, and thus diminishing the tangential stresses without affecting the normal stresses. In the appendix, dated May 1879, Maxwell determined the character of the tangential action on certain hypotheses respecting the nature of the surface of the solid, and the character of the collisions, and concluded that the gas may slide over the surface of the solid with a finite velocity, and that inequalities of temperature at the surface " give rise to a force tending to make the gas slide along the surface from colder to hotter places."

Most of the more elementary theorems respecting the kinetic theory of gases are given in a very concise form by Maxwell in his *Theory of Heat*, the more recent editions of which also give an account of Professor Willard Gibbs'

Thermodynamic Surface. This surface, in which the co-ordinates represent respectively the energy, entropy, and volume of the substance to which it corresponds, was modelled in clay by Maxwell's own hands in the Cavendish Laboratory. From the clay model a number of plaster casts were taken. These casts Maxwell placed in the sunshine in particular positions, and drew upon them in water-colours the boundary-lines between the light and shadow, which correspond to constant pressure or constant temperature, on the part of the substance. An account of this surface, and many of the properties which it represents, will be found in the work referred to.

We have now presented to the reader a scanty selection from the results of Maxwell's scientific work. The mere enumeration of his original papers would occupy several pages of this book, and those who are desirous of forming any approach to a true conception of his contributions to science should consult the memorial edition of Maxwell's papers, edited by Mr. W. D. Niven, F.R.S., and about to be published by the Cambridge University press.

END OF PART II.

PART III.

POEMS.

I. JUVENILE VERSES AND TRANSLATIONS.

TRANSLATION OF VIRGIL—ÆNEID, I. 159 TO 169.

School Exercise, 10th May 1844. *Æt.* 12.

THERE lies within a long recess a bay,
An isle with gulfing sides restrains the sea,
The waves, divided ere they reach the shore,
Run through the winding bay, and cease to roar;
On this side and on that vast rocks arise,
And two twin crags ascending threat the skies,
Beneath whose shade the water silent lies;
Above, with waving branches, stands a wood,
A grove with awful shade o'erhangs the flood,
And on the further side a cave is shown,—
Within, fresh springs, and seats of living stone—
The nymphs' abode; no chains or anchors bind
The worn-out ships, secure from waves and wind.

(SCHOOL RHYMES). [Nov.] 1844.[1] *Æt.* 13.

O ACADEMIC muse that hast for long
Charmed all the world with thy disciples' song,

[1] See p. 66.

As myrtle bushes must give place to trees,
Our humbler strains can now no longer please.
Look down for once, inspire me in these lays
In lofty verse to sing our Rector's praise.

The mighty wheel of Time to light has rolled
That golden age by ancient bards foretold.
Minerva now descends upon our land,
And scatters knowledge with unsparing hand;
Long since Ulysses saw the heavenly maid,
In Mentor's form and Mentor's dress arrayed,
But now to Cambrian lands the goddess flies,
And drops in Williams' form from out the skies;
And as at dawn the brilliant orb of light,
With his bright beams dispels the gloomy night,
So sunk in ignorance our land he finds,
But with his learning drives it from our minds,
And he, a hero, shall with joyful eyes
See crowds of heroes all around him rise;
With great Minerva's wisdom he shall rule
Those boisterous youths—the rector's class at school,
And when in the fifth class begins his power,
And he begins to teach us, from that hour
Dame Poetry begins to show her face,
And witty epigrams the plaster grace;
There growing wild are often to be seen
The names of boys that Duxes erst have been,
And at the chimney-piece is seen the same
All thickly scribbled with the boobie's name.

.　　　.　　　.　　　.

Ne'er shall the dreadful tawse be heard again,
The lash resounding, and the cry of pain;
Carmichael's self will change (O that he would!)
From the imperative to wishing mood;
Ye years roll on, and haste the expected time
When flogging boys shall be accounted crime.

But come, thy real nature let us see,
No more the rector but the goddess be,

Come in thy might and shake the deep profound,
Let the Academy with shouts resound,
While radiant glory all thy head adorns,
And slippers on thy feet protect thy corns ;
O may I live so long on earth below,
That I may learn the things that thou dost know!
Then will I praise thee in heroic verse
So good that Linus' will be counted worse ;
The Thracian Orpheus never will compare
With me, nor Dods that got the prize last year.
But stay, O stay upon this earth a while,
Even now thou seest the world's approving smile,
And when thou goest to taste celestial joys,
Let thy great nephew[1] teach the mourning boys,
Then mounting to the skies upon the wind,
Lead captive ignorance in chains behind.

TORTO VOLITANS SUB VERBERE TURBO
QUEM PUERI MAGNO IN GYRO VACUA ATRIA CIRCUM
INTENTI LUDO EXERCENT.

Virgil, Æn. vii. 378.

Nov. 1844.

OF pearies[2] and their origin I sing :
How at the first great Jove the lord of air
Impelled the planets round the central sun
Each circling within each, until at last
The winged Mercury moves in molten fire.
And which of you, ye heavenly deities,
That hear the endless music of the spheres,
Hast given to man the secret of the Top ?
Say, was it thou, O Fun, that dost prefer,
Before all temples, liberty and play ?
Yes, yes, 'twas only thou, thou from the first
Wast present when the Roman children came
To the smooth pavement, where with heavy lash
They chased the wooden plaything without end.

[1] Mr. Theodore Williams, English Master in the Academy
[2] See p. 51.

But not to tell of these is now my task,
Nor yet of humming-tops, whose lengthened neck,
With packthread bound, and handle placed above,
Amuses little children. Not of these,
But of the pearie, chief of all his tribe,
Do I now sing. He with a sudden bound
From out his station in the player's hand
Descends like Maia's son, on one foot poised,
And utters gentle music circling round,
Till in the centre of the ring it sleeps.
    When lo, as in the bright blue vault of heaven
A falcon, towering in his pride of place
Perceives from far a partridge on the wing,
And stoops to seize him, even so comes down
Another pearie, and as when the sword
Of faithful Abdiel struck the apostate's crest
And " sent him reeling back ten paces huge,"
So reeled the former pearie, nor can stand
The latter's iron peg, and more come down;
Innumerable hosts of pearies, armed
With dire destructive steel. The players shout;
It is the shout of battle; the loud cry
Of victors rushing to the spoil; the wail
Of ruined boys, their pearie split, and all,
All lost.
        Thus wags this ever-changing world,
And we may morals from the pearie draw.

(NATHALOCUS.)

I.   *Jan. 1845.*

BLEAK was the pathway and barren the mountain,
    As the traveller passed on his wearisome way;
Sealed by the frost was each murmuring fountain,
    And the sun shone through mist with a blood-coloured ray.
But neither the road nor the danger together,
Could alter his purpose, nor yet the rough weather;
So on went the wayfarer through the thick heather,
    Till he came to the cave where the dread witches stay.

II.

Hewn from the rock was that cavern so dreary,
    And the entrance by bushes was hid from the sight,
But he found his way in, and with travelling weary,
    With joy he beheld in the darkness a light.
And in a recess of that wonderful dwelling,
He heard the strange song of the witch wildly swelling,
In magical numbers unceasingly telling
    The fortunes of kingdoms, the issue of fight.

III.

Up rose the witch as the traveller entered,
    "Welcome," she said, "and what news from the king;
And why to inquire of me thus has he ventured,
    When he knows that the answer destruction will bring?
Sit here and attend." Then her pale visage turning
To where the dim lamp in the darkness was burning,
She took up a book of her magical learning,
    And prepared in prophetical numbers to sing.

IV.

Now she is seated, the curtain is o'er her,
    The god is upon her; attend then and hear!
The vapour is rising in volumes before her,
    And forms of the future in darkness appear.
Hark, now the god inspiration is bringing,
'Tis not her voice through the cavern is ringing;
No, for the song her familiar is singing,
    And these were the words of the maddening seer.

V.

"Slave of the monarch, return to thy master,
    Whisper these words in Nathalocus' ear;
Tell him, from me, that Old Time can fly faster
    Than he is aware, for his death hour is near;
Tell him his fate with the mystery due it,
But let him not know of the hand that shall do it;"

"Tell me, vile witch, or I swear thou shalt rue it!"
"Thou art the murderer," answered the seer.

VI.

"Am I a dog that I'd do such an action!"
　　Answered the chief as in anger he rose,
"Would I, ungrateful, be head of a faction,
　　And call myself one of Nathalocus' foes?"
"No more," said the witch, "the enchantment is ended,
I brave not the wrath of the demon offended,
Whatever thy fate, 'tis not now to be mended."
　　So the stranger returned through the thick - driving
　　　　snows.

VII.

High from his eyrie the eagle was screaming,
　　Pale sheeted spectres stalked over the heath;
Bright in his mind's eye a dagger was gleaming,
　　Waiting the moment to spring from its sheath.
Hoarse croaked the raven that eastward was flying;
Well did he know of the king that was dying;
Down in the river the Kelpie was sighing,
　　Mourning the king in the water beneath.

VIII.

His mind was confused with this terrible warning,
　　Horrible spectres were with him by night;
Still in his sorrow he wished for the morning,
　　Cursing the day when he first saw the light.
He said in his raving, "The day that she bore me,
Would that my mother in pieces had tore me;
See there is Nathalocus' body before me;
　　Hence, ye vain shadows, depart from my sight!"

IX.

And when from the palace the king sent to meet him,
　　To ask what response from the witch he might bear;
When the messenger thought that the stranger would greet him,
　　He answered by nought but a meaningless stare.

On his face was a smile, but it was not of gladness,
For all was within inconsolable sadness.
And aye in his eye was the fixt glare of madness,—
 " In the king's private chamber, I'll answer him there."

X.

" Tell me, my sovereign, have I been unruly ;
 Have I been ever found out of my place ;
Have not I followed thee faithfully, truly,
 Though danger and death stared me full in the face ?
Have I been seen from the enemy flying,
Have I been wanting in danger most trying ?
Oh, if I have, judge me worthy of dying,
 Let me be covered with shame and disgrace !

XI.

" Couldst thou imagine that I should betray thee,
 I whom thy bounty with friendship has blessed ?
But the witch gave for answer that my hand should slay thee,
 'Tis this that for long has deprived me of rest,
Ever since then have my slumbers been broken,
But true are the words that the prophet has spoken,
Nathalocus, now receive this as a token,"
 So saying the dagger he plunged in his breast.

THE DEATH OF SIR JAMES, LORD OF DOUGLAS.

Prize Poem, July 1845.[1] *Æt.* 14.

 "Men may weill wyt, thouch nane thaim tell,
  How angry for sorow, and how fell,
  Is to tyne sic a Lord as he
  To thaim that war off hys mengye."
             Barbour's *Bruce,* B. XX. i. 507.

WHERE rich Seville's proud turrets rise
A foreign ship at anchor lies ;

---

[1] P. 69.

The pennons, floating in the air,
Proclaim that one of rank is there—
The Douglas, with a gallant band
Of warriors, seeks the Holy Land.
But wherefore now the trumpet's bray,
The clang of arms and war's array,
The atabal and martial drum ?
The Moor—the infidel is come;
And there is Sultan Osmyn—see !
With all his Paynim chivalry ;
And they have sworn to glut their steel
With the best blood of fair Castile.
" And do we here inactive stand ? "
The Douglas cries ; " Land ! comrades, land !"
Then for the Christian camp he makes,
When thus Alphonso silence breaks :
" What news from Scotland do you bring ;
And where is now your patriot king ? "
" Alas ! within this casket lies
The heart so valiant, good, and wise,
This to the Holy Land we bear,
For we have sworn to lay it there.
But let us forward to the fight,
And God protect the Christian right !"
To whom Alphonso—" Scottish lord,
That now for Spain dost draw that sword,
The terror of thy English foes,
When for her freedom Scotland rose ;
With knights like thee and thy brave band
We'll drive the Moslem from the land."
The Douglas thus his comrades cheers—
" Be brave ! and as for him that fears,
Let the base coward turn and fly,
For we will gain the day, or die.
Now couch the trusty Scottish spear,
And think King Robert's heart is here,
And boldly charge—already, see
The dogs of Moslems turn and flee."

At the first onset, with the slain
Those valiant warriors strew the plain ;
But, hark ! the Allah Hu ! the foes
Rally, and hot the combat grows,
For here the Spaniards yield, and there
The Moors have slain the brave St. Clair.
Then, midst the thickest of his foes,
The precious casket Douglas throws—
" Pass on before us," hear him cry,
" For I will follow thee, or die."
He rushes on—but all in vain,
For thicker comes the arrowy rain ;
And now, by multitudes opprest,
With many a wound upon his breast,
Where 'midst the slain the casket lies,
A noble death the Douglas dies.

THE VAMPYRE.[1]

Compylt into Meeter by James Clerk Maxwell   [1845.]

THAIR is a knichte rydis through the wood,
    And a douchty knichte is hee,
And sure hee is on a message sent,
    He rydis sae hastilie.
Hee passit the aik, and hee passit the birk,
    And hee passit monie a tre,
Bot plesant to him was the saugh sae slim,
    For beneath it hee did see
The boniest ladye that ever he saw,
    Scho was sae schyn and fair.
And there scho sat, beneath the saugh,
    Kaiming hir gowden hair.
And then the knichte—" Oh ladye brichte,
    What chance hes broucht you here,
But say the word, and ye schall gang
    Back to your kindred dear."

[1] P. 69.

Then up and spok the Ladye fair—
  "I have nae friends or kin,
Bot in a littel boat I live,
  Amidst the waves' loud din."
Then answered thus the douchty knichte—
  "I'll follow you through all,
For gin ye bee in a littel boat,
  The world to it seemis small."
They gaed through the wood, and through the wood
  To the end of the wood they came :
And when they came to the end of the wood
  They saw the salt sea faem.
And then they saw the wee, wee boat,
  That daunced on the top of the wave,
And first got in the ladye fair,
  And then the knichte sae brave ;
They got into the wee, wee boat,
  And rowed wi' a' their micht;
When the knichte sae brave, he turnit about,
  And lookit at the ladye bricht ;
He lookit at her bonie cheik,
  And hee lookit at hir twa bricht eyne,
Bot hir rosie cheik growe ghaistly pale,
  And scho seymit as scho deid had been.
The fause fause knichte growe pale wi frichte,
  And his hair rose up on end,
For gane-by days cam to his mynde,
  And his former luve he kenned.
Then spake the ladye,—" Thou, fause knichte,
  Hast done to mee much ill,
Thou didst forsake me long ago,
  Bot I am constant still ;
For though I ligg in the woods sae cald,
  At rest I canna bee
Until I sucke the gude lyfe blude
  Of the man that gart me dee."
Hee saw hir lipps were wet wi' blude,
  And hee saw hir lyfelesse eyne,

And loud hee cry'd, " Get frae my syde,
　Thou vampyr corps uncleane ! "
Bot no, hee is in hir magic boat,
　And on the wyde wyde sea ;
And the vampyr suckis his gude lyfe blude,
　Sho suckis hym till hee dee.
So now beware, whoe're you are,
　That walkis in this lone wood ;
Beware of that deceitfull spright,
　The ghaist that suckis the blude.

Seventh Ode of the Fourth Book of Horace. 1846.

(In the metre of the original.)

ALL the snows have fled, and grass springs up on the meadows,
　And there are leaves on the trees ;
Earth has changed her looks, and turbulent rivers decreasing,
　Slowly meander along ;
Now, with the naked nymphs and her own twin sisters, Aglaïa
　Gracefully dances in time.
But the Year, and the Hours which hurry along our existence,
　Solemnly warn us to die.
Zephyr removes the frost, and Summer, soon destined to perish,
　Treads in the footsteps of Spring,
After the joyous reign of Autumn, abounding in apples,
　Shivering Winter returns.
Heavenly waste is repaired by the moon in her quick revolutions,
　But when we go to the grave,
Beside the pious Æneas, and rich old Tullus, and Ancus,
　We are but dust and a shade.
Who knows if the gods above have determined whether tomorrow
　We shall be living or dead.

Nothing will come to the greedy hands of your spendthrift
successor
    Which you have given away.
When you are gone to the grave, and Minos, sitting in judg-
ment,
    Utters your terrible doom,
Neither your rank nor your talents will bring you to life, O
Torquatus,
    Nor will affection avail;
Even the chaste Hippolytus was not released by Diana
    From the infernal abyss,
Nor could Theseus break from his friend the rewards of
presumption
    Which the stern monarch imposed.

HOR., OD. III. 9.   1846.

(In the metre of the original.)

Horace.

WHILE I was your beloved one,
And while no other youth threw his fond arms around
    Your white neck so easily,
Than the King of the world I was far happier.

Lydia.

While you loved not another one,
While you did not prefer Chloë to Lydia,
    I then thought myself happier
Than the mother of Rome, great Rhea Silvia.

Horace.

Thracian Chloë now governs me,
She can merrily sing, playing the cithara;
    I'd not scruple to die for her,
If the Implacable spared Chloë, the auburn haired.

Lydia.

I now love and am loved again,
By my Calaïs, son of the old Ornytus;
 Twice I'd die for him willingly,
If the terrible fates spared but my Calaïs.

Horace.

What if love should return again,
And unite us by ties more indissoluble?
 What if Chloë were cast away,
And the long-closèd door open to Lydia?

Lydia.

My love's brighter than any star;
You, too, lighter than cork, tossed on the waves of the
 Hadriatic so terrible;
Still I'd live but with thee, and I could die with thee.

HORACE, SEVENTH EPODE. 1846.

WHITHER, whither, reckless Romans,
 Are you rushing, sword in hand?
Has not yet the blood of brothers,
 Fully stained the sea and land?

Not that raging conflagration
 Should o'er fallen Carthage play;
Not that the unconquered Briton
 Should descend the sacred way.

" Rome," exclaims the joyful Parthian,
 " Ruin for herself prepares;
Wolves with wolves are never savage,
 Lion lion never tears."

Is this fury? is it madness?
 Speedy answer I demand;
Foolish, blinded, guilty Romans,
 Silent, stupefied you stand.

Thus 'tis fated, blood of brothers
Must atone for brothers' guilt,
Since the blood of injured Remus
Romulus in anger spilt.

(AN ONSET.)   [1848.]

HALLO ye, my fellows! arise and advance,
See the white-crested waves how they stamp and they
    dance!
High over the reef there in anger and might,
So wildly we dance to the bloody red fight.
Than gather, now gather, come gather ye all,
Each thing that hath legs and arms, come to our call;
Like reeds on the moor when the whirlwinds vie
Our lances and war-axes darken the sky;
Sharp, sharp, as the tooth of the sea-hound and shark,
They'll tear ye, they'll split ye, fly lance to the mark,
Home, home to the heart, and thou battle-axe grim,
Split, splintring and shivering through brain-pan and limb;
To-day we ask vengeance, to-day we ask blood,
We ask it; we're coming to make our words good;
The storm flinches not tho' the woods choke its path,
We ask it; we're coming, beware of our wrath.
At home wives and children a hearth for us lay,
A savoury flesh-feast awaits us to-day;
Behind yonder mountains e'en now the smoke streams,
And the blaze of the bush fire crackles and gleams.
Long, long have we hungered and thirsted for you,
At home the dogs bark round the clean table too,
Loud shouting we'll eat you to-night every one,
Devour you clean to the white sinewy bone.
Rush, rush ye my fellows, rush on them like hail,
Soon, soon shall their roasting your nostrils regale,
The fire is flaring, the oven's a glow,
Heave to now, hew thro' now, Holloa, Hollo.

(SONG OF THE EDINBURGH ACADEMICIAN.)   [1848.]

IF ony here has got an ear,
    He'd better tak' a haud o' me,
Or I'll begin, wi' roarin' din,
    To cheer our old Academy.

        Dear old Academy,
        Queer old Academy,
    A merry lot we were, I wot,
        When at the old Academy.

There's some may think me crouse wi' drink,
    And some may think it mad o' me,
But ither some will gladly come
    And cheer our old Academy.

Some set their hopes on Kings and Popes,
    But, o' the sons of Adam, he
Was first, without the smallest doubt,
    That built the first Academy.

Let Pedants seek for scraps of Greek,
    Their lingo to Macadamize ;
Gie me the sense, without pretence,
    That comes o' Scots Academies.

Let scholars all, both grit and small,
    Of Learning mourn the sad demise ;
That's as they think, but we will drink
    Good luck to Scots Academies.

SPECIMEN OF TRANSLATION FROM THE AJAX OF SOPHOCLES.
1852.

(*Lines* 1192-1222.)

O HAD he first been swept away,
    Through air by wild winds tossed,

Or sunk from Heaven's ethereal ray,
    To Pluto's dreary coast.
Who trained the Grecians to the field,
Taught them the sword, the spear to wield,
    And steeled the gentle mind !
Hence toil gives birth to toil again,
Hence carnage stains the ensanguined plain,
    For he destroyed mankind.

Nor the brow with chaplets bound,
Breathing balmy odours round,
Nor the social glow of soul,
Kindling o'er the generous bowl,
Nor the dulcet strain that rings
Jocund from the sounding strings,
Nor endearing love's delight,
Which with rapture fills the night,
Me will he permit to prove,
He, alas ! hath murdered love.
But neglected here I lie,
Open to the inclement sky ;
And my rough and matted hair
Drinks the dews of night's moist air,
    Memorials sad of Troy.
Yet till now, when pale affright
Rolled her hideous form through night
Great in arms, thy shield to oppose,
Ajax at his rampire rose,
And my terror was no more.
Now the hero I deplore,
To the gloomy god consigned,
Now, what joy can touch the mind ?
O that on the pine-clad brow,
Darkening o'er the sea below,
Where the cliffs of Sunium rise,
Rocky bulwarks to the skies,
I were placed—with sweet address
Sacred Athens would I bless,
    And feel a social joy.

II. OCCASIONAL PIECES.

REFLEX MUSINGS:

REFLECTION FROM VARIOUS SURFACES.

18th April 1853.

In the dense entangled street,
  Where the web of Trade is weaving,
Forms unknown in crowds I meet
  Much of each and all believing;
  Each his small designs achieving
Hurries on with restless feet,
  While, through Fancy's power deceiving,
Self in every form I greet.

Oft in yonder rocky dell
  Neath the birches' shadow seated,
I have watched the darksome well,
  Where my stooping form, repeated,
  Now advanced and now retreated
With the spring's alternate swell,
  Till destroyed before completed
As the big drops grew and fell.

By the hollow mountain-side
  Questions strange I shout for ever,
While the echoes far and wide
  Seem to mock my vain endeavour;
  Still I shout, for though they never
Cast my borrowed voice aside,
  Words from empty words they sever—
Words of Truth from words of Pride.

Yes, the faces in the crowd,
  And the wakened echoes, glancing
From the mountain, rocky browed,
  And the lights in water dancing—

2 Q

Each, my wandering sense entrancing,
Tells me back my thoughts aloud,
All the joys of Truth enhancing
Crushing all that makes me proud.

A STUDENT'S EVENING HYMN.

Cambridge, April 25, 1853.

I.

Now no more the slanting rays
    With the mountain summits dally,
Now no more in crimson blaze
    Evening's fleecy cloudlets rally,
    Soon shall Night from off the valley
Sweep that bright yet earthly haze,
    And the stars most musically
Move in endless rounds of praise.

II.

While the world is growing dim,
    And the Sun is slow descending
Past the far horizon's rim,
    Earth's low sky to heaven extending,
    Let my feeble earth-notes, blending
With the songs of cherubim,
    Through the same expanse ascending,
Thus renew my evening hymn.

III.

Thou that fill'st our waiting eyes
    With the food of contemplation,
Setting in thy darkened skies
    Signs of infinite creation,
    Grant to nightly meditation
What the toilsome day denies—
    Teach me in this earthly station
Heavenly Truth to realise.

IV.

Give me wisdom so to use
  These brief hours of thoughtful leisure,
That I may no instant lose
  In mere meditative pleasure,
  But with strictest justice measure
All the ends my life pursues,
  Lies to crush and truths to treasure,
Wrong to shun and Right to choose.

V.

Then, when unexpected Sleep,
  O'er my long-closed eyelids stealing,
Opens up that lower deep
  Where Existence has no feeling,
  May sweet Calm, my languor healing,
Lend me strength at dawn to reap
  All that Shadows, world-concealing,
For the bold enquirer keep.

VI.

Through the creatures Thou hast made
  Show the brightness of Thy glory,
Be eternal Truth displayed
  In their substance transitory,
  Till green Earth and Ocean hoary,
Massy rock and tender blade
  Tell the same unending story—
" We are Truth in Form arrayed."

VII.

When to study I retire,
  And from books of ancient sages
Glean fresh sparks of buried fire
  Lurking in their ample pages—
  While the task my mind engages
Let old words new truths inspire—
  Truths that to all after-ages
Prompt the Thoughts that never tire.

VIII.

Yet if, led by shadows fair
I have uttered words of folly,
Let the kind absorbing air
Stifle every sound unholy.
So when Saints with Angels lowly
Join in heaven's unceasing prayer,
Mine as certainly, though slowly,
May ascend and mingle there.

*Two stanzas omitted, the Author knows where, but not
to be inserted till he knows how.*

Teach me so Thy works to read
That my faith,—new strength accruing,—
May from world to world proceed,
Wisdom's fruitful search pursuing ;
Till, thy truth my mind imbuing,
I proclaim the Eternal Creed,
Oft the glorious theme renewing
God our Lord is God indeed.

Give me love aright to trace
Thine to everything created,
Preaching to a ransomed race
By Thy mercy renovated,
Till with all thy fulness sated
I behold thee face to face
And with Ardour unabated
Sing the glories of thy grace.

(ON ST. DAVID'S DAY.[1]   TO MRS. E. C. MORRIESON.)

1st March 1854.

'TWAS not chance but deep design,
Tho' of whom I can't divine
Made the courtly Valentine
(Corpulent saint and bishop)

[1] See p. 207.

Such a time with Bob to stay :—
Let me now in bardish way
On your own St. David's day
    Toss you a simple dish up.

'Tis a tale we learnt at school,—
Oft we broke domestic rule,
Standing till our brows were cool
    In the forbidden lobby.[1]
There we talked and there we laughed,
Till the townsfolk thought us daft,
What of that ? a thorough draft
    Was and is still my hobby.

To my tale : In ancient days,
Ere men left the good old ways,
Lived a lady whose just praise
    Passes all fancied glory.
Rich was she in field and store,
Richer in the sons she bore,
How could she be honoured more ?
    Listen and hear the story.

On a high and festive day
When the chariots bright and gay
To the temple far away
    Passed in majestic order,—
When the hour was nigh at hand,
She who should have led the band
Found no oxen at command,
    Searching through all her border.[2]

Then her two sons brave and strong
Girt their limbs with band and thong,
And before the wondering throng
    Drew their exulting mother.

[1] P. 68.   [2] Herodotus, i. 31.

Swift and steady, on they came ;
At the temple loud acclaim
Greeted that illustrious dame,
    Blest above every other.

Then, while triumph filled her breast,
Loud she prayed above the rest,
Give my sons whatever best
    Man may receive from heaven.
To the shrine the brothers stept,
Low they bowed, they sunk, they slept,
Stillness o'er their brave limbs crept :—
    Rest was the guerdon given.

Such the simple story told,
By a sage renowned of old, [1]
To a king [2] whose fabled gold
    Could not procure him learning.
Heathen was the sage indeed,
Yet his tale we gladly read,
Thro' his dark and doubtful creed
    Glimpses of Truth discerning.

Now no more the altar's blaze
Glares athwart our worldly haze,
Warning men how evil ways
    Lead to just tribulation.
Now no more the temple stands,
Pointing out to godless lands
That which is not made with hands,
    Even the whole Creation.

Ask no more, then, " what is best,
How shall those you love be blest,"
Ask at once, eternal Rest,
    Peace and assurance giving.

----

[1] Solon.   [2] Crœsus.

Rest of Life and not of death,
Rest in Love and Hope and Faith,
Till the God who gives their breath
Calls them to rest from living.

RECOLLECTIONS OF DREAMLAND.

Cambridge, June ? 1856.[1]

ROUSE ye! torpid daylight-dreamers, cast your carking cares
    away!
As calm air to troubled water, so my night is to your day;
All the dreary day you labour, groping after common sense,
And your eyes ye will not open on the night's magnificence.
Ye would scoff were I to tell you how a guiding radiance
    gleams
On the outer world of action from my inner world of dreams.

When, with mind released from study, late I lay me down
    to sleep,
From the midst of facts and figures, into boundless space I
    leap;
For the inner world grows wider as the outer disappears,
And the soul, retiring inward, finds itself beyond the spheres.
Then, to this unbroken sameness, some fantastic dream
    succeeds,
Vague emotions rise and ripen into thoughts and words and
    deeds.
Old impressions, long forgotten, range themselves in Time
    and Space,
Till I recollect the features of some once familiar place.
Then from valley into valley in my dreaming course I roam,
Till the wanderings of my fancy end, where they began, at
    home.
Calm it lies in morning twilight, while each streamlet far
    and wide
Still retains its hazy mantle, borrowed from the mountain's
    side;

[1] See p. 249.

Every knoll is now an island, every wooded bank a shore,
To the lake of quiet vapour that has spread the valley o'er.
Sheep are couched on every hillock, waiting till the morning
    dawns,
Hares are on their early rambles, limping o'er the dewy
    lawns.
All within the house is silent, darkened all the chambers
    seem,
As with noiseless step I enter, gliding onwards in my
    dream.

What! has Time run out his cycle, do the years return
    again?
Are there treasure-caves in Dreamland where departed days
    remain?
I have leapt the bars of distance—left the life that late I
    led—
I remember years and labours as a tale that I have read;
Yet my heart is hot within me, for I feel the gentle power
Of the spirits that still love me, waiting for this sacred hour.
Yes,—I know the forms that meet me are but phantoms of
    the brain,
For they walk in mortal bodies, and they have not ceased
    from pain.
Oh! those signs of human weakness, left behind for ever
    now,
Dearer far to me than glories round a fancied seraph's brow.
Oh! the old familiar voices! Oh! the patient waiting eyes!
Let me live with them in dreamland, while the world in
    slumber lies!
For by bonds of sacred honour will they guard my soul in
    sleep
From the spells of aimless fancies, that around my senses
    creep.
They will link the past and present into one continuous life,
While I feel their hope, their patience, nerve me for the
    daily strife.

For it is not all a fancy that our lives and theirs are one,
And we know that all we see is but an endless work begun.
Part is left in Nature's keeping, part is entered into rest,
Part remains to grow and ripen, hidden in some living breast.
What is ours we know not, either when we wake or when
    we sleep,
But we know that Love and Honour, day and night, are ours
    to keep.
What though Dreams be wandering fancies, by some lawless
    force entwined,
Empty bubbles, floating upwards through the current of the
    mind ?
There are powers and thoughts within us, that we know not,
    till they rise
Through the stream of conscious action from where Self in
    secret lies.
But when Will and Sense are silent, by the thoughts that
    come and go,
We may trace the rocks and eddies in the hidden depths
    below.

Let me dream my dream till morning; let my mind run slow
    and clear,
Free from all the world's distraction, feeling that the Dead
    are near,
Let me wake, and see my duty lie before me straight and
    plain.
Let me rise refreshed, and ready to begin my work again.

To the Air of " Lörelei."

Aberdeen, January 1858.

I.

Alone on a hillside of heather,
    I lay with dark thoughts in my mind,
In the midst of the beautiful weather
    I was deaf, I was dumb, I was blind.

I knew not the glories around me,
　　I counted the world as it seems,
Till a spirit of melody found me,
　　And taught me in visions and dreams.

II.

For the sound of a chorus of voices
　　Came gathering up from below,
And I heard how all Nature rejoices,
　　And moves with a musical flow.
O strange! we are lost in delusion,
　　Our ways and doings are wrong,
We are drowning in[1] wilful confusion,
　　The notes of that[2] wonderful song.

III.

But listen, what harmony holy
　　Is mingling its notes with our own!
The discord is vanishing slowly,
　　And melts in that dominant tone.
And they that have heard it can never
　　Return to confusion again,
Their voices are music for ever,
　　And join in the mystical strain.

IV.

No mortal can[3] utter the beauty
　　That dwells[4] in the song that they sing;
They move in the pathway of duty,
　　They follow the steps of their King.
I would barter the world and its glory,
　　That vision of joy to prolong,
Or to hear and remember the story
　　That lies in the heart of their song.

[1] v.r. marring with.　　　　　　　　[2] v.r. the.
[3] v.r. may.　　　　　　　　　　　　[4] v.r. breathes through.

"I'VE HEARD THE RUSHING."

Aberdeen, 1858.

I'VE heard the rushing of mountain torrents, gushing
  Down through the rocks, in a cataract of spray,
    Onward to the ocean;
    Swift seemed their motion,
Till, lost in the desert, they dwindled away.

I've learnt the story of all human glory,
  I've felt high resolves growing weaker every day,
    Till cares, springing round me,
    With creeping tendrils bound me,
And all I once hoped for was wearing fast away.

I've seen the river rolling on for ever,
  Silent and strong, without tumult or display.
    In the desert arid,
    Its waters never tarried,
Till far out at sea we still found them on their way.

Now no more weary we faint in deserts dreary,
  Toiling alone till the closing of the day;
    All now is righted,
    Our souls flow on united,
Till the years and their sorrows have all died away.

"WILL YOU COME ALONG WITH ME?"

Aberdeen, 1858.

I.

WILL you come along with me,
  In the fresh spring-tide,
My comforter to be
  Through the world so wide?
Will you come and learn the ways
A student spends his days,
On the bonny, bonny braes
  Of our ain burnside?

II.

For the lambs will soon be here,
　In the fresh spring-tide ;
As lambs come every year
　On our ain burnside.
Poor things, they will not stay,
But we will keep the day
When first we saw them play
　On our ain burnside.

III.

We will watch the budding trees
　In the fresh spring-tide,
While the murmurs of the breeze
　Through the branches glide.
Where the mavis builds her nest,
And finds both work and rest,
In the bush she loves the best,
　On our ain burnside.

IV.

And the life we then shall lead
　In the fresh spring-tide,
Will make thee mine indeed,
　Though the world be wide.
No stranger's blame or praise
Shall turn us from the ways
That brought us happy days
　On our ain burnside.

"WHY, WHEN OUR SUN SHINES CLEAREST."

1858.

WHY, when our sun shines clearest,
Why, when our hopes seem nearest,
Why, when our life feels dearest,
　Rises a secret pain—

Hope's perfect mirror broken—
Shadows of things unspoken—
Why will not some sure token
    Calm us to rest again ?

Mixed with all earthly blessing
Lingers the fear distressing—
Conscience within confessing
    Nothing of ours is pure.
Still must such thoughts upbraid us,
Seeking our own to aid us;
God, not ourselves, hath made us ;
    Trusting in Him we're sure.

Thus, from our sorrows gleaning
Thoughts of the world's deep meaning,
Let us rejoice while leaning
    Firm on our Father's arm.
Now are we one for ever,
Joined so that none may sever,
Souls, so united, never
    Faint through mischance or harm.

To K. M. D.

Aberdeen, 1858.

IN the buds, before they burst,
    Leaves and flowers are moulded ;
Closely pressed they lie at first,
    Exquisitely folded.

Though no hope of change they felt,
    Folded hard together,
Soon their sap begins to melt
    In the warmer weather.

Till, when Life returns with Spring,
    Through them softly stealing,
All their freshness forth they fling,
    Hidden forms revealing.

Who can fold those flowers again,
In the way he found them?
Or those spreading leaves restrain,
In the buds that bound them?

Trust me, Spring is very near,
All the buds are swelling;
All the glory of the year
In those buds is dwelling.

What the opened buds reveal
Tells us—Life is flowing;
What the buds, still shut, conceal,
We shall end in knowing.

Long I lingered in the bud,
Doubting of the season,
Winter's cold had chilled my blood—
I was ripe for treason.

Now no more I doubt or wait,
All my fears are vanished,
Summer's coming, dear, though late,
Fogs and frosts are banished.

Tune, *Il Segreto per esser felice.*

24th March 1858.

I.

THERE are some folks that say,
They have found out a way,
      To be healthy and wealthy and wise—
" Let your thoughts be but few,
Do as other folk do,
      And never be caught by surprise.
Let your motto be—Follow the fashion,
      But let other people alone;

Do not love them, nor hate them, nor care for their fate,
    But keep a look out for your own.
Then what though the world may run riot,
    Still playing at catch who catch can ;
You may just eat your dinner in quiet,
    And live like a sensible man."

II.

'Twere a beautiful thing,
Thus to sit like a king,
    And talk of the world turning round,
If it were not that we,
Like all things that we see,
    Are standing on moveable ground.
While we boast of our tranquil enjoyments,
    The means of enjoyment are flown,
Both our joys and our pains, till there's nothing remains,
    But the tranquil repose of a stone.
The world may be utterly crazy,
    And life may be labour in vain ;
But I'd rather be silly than lazy,
    And would not quit life for its pain.

III.

In Nature I read
Quite a different creed,
    There everything lives in the rest ;
Each feels the same force,
As it moves in its course,
    And all by one blessing are blest.
The end that we live for is single,
    But we labour not therefore alone,
For together we feel how by wheel within wheel,
    We are helped by a force not our own.
So we flee not the world and its dangers,
    For He that has made it is wise,
He knows we are pilgrims and strangers,
    And He will enlighten our eyes.

(To his Wife.)

1858.

Oft in the night, from this lone room
   I long to fly o'er land and sea,
To pierce the dark, dividing gloom,
   And join myself to thee.

And thou to me wouldst gladly fly,
   I know thee well, my own true wife!
We feel, that when we live not nigh,
   We lose the crown of life.

Yet soon I hope, at dead of night,
   To meet where all is strange beside,
And mid the train's resounding flight
   To have thee by my side.

Then shall I feel that thou art near,
   Joined hand to hand and soul to soul;
Short will that happy night appear,
   As through the dark we roll.

Then shall the secret of the will,
   That dares not enter into bliss;
That longs for love, yet lingers still,
   Be solved in one long kiss.

I, drinking deep of thy rich love,
   Thou feeling all the strength of mine,
Our souls will rise in faith above
   The cares which make us pine.

Till I give thee, thou giving me,
   As that which either loves the best,
To Him that loved us both, that He
   May take us to His rest.

Wandering and weak are all our prayers,
 And fleeting half the gifts we crave ;
Love only, cleansed from sins and cares,
 Shall live beyond the grave.

Strengthen our love, O Lord, that we
 May in Thine own great love believe
And, opening all our soul to Thee,
 May Thy free gift receive.

All powers of mind, all force of will,
 May lie in dust when we are dead,
But love is ours, and shall be still,
 When earth and seas are fled.

III. SERIO-COMIC VERSE.

" NUMA POMPILIUS."

A Lay of Ancient Rome.

Junior Sophs' Exam.,
Cambridge, December 1851.

O WELL is thee ! King Numa,
 Within thy secret cave,
Where thy bones are ever moistened
 By sad Egeria's wave ;
None now have power to pilfer
 The treasure of thy tomb,
And reveal the institutions
 And secret Rites of Rome.
O blessed be the Senate
 That stowed those books away,
Curst be the attempt of Niebuhr
 To drag them into day ;
Light be the pressure, Numa,
 Around thy watery bed,
May no perplexing problems
 Infest thy kingly head !

As thus I blessed King Numa
  And struggled hard with sleep,
I felt unwonted chillness
  O'er all my members creep;
Before mine eyes in fragments
  The fireplace seemed to roll,
The chillness left my body
  And slid into my soul.
Deep in Egeria's grotto
  I saw the darksome well;
I slowly sunk to Numa,
  But *why* I cannot tell.

" What !   Livest thou still, old Sabine,
  With thy mysterious wife ? "
" Yes, here beneath the surface,
  We lead a torpid life.
But little think the Critics
  Who nullify old Rome,
That in these benumbing waters
  I always lived at home.
Never was I a Sabine,
  Or lived like men above;
No mortal wight was Numa,
  Who quelled the fear of Jove.
Before my day the Romans
  Served gods of wood and stone,
But what each man had fashioned
  That worshipped he alone;
With care he saved the silver,
  With pains the mould designed,
He loved and feared the offspring
  Of his pocket and his mind.
To him he went for counsel
  And then to Common Sense;
When both of these had failed him
  He took to tossing pence;

But I forbade all tossing,
　Made men enquire of beasts,
Pulled down all private idols
　And set up public priests.
'Birds, too,' said I, 'are holy,
　They show us things to come,
They have more subtle spirits
　Than wooden idols dumb.
No longer burn your incense
　Before your private shrine,
My Vestals are most careful
　To feed the flame divine;
Dismiss all fear of idols,
　Of demons, and of gods,
My Augurs will protect you
　With their long crooked rods.
(With such the careful shepherd
　Drags lambs from ditches deep;
With such he points to heaven
　When they are fast asleep.)
O, trust me, those same Augurs
　Know more about the stars
Than you whose only business
　Is everlasting wars.
How can you be religious,
　How can they work for bread?
You sinners must be shriven,
　My Augurs must be fed.
You know dividing labour
　To nations riches brings,
So let my Augurs shrive you
　While you mind earthly things.
Your case I've set before you,
　You see the thing to do,
If you fork out the needful,
　They do your job for you.'
With this and other speeches
　I brought the people round,

Till not a single Roman
  In Jove's house can be found.
For well he knows each evening
  When bells in steeples toll,
'Tis a sign that well-paid Augurs
  Are helping on his soul.
'Twas this that kept 'em quiet
  Through all my fabled reign,
Till quarrelsome young Tullus
  Brought battles back again.
Thus my cold-blooded doctrines
  The fear of Jove could quell,
Wonder not then to find me
  Alive here in a well."

A VISION

Of a Wrangler, of a University, of Pedantry,
and of Philosophy.

10th November 1852.

DEEP St. Mary's bell had sounded,
And the twelve notes gently rounded
Endless chimneys that surrounded
    My abode in Trinity.
(Letter G, Old Court, South Attics),
I shut up my mathematics,
That confounded hydrostatics—
    Sink it in the deepest sea !

In the grate the flickering embers
Served to show how dull November's
Fogs had stamped my torpid members,
    Like a plucked and skinny goose.
And as I prepared for bed, I
Asked myself with voice unsteady,
If of all the stuff I read, I
    Ever made the slightest use.

Late to bed and early rising,
Ever luxury despising,
Ever training, never " sizing,"
　　　I have suffered with the rest.
Yellow cheek and forehead ruddy,
Memory confused and muddy,
These are the effects of study
　　　Of a subject so unblest.

Look beyond, and see the wrangler,
Now become a College dangler,
Court some spiritual angler,
　　　Nibbling at his golden bait.
Hear him silence restive Reason,
Her advice is out of season,
While her lord is plotting treason
　　　Gainst himself, and Church or State.

See him next with place and pension,
And the very best intention
Of upholding that Convention
　　　Under which his fortunes rose.
Every scruple is rejected,
With his cherished schemes connected,
" Higher Powers may be neglected—
　　　His result no further goes."

Much he lauds the education
Which has raised to lofty station,
Men, whose powers of calculation
　　　Calculation's self defied.
How the learned fool would wonder
Were he now to see his blunder,
When he put his reason under
　　　The control of worldly Pride.

Thus I muttered, very seedy,
Husky was my throat, and reedy ;
And no wonder, for indeed I
　　　Now had caught a dreadful cold.

Thickest fog had settled slowly
Round the candle, burning lowly,
Round the fire, where melancholy
        Traced retreating hills of gold.

Still those papers lay before me—
Problems made express to bore me,
When a silent change came o'er me,
        In my hard uneasy chair.
Fire and fog, and candle faded,
Spectral forms the room invaded,
Little creatures, that paraded
        On the problems lying there.

Fathers there, of every college,
Led the glorious ranks of knowledge,
Men, whose virtues all acknowledge
        Levied the proctorial fines ;
There the modest Moderators,
Set apart as arbitrators
'Twixt contending calculators,
        Scrutinised the trembling lines.

All the costly apparatus,
That is meant to elevate us
To the intellectual status
        Necessary for degrees—
College tutors—private coaches—
Line the Senate-house approaches.
If our Alma Mater dote, she's
        Taken care of well by these.

Much I doubted if the vision
Were the simple repetition
Of the statements of Commission,
        Strangely jumbled, oddly placed.
When an awful form ascended,
And with cruel words defended
Those abuses that offended
        My unsanctioned private taste.

Angular in form and feature,
Unlike any earthly creature,
She had properties to meet your
        Eye whatever you might view.
Hair of pens and skin of paper ;
Breath, not breath but chemic vapour ;
Dress,—such dress as College Draper
        Fashions with precision due.

Eyes of glass, with optic axes
Twisting rays of light as flax is
Twisted, while the Parallax is
        Made to show the real size.
Primary and secondary
Focal lines in planes contrary,
Sum up all that's known to vary
        In those dull, unmeaning eyes.

Such the eyes, through which all Nature
Seems reduced to meaner stature.
If you had them you would hate your
        Symbolising sense of sight.
Seeing planets in their courses
Thick beset with arrowy " forces,"
While the common eye no more sees
        Than their mild and quiet light.

" Son," she said (what could be queerer
Than thus tête-à-tête to hear her
Talk, in tones approaching nearer
        To a saw's than aught beside ?
For the voice the spectre spoke in
Might be known by many a token
To proceed from metal, broken
        When acoustic tricks were tried.

Little pleased to hear the Siren
" Own " me thus with voice of iron,
I had thoughts of just retiring
        From a mother such a fright).

" No," she said, " the time is pressing,
So before I give my blessing,
I'll excuse you from confessing
        What you thought of me to-night.

" Powers ! " she cried, with hoarse devotion,
" Give my son the clearest notion
How to compass sure promotion,
        And take care of Number One.
Let his college course be pleasant,
Let him ever, as at present,
Seem to have read what he hasn't,
        And to do what can't be done.

" Of the Philosophic Spirit
Richly may my son inherit ;
As for Poetry, inter it
        With the myths of other days.
" Cut the thing entirely, lest yon
College Don should put the question,
Why not stick to what you're best on ?
        Mathematics always pays."

As the Hag was thus proceeding
To prescribe my course of reading,
And as I was faintly pleading,
        Hardly knowing what to say,
Suddenly, my head inclining
I beheld a light form shining ;
And the withered beldam, whining,
        Saw the same and slunk away.

Then the vision, growing brighter,
Seemed to make my garret lighter ;
As when noisome fogs of night are
        Scattered by the rising sun.
Nearer still it grew and nearer,
Till my straining eyes caught clearer
Glimpses of a being dearer,
        Dearer still than Number One.

In that well-remembered Vision
I was led to the decision
Still to hold in calm derision
       Pedantry, however draped ;
Since that artificial spectre
Proved a paltry sub-collector,
And had nothing to connect her
       With the being whom she aped.

I could never finish telling
You of her that has her dwelling
Where those springs of truth are welling,
       Whence all streams of beauty run.
She has taught me that creation
Bears the test of calculation,
But that Man forgets his station
       If he stops when that is done.

Is our algebra the measure
Of that unexhausted treasure
That affords the purest pleasure,
       Ever found when it is sought ?
Let us rather, realising
The conclusions thence arising
Nature more than symbols prizing,
       Learn to worship as we ought.

Worship ?   Yes, what worship better
Than when free'd from every fetter
That the uninforming letter
       Rivets on the tortured mind,
Man, with silent admiration
Sees the glories of Creation,
And, in holy contemplation,
       Leaves the learned crowd behind !

To F. W. F.

6th April 1853.

FARRAR, when o'er Goodwin's page
    Late I found thee poring,
From the hydrostatic Sage
    Leaky Memory storing,
Or when groaning yesterday
    Needlessly distracted
By some bright erratic ray,
    Through a sphere refracted,—

Then the quick words, oft suppressed,
    In my fauces fluttered;
Thoughts not yet in language drest
    Pleasing to be uttered.
He that neatly gilds the pill
    Hides the drug but vainly,
So, in chance-sown words, I will
    Speak the matter plainly.

Men there are, whose patient minds,
    In one object centred,
Wait, till through their darkened blinds
    Truth has burst and entered.
Then, that ray so barely caught
    Joyfully absorbing,
They behold the realms of Thought
    Into Science orbing.

Thus they wait, and thus they toil,
    Thus they end in knowing,
Like good seed in kindly soil
    Taking root and growing.
Men there are whose ambient souls,
    In rapt Intuition,
Seize Creation as it rolls,
    Whole, without partition.

Not for them the darkened room,
    Lens, and perforation ;
Enemies are they to gloom,
    Foes to Insulation.
Theirs the light of perfect Day,
    Theirs the sense of Freedom ;
Dungeons, and the tortured ray,
    Serve for those that need 'em.

Song to them of right belongs,
    Eloquently flowing ;
Sweeping down time-honoured wrongs,
    Surging, burning, glowing.
Songs in which all hearts rejoice,
    Songs of ancient story ;
Songs that fill a People's voice
    Marching on.to glory.

Thus they live, and thus they love,
    Thus they soar in singing ;
Like glad larks in heaven above,
    Dazzling courses winging.—
Here, I prithee, turn thy mind
    To a little fable
Of the fledged and rooted kind,
    Bird and vegetable.

Pensive in his lowly nest
    Once a Lark was lying ;
Often did he heave his breast
    Querulously sighing.
For he saw with envious eyes,
    Pampered vegetation—
Cabbages of goodly size,
    Swoll'n with emulation.

Till their self-infolded green
    Tight crammed, wide distended,
Seemed in spheréd pomp to mean
    All that it pretended.

Long he sought to win their place
In the Gardener's favour;
Well he caught the silent grace
Of a plant's behaviour.

All was useless, he confest,
Earth for him unsuited;
Terror seized upon him, lest
He should there be rooted.
" Cabbages are cabbages,
Larks are larks," he muttered;
Then, light springing in the breeze,
Through the sky he fluttered.

Farrar, mark my fable well,
Fling away Ambition;
By that sin the angels' fell
Into black perdition.
Cut the Calculus, and stop
Paths that lead to error;
Think—below the Junior Op.,[1]
Gapes the Gulph's grim terror.

[1] The occasion of this and of the following poem is fully explained
in a note from the Rev. Canon Farrar, which is here subjoined :—

To Prof. Campbell.

30th March 1882.

My dear Sir—I am sorry that the playful nonsense of my undergraduate days should be printed, but if the lines will in the slightest
degree help to illustrate the extreme goodness and kind-heartedness of
Maxwell, I am content.

The circumstances were these. I went up to Cambridge very ill
prepared in mathematics, and my classical work was much hindered
by the considerable amount of mathematics required for a "Senior
Optime," at which I aimed. In despair I used to say to Maxwell
sometimes that I should be "plucked" in mathematics, which in those
days would have prevented my taking classical honours. He, in his
ready sympathy, and the kind interest which he showed towards me,
took my words *d'un trop grand sérieux*, and wrote me the poetic apologue,

Then your Mathematic wings,
　　Plucked from off your shoulder,
Will express what Horace sings
　　Of that rash youth, bolder
Than his waxen wings allowed,
　　Or his cautious father.
Fall not *thou* from out thy cloud
　　Algebraic, rather

Try the Poll, for none but fools,—
　　Fools, I mean, at College,
Reach the earth between two stools,
　　Triposes of Knowledge.
Better in poetic rage
　　Sing, through heaven soaring,
Than disfigure Goodwin's page
　　By incessant poring.

REPLY TO THE ABOVE, BY F. W. F.

" *Te quoque vatem dicunt pastores.*"—VIRGIL.

O MAXWELL, if by reason's strength
　　And studying of Babbage,
You have transformed yourself at length
　　Into a mental cabbage ;
And if I've proved myself a lark
　　At morn and blushing even,
By soaring like a music-spark
　　Thro' sapphire fields of Heaven,

in which he advised me to aim at poetry like a lark, while I left mathematicians to the distinction of being what Carlyle calls the "quickest
and completest of vegetables." The point of his advice was that I should
give up the effort to win honour in the Mathematical Tripos, and
should be content with the Poll, or ordinary mathematical degree,
reserving my strength for the Classical Tripos. I sent him my reply
the same day, and I remember he pressed me to let him read it at one
of our social meetings—which I absolutely refused to allow him to do.
Without this word of explanation the verses would seem *too* nonsensical.
—Yours very truly,　　　　　　　　　　　　F. W. FARRAR.

Our diverse fates are now reversed
By strange metempsychosis,
Into a cabbage I have burst
And scorn poetic posies ;
But you a lark with twinkling wings
O'er violet-banks are soaring ;
Your voice the dewy rose-cloud rings
While Statics me are boring.

Yet *cabbage* as I will—on earth
My roots I cannot anchor,
For at my mathematic birth
Was also born a canker !
It soon will gnaw my roots away—
But when I weigh a chœnix
I'll freely soar to realms of day
An emerald cabbage-Phœnix.

Then talk not of the Poll to me,
I hate, detest, and scorn it ;
I am as earnest as a bee,
But savage as a hornet.
And if they pluck me I will drown
Each pedant in a sonnet,
And of their pluckings make a crown
With golden plumes upon it.

So if my cabbage growth be slow
I'll try to be a carrot,
Or still remain a lark—but know
I'll not be Poll, or Parrot.
Then if I fall beneath the mark,
I'll shout with accent savage,
" It is a lark to be a lark,
'Tis green to be a cabbage."

LINES *written under the conviction that it is not wise to read Mathematics in November after one's fire is out.*

10th Nov. 1853.

IN the sad November time,
When the leaf has left the lime,
And the Cam, with sludge and slime,
Plasters his ugly channel,

While, with sober step and slow,
Round about the marshes low,
Stiffening students stumping go
    Shivering through their flannel.

Then to me in doleful mood
Rises up a question rude,
Asking what sufficient good
    Comes of this mode of living?
Moping on from day to day,
Grinding up what will not " pay,"
Till the jaded brain gives way
    Under its own misgiving.

Why should wretched Man employ
Years which Nature meant for joy,
Striving vainly to destroy
    Freedom of thought and feeling?
Still the injured powers remain
Endless stores of hopeless pain,
When at last the vanquished brain
    Languishes past all healing.

Where is then his wealth of mind—
All the schemes that Hope designed?
Gone, like spring, to leave behind
    Indolent melancholy.
Thus he ends his helpless days,
Vex't with thoughts of former praise—
Tell me, how are Wisdom's ways
    Better than senseless Folly?

Happier those whom trifles please,
Dreaming out a life of ease,
Sinking by unfelt degrees
    Into annihilation.
Or the slave, to labour born,
Heedless of the freeman's scorn,
Destined to be slowly worn
    Down to the brute creation.

Thus a tempting spirit spoke,
As from troubled sleep I woke
To a morning thick with smoke,
    Sunless and damp and chilly.
Then to sleep I turned once more,
Eyes inflamed and windpipe sore,
Dreaming dreams I dreamt before,
    Only not quite so silly.

In my dream methought I strayed
Where a learned-looking maid
Stores of flimsy goods displayed,
    Articles not worth wearing.
" These," she said, with solemn air,
" Are the robes that sages wear,
Warranted, when kept with care,
    Never to need repairing."

Then unnumbered witlings, caught
By her wiles, the trappings bought,
And by labour, not by thought,
    Honour and fame were earning.
While the men of wiser mind
Passed for blind among the blind ;
Pedants left them far behind
    In the career of learning.

" Those that fix their eager eyes
Ever on the nearest prize
Well may venture to despise
    Loftier aspirations.
Pedantry is in demand !
Buy it up at second-hand,
Seek no more to understand
    Profitless speculations."

Thus the gaudy gowns were sold,
Cast off sloughs of pedants old ;
Proudly marched the students bold
    Through the domain of error,

Till their trappings, false though fair,
Mouldered off and left them bare,
Clustering close in blank despair,
  Nakedness, cold, and terror.

Then, I said, " These haughty Schools
Boast that by their formal rules
They produce more learned fools
  Than could be well expected.
Learned fools they are indeed,
Learned in the books they read ;
Fools whene'er they come to need
  Wisdom, too long neglected.

" Oh ! that men indeed were wise,
And would raise their purblind eyes
To the opening mysteries
  Scattered around them ever.
Truth should spring from sterile ground,
Beauty beam from all around,
Right should then at last be found
  Joining what none may sever."

A PROBLEM IN DYNAMICS.

19th Feb. 1854.

AN inextensible heavy chain
Lies on a smooth horizontal plane,
An impulsive force is applied at A,
Required the initial motion of K.

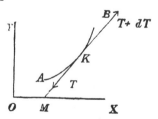

Let *ds* be the *infinitesimal* link,
Of which for the present we've only to think ;

2 s

Let T be the tension, and T + dT
The same for the end that is nearest to B.
Let a be put, by a common convention,
For the angle at M 'twixt OX and the tension ;
Let V_t and V_n be ds's velocities,
Of which V_t along and V_n across it is ;
Then $\dfrac{V_n}{V_t}$ the tangent will equal,
Of the angle of starting worked out in the sequel.

In working the problem the first thing of course is
To equate the impressed and effectual forces.
K is tugged by two tensions, whose difference dT
(1) Must equal the element's mass into V_t.
V_n must be due to the force perpendicular
To ds's direction, which shows the particular
Advantage of using da to serve at your
Pleasure to estimate ds's curvature.
For V_n into mass of a unit of chain
(2) Must equal the curvature into the strain.

Thus managing cause and effect to discriminate,
The student must fruitlessly try to eliminate,
And painfully learn, that in order to do it, he
Must find the Equation of Continuity.
The reason is this, that the tough little element,
Which the force of impulsion to beat to a jelly meant,
Was endowed with a property incomprehensible,
And was " given," in the language of *Shop*, " inexten-
        sible."
It therefore with such pertinacity odd defied
The force which the length of the chain would have
        modified,
That its stubborn example may possibly yet recall
These overgrown rhymes to their prosody metrical.
The condition is got by resolving again,
According to axes assumed in the plane.
If then you reduce to the tangent and normal,
(3) You will find the equation more neat tho' less formal.

(4) The condition thus found after these preparations,
When duly combined with the former equations,
Will give you another, in which differentials
(5) (When the chain forms a circle), become in essentials
No harder than those that we easily solve
(6) In the time a T totum would take to revolve.

Now joyfully leaving ds to itself, a-
Ttend to the values of T and of a.
The chain undergoes a distorting convulsion,
Produced first at A by the force of impulsion.
In magnitude R, in direction tangential,
(7) Equating this R to the form exponential,
Obtained for the tension when a is zero,
It will measure the tug, such a tug as the " hero
Plume-waving " experienced, tied to the chariot.
But when dragged by the heels his grim head could not
    carry aught,
(8) So give a its due at the end of the chain,
And the tension ought there to be zero again.
From these two conditions we get three equations,
Which serve to determine the proper relations
Between the first impulse and each coefficient
In the form for the tension, and this is sufficient
To work out the problem, and then, if you choose,
You may turn it and twist it the Dons to amuse.

Equations referred to.

(1) $$dT = mV_t\, ds$$

(2) $$mV_n = T\frac{da}{ds}$$

(3) $$\frac{dV_t}{ds} = V_n\frac{da}{ds}$$

(4) $$\frac{d^2T}{ds^2} - T\left(\frac{da}{ds}\right)^2 = 0$$

(5) $$\frac{d^2T}{da^2} - T = 0$$

$$(6) \quad T = C_1 e^a + C_2 e^{-a}$$

$$(7) \quad R = C_1 + C_2$$

$$(8) \quad 0 = C_1 e^{a_1} + C_2 e^{-a_1}$$

$$\frac{V_n}{V_t} = \tan \beta = - \frac{e^{(a_1 - a)} - e^{-(a_1 - a)}}{e^{(a_1 - a)} + e^{-(a_1 - a)}}$$

VALEDICTORY ADDRESS TO THE D—N.

May 1854.

JOHN Alexander Frere, John,
　　When we were first acquent,
You lectured us as Freshmen
　　In the holy term of Lent;
But now you're gettin' bald, John,
　　Your end is drawing near,
And I think we'd better say " Goodbye,
　　John Alexander Frere."

John Alexander Frere, John,
　　How swiftly Time has flown!
The weeks that you refused us
　　Are now no more your own;
Tho' Time was in your hand, John,
　　You lingered out the year,
That Grace might more abound unto
　　John Alexander Frere.[1]

There's young Monro of Trinity,
　　And Hunter bold of Queen's,
Who spurn the chapel system,
　　And " vex the souls of Deans."
But all their petty squabbles
　　More ludicrous appear,
When we muse on thy departed form,
　　John Alexander Frere.

---

[1] Mr. Frere had accepted the living of Shillington, but retained his fellowship for the customary " year of grace."

There's many a better man, John,
　That scorns the scoffing crew,
But keeps with fond affection
　The notes he got from you—
" Why he was out of College,
　Till two o'clock or near,
The Senior Dean requests to know,
　Yours truly, J. A. Frere."

John Alexander Frere, John,
　I wonder what you mean
By mixing up your name so
　With me, and with " The Dean."
Another Don may dean us,
　But ne'er again, we fear,
Shall we receive such notes as yours,
　John Alexander Frere.

The Lecture Room no more, John,
　Shall hear thy drowsy tone,
No more shall men in Chapel
　Bow down before thy throne.
But Shillington with meekness,
　The oracle shall hear,
That set St. Mary's all to sleep—
　John Alexander Frere.

Then once before we part, John,
　Let all be clean forgot,
Our scandalous inventions,
　[Thy note-lets, prized or not].
For under all conventions,
　The small man lived sincere,
The kernel of the Senior Dean,
　John Alexander Frere.[1]

---

[1] The genuine esteem expressed in the concluding words, which alone in this youthful pasquinade are to be taken seriously, must be the apology for inserting what Maxwell himself would not have printed.

IN MEMORY OF EDWARD WILSON,
*Who repented of what was in his mind to write after
section.*

Rigid Body (sings).

GIN a body meet a body
Flyin' through the air,
Gin a body hit a body,
Will it fly ? and where ?
Ilka impact has its measure,
Ne'er a ane hae I,
Yet a' the lads they measure me,
Or, at least, they try.

Gin a body meet a body
Altogether free,
How they travel afterwards
We do not always see.
Ilka problem has its method
By analytics high ;
For me, I ken na ane o' them,
But what the waur am I ?

VALENTINE BY A TELEGRAPH CLERK ♂ TO A TELEGRAPH
CLERK ♀.

" THE tendrils of my soul are twined
With thine, though many a mile apart,
And thine in close-coiled circuits wind
Around the needle of my heart.

" Constant as Daniell, strong as Grove,
Ebullient through its depths like Smee,
My heart pours forth its tide of love,
And all its circuits close in thee.

"O tell me, when along the line
From my full heart the message flows,
What currents are induced in thine ?
One click from thee will end my woes."

Through many an Ohm the Weber flew,
And clicked this answer back to me,—
"I am thy Farad, staunch [1] and true,
Charged to a Volt with love for thee."

LECTURES TO WOMEN ON PHYSICAL SCIENCE.

I.

PLACE.—*A small alcove with dark curtains.*

The Class consists of one member.

SUBJECT.—*Thomson's Mirror Galvanometer.*

THE lamp-light falls on blackened walls,
And streams through narrow perforations,
The long beam trails o'er pasteboard scales,
With slow-decaying oscillations.
Flow, current, flow, set the quick light-spot flying,
Flow current, answer light-spot, flashing, quivering, dying,

O look ! how queer ! how thin and clear,
And thinner, clearer, sharper growing
The gliding fire ! with central wire,
The fine degrees distinctly showing.
Swing, magnet, swing, advancing and receding,
Swing magnet ! Answer dearest, What's your final reading ?

O love ! you fail to read the scale
Correct to tenths of a division.
To mirror heaven those eyes were given,
And not for methods of precision.
Break contact, break, set the free light-spot flying ;
Break contact, rest thee, magnet, swinging, creeping, dying.

[1] *v.r.* Stout.

LECTURES TO WOMEN ON PHYSICAL SCIENCE.

July 1874.

II.

*Professor Chrschtschonovitsch, Ph.D., " On the C. G. S.[1]
system of Units."*

Remarks submitted to the Lecturer by a student.

PRIM Doctor of Philosophy
From academic Heidelberg !
Your sum of vital energy
Is not the millionth of an erg.[2]
Your liveliest motion might be reckoned
At one-tenth metre [3] in a second.

" The air," you said, in language fine,
    Which scientific thought expresses,
" The air—which with a megadyne,[4]
    On each square centimetre presses—
The air, and I may add the ocean,
Are nought but molecules in motion."

Atoms, you told me, were discrete,
    Than you they could not be discreter,
Who know how many Millions meet
    Within a cubic millimetre.
They clash together as they fly,
But *you !*—you cannot tell me why.

---

[1] C. G. S. system—the system of units founded on the centimetre, gramme, and second. See report of Committee on units. *Brit. Ass. Report* for 1873, p. 222.

[2] Erg—the energy communicated by a dyne, acting through a centimetre. See p. 60, note 2.   [3] Tenth-metre $= 1$ metre $\times 10^{-10}$.

[4] Megadyne $= 1$ dyne $\times 10^6$. It is somewhat more than the weight of a kilogramme.

And when in tuning my guitar
The interval would *not* come right,
" This string," you said, " is strained too far,
'Tis forty dynes,[1] at least too tight !"
And then you told me, as I sang,
What overtones were in my clang.[2]

You gabbled on, but every phrase
Was stiff with scientific shoddy,
The only song you deigned to praise
Was " Gin a body meet a body,"
" And even there," you said, " collision
Was not described with due precision."

" In the invariable plane,"
You told me, " lay the impulsive couple."[3]
You seized my hand—you gave me pain,
By torsion of a wrist so supple ;
You told me what that wrench would do,—
" 'Twould set me twisting round a screw."[4]

Were every hair of every tress
(Which you, no doubt, imagine mine),
Drawn towards you with its breaking stress—
A stress, say, of a megadyne,
That tension I would sooner suffer
Than meet again with such a duffer !

[1] Dyne—the force which, acting on a gramme for a second, would give a velocity of a centimetre per second. The weight of a gramme is about 980 dynes.

[2] See *Sound and Music*, by Sedley Taylor, p. 89.

[3] See Poinsot, *Théorie nouvelle de la rotation des corps.*

[4] See Prof. Ball on the Theory of Screws, *Phil. Trans.*, 1873.

To the Chief Musician upon Nabla.[1]

A Tyndallic Ode.

I.

I come from fields of fractured ice,
  Whose wounds are cured by squeezing,
Melting [2] they cool, but in a trice,
  Get [3] warm again by freezing.
Here, in the frosty air, the sprays
  With fern-like hoar-frost bristle,
There, liquid stars their watery rays
  Shoot through the solid crystal.

II.

I come from empyrean fires—
  From microscopic spaces,
Where molecules with fierce desires,
  Shiver in hot embraces.
The atoms clash, the spectra flash,
  Projected on the screen,
The double D, magnesian *b*,
  And Thallium's living green.

III.

We place our eye where these dark rays
  Unite in this dark focus,
Right on the source of power we gaze,
  Without a screen to cloak us.
Then where the eye was placed at first,
  We place a disc [4] of platinum,
It glows, it puckers! will it [5] burst?
  How ever shall we [6] flatten him!

---

[1] Nabla was the name of an Assyrian harp of the shape $\triangledown$. $\triangledown$ is a quaternion operator $\left(i\dfrac{d}{dx} + j\dfrac{d}{dy} + k\dfrac{d}{dz} \right)$ invented by Sir W. R. Hamilton, whose use and properties were first fully discussed by Professor Tait, who is therefore called the "Chief Musician upon Nabla."

[2] *v.r.* They melt. [3] *v.r.* Grow. [4] *v.r.* dish.
[5] *v.r.* like to. [6] *v.r* By Jove, I'll have to.

IV.

This crystal tube the electric ray
    Shows optically clean,
No dust or haze within, but stay !
    All has not yet been seen.
What gleams are these of heavenly blue ?
    What air-drawn form appearing,
What mystic fish, that, ghostlike, through [1]
    The empty [2] space is steering ?

V.

I light this sympathetic flame,
    My faintest wish that answers,
I sing, it sweetly sings the same,
    It dances with the dancers.
I shout, I whistle, clap my hands,
    And [3] stamp upon [4] the platform,
The flame responds [5] to my commands,
    In this form and in that form.

VI.

What means that thrilling, drilling scream,
    Protect me ! 'tis the siren :
Her heart is fire, her breath is steam,
    Her larynx is of iron.
Sun ! dart thy beams ! in tepid streams,
    Rise, viewless exhalations !
And lap me round, that no rude sound
    May mar my meditations.

VII.

Here let me pause.——These transient facts,
    These fugitive impressions,
Must be transformed by mental acts,
    To permanent possessions.

---

[1] *v.r.* What fish, what whale is this, that through.   [2] *v.r.* vacuous.
[3] *v.r.* I.   [4] *v.r.* about.   [5] *v.r.* bows down.

Then summon up your grasp of mind,
　Your fancy scientific,
Till [1] sights and sounds with thought combined,
　Become [2] of truth prolific.

VIII.

Go to! prepare your mental bricks,
　Fetch them from every quarter,
Firm on the sand your basement fix
　With best sensation mortar.
The top [3] shall rise to heaven on high—
　Or such an elevation,
That the swift whirl with which we fly
　Shall conquer gravitation.

To the Committee of the Cayley Portrait Fund.

1874.

O WRETCHED race of men, to space confined!
What honour can ye pay to him, whose mind
To that which lies beyond hath penetrated?
The symbols he hath formed shall sound his praise,
And lead him on through unimagined ways
To conquests new, in worlds not yet created.

First, ye Determinants! in ordered row
And massive column ranged, before him go,
To form a phalanx for his safe protection.
Ye powers of the nth roots of -1!
Around his head in ceaseless [4] cycles run,
As unembodied spirits of direction.

And you, ye undevelopable scrolls!
Above the host wave your emblazoned rolls,
Ruled for the record of his bright inventions.
Ye Cubic surfaces! by threes and nines
Draw round his camp your seven-and-twenty lines—
The seal of Solomon in three dimensions.

[1] *v.r.* That.　　[2] *v.r.* May be.　　[3] *v.r.* tower.　　[4] *v.r.* endless.

March on, symbolic host! with step sublime,
Up to the flaming bounds of Space and Time!
There pause, until by Dickenson depicted,
In two dimensions, we the form may trace
Of him whose soul, too large for vulgar space,
In n dimensions flourished unrestricted.

MOLECULAR EVOLUTION.

Belfast, 1874.

AT quite uncertain times and places,
 The atoms left their heavenly path,
And by fortuitous embraces,
 Engendered all that being hath.
And though they seem to cling together,
 And form " associations " here,
Yet, soon or late, they burst their tether,
 And through the depths of space career.

So we who sat, oppressed with science,
 As British asses, wise and grave,
Are now transformed to wild Red Lions,[1]
 As round our prey we ramp and rave.
Thus, by a swift metamorphōsis,
 Wisdom turns wit, and science joke,
Nonsense is incense to our noses,
 For when Red Lions speak, they smoke.

Hail, Nonsense! dry nurse of Red Lions,[2]
 From thee the wise their wisdom learn,
From thee they cull those truths of science,
 Which into thee again they turn.
What combinations of ideas,
 Nonsense alone can wisely form!
What sage has half the power that she has,
 To take the towers of Truth by storm?

[1] The "Red Lions" are a club formed by Members of the British Association, to meet for relaxation after the graver labours of the day.
[2] " Leonum arida nutrix."—*Horace.*

Yield, then, ye rules of rigid reason!
  Dissolve, thou too, too solid sense!
Melt into nonsense for a season,
  Then in some nobler form condense.
Soon, all too soon, the chilly morning,
  This flow of soul will crystallize,
Then those who Nonsense now are scorning,
  May learn, too late, where wisdom lies.

Molecular Evolution.

SONG OF THE CUB.

Belfast, 1874.

I KNOW not what this may betoken,
  That I feel so wondrous wise;
My dream of existence is broken
  Since science has opened my eyes.
At the British Association
  I heard the President's speech,
And the methods and facts of creation
  Seemed suddenly placed in my reach.

My life's undivided devotion
  To Science I solemnly vowed,
I'd dredge up the bed of the ocean,
  I'd draw down the spark from the cloud.
To follow my thoughts as they go on,
  Electrodes I'd place in my brain;
Nay, I'd swallow a live entozöon,
  New feelings of life to obtain.

O where are those high feasts of Science?
  O where are those words of the wise?
I hear but the roar of Red Lions,
  I eat what their Jackal supplies.
I meant to be so scientific,
  But science seems turned into fun;
And this, with his roaring terrific,
  That old red lion hath done.

BRITISH ASSOCIATION, 1874.

Notes of the President's Address.

IN the very beginnings of science, the parsons, who managed
  things then,
Being handy with hammer and chisel, made gods in the
  likeness of men ;
Till Commerce arose, and at length some men of exceptional
  power
Supplanted both demons and gods by the atoms, which last
  to this hour.
Yet they did not abolish the gods, but they sent them well
  out of the way,
With the rarest of nectar to drink, and blue fields of nothing
  to sway.
From nothing comes nothing, they told us, nought happens
  by chance, but by fate ;
There is nothing but atoms and void, all else is mere whims
  out of date !
Then why should a man curry favour with beings who cannot exist,
To compass some petty promotion in nebulous kingdoms of
  mist ?
But not by the rays of the sun, nor the glittering shafts of
  the day,
Must the fear of the gods be dispelled, but by words, and
  their wonderful play.
So treading a path all untrod, the poet-philosopher sings
Of the seeds of the mighty world—the first-beginnings of
  things ;
How freely he scatters his atoms before the beginning of
  years ;
How he clothes them with force as a garment, those small
  incompressible spheres !
Nor yet does he leave them hard-hearted—he dowers them
  with love and with hate,
Like spherical small British Asses in infinitesimal state ;

Till just as that living Plato, whom foreigners nickname
Plateau,[1]
Drops oil in his whisky-and-water (for foreigners sweeten
it so),
Each drop keeps apart from the other, enclosed in a flexible
skin,
Till touched by the gentle emotion evolved by the prick of
a pin:
Thus in atoms a simple collision excites a sensational thrill,
Evolved through all sorts of emotion, as sense, understanding,
and will;
(For by laying their heads all together, the atoms, as coun-
cillors do,
May combine to express an opinion to every one of them
new).
There is nobody here, I should say, has felt true indignation
at all,
Till an indignation meeting is held in the Ulster Hall;
Then gathers the wave of emotion, then noble feelings arise,
Till you all pass a resolution which takes every man by
surprise.
Thus the pure elementary atom, the unit of mass and of
thought,
By force of mere juxtaposition to life and sensation is
brought;
So, down through untold generations, transmission of struc-
tureless germs
Enables our race to inherit the thoughts of beasts, fishes,
and worms.
We honour our fathers and mothers, grandfathers and grand-
mothers too;
But how shall we honour the vista of ancestors now in our
view?
First, then, let us honour the atom, so lively, so wise, and
so small;
The atomists next let us praise, Epicurus, Lucretius, and all;

[1] *Statique Expérimentale et Théorique des Liquides soumis aux seules
Forces Moléculaires.* Par J. Plateau, Professeur à l'Université de Gaud.

Let us damn with faint praise Bishop Butler, in whom many
    atoms combined
To form that remarkable structure, it pleased him to call
    —his mind.
Last, praise we the noble body to which, for the time, we
    belong,
Ere yet the swift whirl of the atoms has hurried us, ruth-
    less, along,
The British Association—like Leviathan worshipped by
    Hobbes,
The incarnation of wisdom, built up of our witless nobs,
Which will carry on endless discussions, when I, and prob-
    ably you,
Have melted in infinite azure—in English, till all is blue.

*Translation of " Notes on the President's Address, British
Association, 1874." By R. Shilleto.*

πρῶτ᾽ ἀρχομένης σοφίας, ἱερεῖς, οἷσπερ τότε πάντ᾽ ἐμεμήλει,
δαιδαλοκομψογλυφανορραισταί, θνητῷ θεὸν εἰκότ᾽ ἔτευξαν·
ἔς τ᾽ ἐμπορία 'γένετ᾽, εἶτα βροτὸς πλατύνωτος ὑπεσκέλισέν τις
δαίμονα καὶ θεὸν, ἀντειληφὼς ἀτόμους μέχρι δεῦρο μενούσας.
οὐ μὴν πάντως γ᾽ ἔκτεινε θεούς, ἀλλ᾽ ἐκτοπίους ἀπέπεμψεν
νέκταρ τε πιεῖν σπάνιον, γλαυκοῖς τ᾽ ἐν ἀγροῖς τοῖς μηδὲν ἀνάσσειν.
ἐξ οὐδενὸς ἔστ᾽ οὐδέν, ἔφασκον· συνέβη γὰρ ἅπαντα κατ᾽ αἶσαν.
πλὴν ἄτομοι καὶ χάος οὐδὲν ἔφυ· κρονίων τἄλλ᾽ οὐ καλὸν ὄζει.
τί ἂν οὖν τοῖς μὴ δυνατοῖς εἶναι μελετῴης ἂν χαρίσασθαι,
ἱμειρόμενος τινὸς ἀρχιδίου 'ν κορυφαῖς ὀμίχλης νεφοέσσης;
οὐ μὴν οὐδ᾽ ἂν σέλας ἠελίου γ᾽, οὐδ᾽ ἡμέριον βέλος αὐγῆς,
σκεδάσειε θεῶν δεῖμ᾽, ἀλλὰ λόγοι, χὦ περὶ τούτων στενολεσχῶν.
πρὸς ταῦτα πατῶν στίβον ἄστειπτον, μεγάλου μετεωροποιητὴς
σπέρματα κόσμου πρωτάς[1] τ᾽ ἀρχὰς πάντων ὁπόσ᾽ ἔστιν ἀείδει·
καταχεῖ τ᾽ ἀτόμους μάλ᾽ ἐλευθερίως πρὶν φῦναι τοὺς ἐνιαυτούς,
ὡς ἱματίῳ δυνάμει περιδὺς τὰ σφαιρίδι᾽ οὐχὶ πιεστά·
φιλίαν μέντοι μῖσός τ᾽ ἐπέδωκ᾽, οὐ σκληρόφρονας καταλείπων,
ὡς σμίκρ᾽ ὄντα σφαιρίχ᾽ ὁποστοῦν φιτύματ᾽ ὄνεια Βρετάννων.
κᾆθ᾽ οἷα Πλάτων ὁ ζῶν νυνί, καλέουσι θεοὶ δὲ Πλαταιᾶ,
γάνος οἰναρίῳ "μιξεν ἐλαίου (γλυκὺ γὰρ ποιεῖ θεὸς οὕτω),
ἑτέρας ἑτέρᾳ χωρὶς σταγόσιν δέρμ᾽ αἰόλον ἀμφιέσας τι,

---

[1] Accentum nolim mutare.—C.

2 T

ἐς τ' ἂν μαλακὴ κίνησις ἀφῆς πλησθῇ κέντροισιν ὄπητος·
οὕτως ἀτόμων σύγκρασις ἁπλῆ παρέχει τινὰ γάργαλον ὀξύν,
κατέδειξεν ἅπασ' ὃν κίνησις, νοῦς οἷα φρόνησις ὄρεξις·
οὐδ' ἂν φαίην γνήσιον ὀργὴν τῶν ἐνταῦθ' οἰδέν' ἐπᾶραι
πρὶν ἐποιήθη σύλλογος ὀργῆς ὅδ' Ἱέρνης [1] ἐν πρυτανείῳ
τότε κίνησις κύματ' ἐγείρει, μεγάλη τότε γίγνεται ὁρμή,
θαύματί θ' ὑμεῖς ψῆφον θέμενοι καταπλήττετε πάντας ἅπαντες.
ἄτομος διὰ ταῦθ' ἡ πρώτη, νοῦ πλήθους θ' ἑνὰς, αἰὲν ἄκρατος
τοῦ παρακεῖσθαι δυνάμει βιοτὴν ἔλαβέν τ' αἴσθησιν ἑκάστη.
χοὕτω γενεῶν ἐξ ἀμετρήτων στοιχεῖα δεδέγμεθ' ἄμορφα,
θηρίον ἰχθὺς αὐτός θ' ὁ φθεὶρ ἃ φρονοῦσιν κληρονομοῦντες.
πατέρας πάππους τιμῶμεν ἀεὶ τηθαῖς καὶ μητράσιν αὐταῖς.
πῶς δ' ἂν προγόνων τήνδε παροῦσαν τιμήσειέν τις ἔποψιν;
ἀτόμῳ πρώτη κείσθω τιμή, διερᾷ μάλα σώφρονι λεπτῇ.
τοὺς ἀτομίζοντας ἔπειτ' αἰνῶ Λουκρήτιον ἠδ' Ἐπίκουρον,
καταμεμφόμενος πόλλ' Οἰνοχόον, κεἰ πολὺ πλῆθος συνεπλέχθη
ἀτόμων μέγα θαῦμ' ἐργαζομένων ὃ καλεῖν οἱ ἔδοξε νόημα.
ὕστατα δ' αἰνῶ χορὸν ἀθάνατον, νῦν ἡμετέρους κορυφαίους,
στρόβος οὓς ἀτόμων οὔπω σοβαρὸς φρεσὶν εἵλκυσε τηλόσ' ἀνοίκτοις,
κνώδαλον οἷον τοῦ ψυχολόγου, φιτύματ' ὄνεια Βρετάννων,
ἀσόφου κτισθέντας ὑπ' ἐγκεφάλου, σοφίας ἔμψυχον ἄγαλμα,
οἵπερ διατρίψουσιν ἀπαύστως, ὁπόταν τάχ' ἴσως σύ τε κἀγὼ
κυανῷ τε τακῶμεν ἐν ἀσβέστῳ, καὶ γλαυκὰ πέριξ συνάπαντ' ᾖ.

To The Additional Examiner for 1875. [2]

July 29, 1874.

QUEEN CRAM went straying
Where Tait was swaying,
In just hands weighing,
    With care immense,
Dry proofs made pleasant
By Routh or Besant
For one who hasn't
    Got too much sense.
Nor marked how, quicker
Than mounts the liquor
In brains made thicker

---

[1] Var. l. Ἱερνικὸς suppedidat nescio quis. Mihi quidem haud displicet.
—[ED.]   [2] Compare Mr. Swinburne's *Bothwell*, Act ii. Sc. 17.

By College beer,
The murderous maiden,
Mistake, walks laden
With tips forgotten and slips so queer.

How, like a spider,
She still spreads wider,
O'er bookwork, rider,
    And problem too,
Her flimsy curtain
Of terms uncertain,
Till all seems dirt in
    The marker's view.
For if Cram were not,
Which markers spare not,
Wise men would care not
    To pluck too soon,
Seeing all life's season
Of budding [1] reason
Finds good stiff work for a wooden spoon.

As Tait sat joking,
And marked while smoking,
Still slyly poking
    Where jests might hit,
She came, soft-gliding,
Her false face hiding,
Rich food providing
    For Tait's sharp wit.
Through symbols tangled,
The Wranglers wrangled
Like sweet bells jangled
    And out of tune.
For though their music
Would soon make *you* sick
The tides they measure and guide the moon.

[1] *v.r.* opening.

Cram found no cover
Wherein to hover,
For still above her
    Tait held his pen,
Which, onward creeping,
Might find her sleeping,
But left her weeping
    O'er ruined men.
For, like a blister,
Mistake, Cram's sister,
Would wring and twist her
    In awkward [1] ways,
Till all the knowledge
Acquired at College
Had passed from thought [2] in the last six days.

PROFESSOR TAIT, LOQUITUR.

June 1877.

WILL mounted ebonite disk
    On smooth unyielding bearing,
When turned about with motion brisk
    (Nor excitation sparing),
Affect the primitive repose,
    Of $+$ and $-$ in a wire,
So that while either downward flows,
    The other upwards shall aspire?
Describe the form and size of coil,
    And other things that we may need,
Think not about increase of toil
    Involved in work at double speed.
I can no more, my pen is bad,
    It catches in the roughened page—
But answer us and make us glad,
    THOU ANTI-DISTANCE-ACTION SAGE!

[1] *v.r.* secret.   [2] *v.r.* away.

Yet have I still a thousand things to say,
    But work of other kinds is pressing—
So your petitioner will ever pray
    That your defence be triple *messing*.[1]

Answer to Tait.

The mounted disk of ebonite
    Has whirled before, nor whirled in vain :
Rowland of Troy, that doughty knight,
    Convection currents did obtain [2]
In such a disk, of power to wheedle,
From its loved North the subtle [3] needle.

'Twas when Sir Rowland, as a stage
    From Troy to Baltimore, took rest
In Berlin, there old Archimage,[4]
    Armed him to follow up this quest ;
Right glad to find himself possessor
Of the irrepressible Professor.[5]

But wouldst thou twirl [6] that disk once more,
    Then follow in Childe Rowland's train,
To where in busy Baltimore
    He brews [7] the bantlings of his brain ;
As he may do who still prefers
One Rowland to two Olivers.[8]

But Rowland,—no, nor Oliver,—
    Could get electromotive force,
Which fact and reason both aver,
    Has change of some kind as its source,
Out of a disk in swift rotation
Without the least acceleration.

---

[1] *Illi robur et aes triplex !* (Horace.)   [2] Berlin Monats-berichte.
[3] *v.r.* pensile.   [4] *i.e.* Helmholtz—the chief magician !
[5] See Prof. Sylvester's Address to the Johns Hopkins University.
[6] *v.r.* whirl.   [7] *v.r.* rears.   [8] Heaviside and Lodge, see p. 374.

But with your splendid roundabout
    Of mighty power, new-hung and greasy,
With galvanometer so stout,
    A new research would be as easy ;
A test which might perchance disclose,
Which way the electric current flows.

Take then a coil of copper pure,
    And fix it on your whirling table ;
Place the electrodes firm and sure
    As near the axis as you're able,
And soon you'll learn the way to work it,
With galvanometer in circuit.

Not while the coil in spinning sleeps,
    On her smooth axle swift and steady ;
But when against the stops she sweeps,
    To watch the light-spot then be ready,
That you may learn from its deflexion
The electric current's true direction.

It may be that it does not move,
    Or moves but for some other reason ;
Then let it be your boast to prove
    (Though some may think it out of season,
And worthy of a fossil Druid),
That there is no Electric Fluid.

REPORT ON TAIT'S LECTURE ON FORCE :—B.A., 1876.

YE British Asses, who expect to hear
        Ever some new thing,
I've nothing new to tell, but what, I fear,
        May be a true thing.
For Tait comes with his plummet and his line,
        Quick to detect your
Old bosh new dressed in what you call a fine
        Popular lecture.

Whence comes that most peculiar smattering,
    Heard in our section ?
Pure nonsense, to a scientific swing
    Drilled to perfection ?
That small word "Force," they make [1] a barber's block,
    Ready to put on
Meanings most strange and various, fit to shock
    Pupils of Newton.

Ancient and foreign ignorance they throw
    Into the bargain ;
The shade of Leibnitz [2] mutters from below
    Horrible jargon.
The phrases of last century in this
    Linger to play tricks—
Vis Viva and *Vis Mortua* and *Vis*
    Acceleratrix :—

Those long-nebbed words that to our text books still
    Cling by their titles,
And from them creep, as entozoa will,
    Into our vitals.
But see ! Tait writes in lucid symbols clear
    One small equation ;
And Force becomes of Energy a mere
    Space-variation.

Force, then, is Force, but mark you ! not a thing,
    Only a Vector ;
Thy barbèd arrows now have lost their sting,
    Impotent spectre !
Thy reign, O Force ! is over.   Now no more
    Heed we thine action ;
Repulsion leaves us where we were before,
    So does attraction.

---

[1] *v.r.* is made.   [2] *v.r.* sage of Leipzig.

Both Action and Reaction now are gone.
        Just ere they vanished,
Stress joined their hands in peace, and made them one ;
        Then they were banished.
The Universe is free from pole to pole,
        Free from all forces.
Rejoice ! ye stars—like blessed gods ye roll
        On in your courses.

No more the arrows of the Wrangler race,
        Piercing shall wound you.
Forces no more, those symbols of disgrace,
        Dare to surround you :
But those whose statements baffle all attacks,
        Safe by evasion,—
Whose definitions, like a nose of wax,
        Suit each occasion,—

Whose unreflected rainbow far surpassed
        All our inventions,
Whose very energy appears at last
        Scant of dimensions :—
Are these the gods in whom ye put your trust,
        Lordlings and ladies ?
The hidden [1] potency of cosmic dust
        Drives them to Hades.

While you, brave Tait ! who know so well the way
        Forces to scatter,
Calmly await the slow but sure decay,
        Even of Matter.

(CATS) CRADLE SONG,

By a Babe in Knots.

PETER the Repeater,
        Platted round a platter

----

[1] *v.r.* secret.

Slips of slivered paper,
    Basting them with batter.

Flype 'em, slit 'em, twist 'em,
    Lop-looped laps of paper ;
Setting out the system
    By the bones of Neper.

Clear your coil of kinkings
    Into perfect plaiting,
Locking loops and linkings
    Interpenetrating.

Why should a man benighted,
    Beduped, befooled, besotted,
Call knotful knittings plighted,
    Not knotty but beknotted ?

It's monstrous, horrid, shocking,
    Beyond the power of thinking,
Not to know, interlocking
    Is no mere form of linking.

But little Jacky Horner
    Will teach you what is proper,
So pitch him, in his corner,
    Your silver and your copper.

To HERMANN STOFFKRAFT, Ph.D., the Hero of a recent work
called "Paradoxical Philosophy."

A Paradoxical Ode.

[After Shelley.]

I.

1878.

MY soul is an entangled [1] knot,
Upon a liquid vortex wrought

[1] *v.r.* 's an amphicheiral.

By Intellect, in the Unseen residing,
  And thine doth like a convict sit,
  With marlinspike untwisting it,
Only to find its knottiness abiding;
  Since all the tools for its untying
  In four-dimensioned space are lying,
  Wherein thy fancy intersperses
  Long [1] avenues of universes,
  While Klein and Clifford fill the void
  With one finite, unbounded homaloid,
And think the Infinite is now at last destroyed.

II.

But when thy Science lifts her pinions
  In Speculation's wild dominions,
We treasure every dictum thou emittest,
  While down the stream of Evolution
  We drift, expecting no solution
But that of the survival of the fittest.
  Till, in the twilight of the gods,
  When earth and sun are frozen clods,
  When, all its energy degraded,
  Matter to æther shall have faded;
  We, that is, all the work we've done,
  As waves in æther, shall for ever run
In ever-widening [2] spheres through heavens beyond the sun.

III.

Great Principle of all we see,
  Unending Continuity!
By thee are all our angles sweetly rounded,
  By thee are our misfits adjusted,
  And as I still in thee have trusted,
So trusting, let me never be confounded!

[1] *v.r.* Whole.   [2] *v.r.* swift expanding.

Oh never may direct Creation
Break in upon my contemplation ;
Still may thy causal chain, ascending,
Appear unbroken and unending,
While Residents in the Unseen—
Æons and Emanations—intervene,
And from my shrinking soul the Unconditioned screen.[1]

[1] *v.r.* And where that chain is lost to sight
Let viewless fancies guide my darkling flight,
Through atom-haunted worlds in series infinite.

END OF PART III.

INDEX.

INDEX. 657

Force, suggestions from Faraday as to definition of, 289.
Foucault, his pendulum for showing the rotation of the earth, 153.
his determination of the velocity of light, 545.
Fourier's Theorie de la Chaleur perused by Maxwell while at Edinburgh University, 133 *note*, 134.
method of harmonics, 507.
Fraunhofer's lines, preparation of water prism for showing, to students, 267.
Free will, 305, 306.
Fresnel, 129, 486.
Frog, a subject of childish interest, 34.
transformations ingeniously delineated on magic disc, 37.

Galileo, 501.
Galton, F., 435.
Galvanism (see Electricity).
Gases, experiments on viscosity of, at different pressures and temperatures, 318.
paper on this subject, delivered at Bakerian Lecture 1866, 325.
Geometry, original inquiries suggested to J. C. M. by school exercises in, 69.
Maxwell's investigations in, 465, 496-498.
Germs, speculations upon, 390.
Glacier-markings on Arthur's Seat, J. C. M. points out to L. C., in 1846, 73.
Glaciers, Forbes's investigations on, a subject of interest to J. C. M., 80.
Glenlair, name of farm acquired by Mr. John C. Maxwell, and house built by him, 24-26.
journey from, on J. C. M. going to Edinburgh for schooling, 46.
holidays at, 61-65, etc.
completion of new offices at, 64.
Maxwell's residence in intervals of professorial work, 320-328.
Gloag, J. C. M.'s first mathematical teacher, 68.
Graham's experiments on gases, 315, 332, 564.
Gregory, Professor, at Edinburgh, his chemistry class attended by Maxwell, 107, 114, 126, 127.
Guillemard, Rev. Dr., of Little St. Mary's,[1] Cambridge), his account of Maxwell's last hours, 414-416.

Haidinger's Brushes, 84, 489.

Hamilton, Sir William, Edinburgh, his logic and metaphysic class attended by Maxwell, 90, 107, 116, 126.
influence on Maxwell of his teaching, 108, 165, 227.
Happy Valley, Vale of Urr, in *Coterie-Sprache*, 62.
Harmonic analysis, 374.
Hay, Mr. D. R., his work suggested J. C. M.'s investigations on the subject of ovals, 73.
furnishes colour-patterns, 198.
his book of colours, 212.
Heat, a subject in chair of experimental physics, 350.
treatise on, 374.
Helmholtz, his investigations on colours referred to, 214, 468.
on conversion of energy, 335.
on acoustics, 363.
Heriot Row, "Old 31," house of aunt, Mrs. Wedderburn, and home of J. C. M. while at school and college in Edinburgh — arrival there sketched by J. W., 46.
home occupations, 52.
Herschel, Sir John, his essays, 302.
his style, 305.
Sir William, his observation of Saturn's rings, 502.
Hills and Dales, paper on (British Association 1870), 327, 497.
Hopkins, William, his remark that "Maxwell never made a mistake," 133 *note*.
Maxwell commences to read with him, 154.
reminiscence of a fellow-pupil under, 175.
Hort, Professor F. J. A., his recollections of Maxwell, 417-421.
Hughes, Professor (microphone), 363.
Huyghens' discovery of Saturn's rings, 501.

Impenetrability of matter, doctrine stated in paper on properties of matter, written for Mental Philosophy class in Edinburgh University, 109.
Inertia, as a property of matter, defined in paper written for Mental Philosophy class in University of Edinburgh, 109.
Investigation, early proclivities for, 27.
Irving, Miss Janet, paternal grandmother of J. C. M., 4, 21, 22.

---

[1] N.B.—*not* Trinity, as in 413.

2 U

Maxwell, James Clerk, his father's death, professorship at Aberdeen, marriage, 247-313.
King's College, London—Glenlair 1860-1870, 314-347.
Examiner at Cambridge, 325.
Cambridge — Cavendish laboratory (1871-1879), 348-405.
illness and death 1879, 406-433.
last essays at Cambridge, 434-463.
poems, 577-651.
John Clerk, father of J. C. M., 2.
his place in descent of the Clerks of Penicuik, 22.
inherits remnant of Middlebie under an entail which debarred his elder brother George (Clerk) from holding it with Penicuik, 3.
converts estate into habitation (Glenlair), 5, 25.
inherited and personal traits of character, 5-12.
portrait-sketch by J. W., 38.
genial intercourse between father and son, 53.
interest in son's progress, 72, 73.
places son at Cambridge, 146.
extracts from his letters to J. C. M., 150, 156, 160, 181, 184-187, 193, 194, 196, 207, 213, 214, 217-220.
anxiety about his health, 206, 210.
his death, 248.
Mrs. John Clerk (née Frances Cay), mother of J. C. M., 2.
traits of character, 12.
her death in J. C. M.'s ninth year, 15.
Maxwells of Middlebie, note on their descent, 22.
Meloid, delineation of, 87, 88, 91-97.
Microphone (Hughes'), 363.
Middlebie, the family estate descended from the Maxwells, 3.
Maxwells of, note on their descent, 22.
how the bulk of the old estate was disposed of, 23.
situation of remaining estate, 24.
Milton, Maxwell's early familiarity with, 32.
Moigno's Répertoire d'Optique, read during vacation at Edinburgh University, 129.

Molecular physics, account of Maxwell's investigations in, 465, 559-574.
Molecules, Maxwell's discourse on, British Association at Bradford 1873, 358-360.
his paper on dynamical evidence of the molecular constitution of bodies, 361.
his theory of, 561.
Moral Philosophy, Professor Wilson's (Christopher North), lectures on, 107, 114, 127, 128.
thoughts upon, in correspondence, 140-44.
and in an essay, 234.
Maxwell's comments upon Professor Wilson's lectures, 145.

NECESSITY, 306 (and see Determinism).
Neptune, interest in discovery of planet, 85.
Newton, his optics perused by Maxwell while at Edinburgh University, 133 note.
Principia, etc., referred to, 398, 437.
class at Aberdeen upon, 291, 295, 302.
Newton's rings, engaged attention of J. C. M. (Æt. 15), 84.
Nichol, Professor, his lecture on discovery of planet Neptune, 85.
Nicol, Mr., inventor of the polariscope visited, 84.
letters of J. C. M. referring to this visit, 122, 123.
Nicol's prism described, 486.
Niven, W. D., 402, 575.

OERSTED, 305, 521.
Ohm's Law, 316, 365, 555.
Ophthalmoscope, instrument made by Maxwell for seeing into the eye, 198, 208.
dogs trained to submit to observation by it, 39, 208, 212, 342.
Optics, Maxwell's contributions to, 483.
projected work on, 204.
class at Aberdeen on, 291.
Orr (or Urr), the name of the river by Glenlair, 24.
scenery of, 33.
spelling of, 67.
Ovals, investigations on, by J. C. M., as a boy of 14—74, 75, 76, 88, 91, 98-104.
identified with those of Descartes, but J. C. M.'s method new and simpler, 79, 88.

THE END.